Two-dimensional Inorganic Nanomaterials for Conductive Polymer Nanocomposites

Inorganic Materials Series

Series editors:
Duncan W. Bruce, *University of York, UK*
Dermot O'Hare, *University of Oxford, UK*
Richard I. Walton, *University of Warwick, UK*

Titles in the Series:
1: Pre-combustion Carbon Dioxide Capture Materials
2: Post-combustion Carbon Dioxide Capture Materials
3: Solar Energy Capture Materials
4: Bioinspired Inorganic Materials: Structure and Function
5: Energy Storage and Conversion Materials
6: Two-dimensional Inorganic Nanomaterials for Conductive Polymer Nanocomposites

How to obtain future titles on publication:
A standing order plan is available for this series. A standing order will bring delivery of each new volume immediately on publication.

For further information please contact:
Book Sales Department, Royal Society of Chemistry, Thomas Graham House, Science Park, Milton Road, Cambridge, CB4 0WF, UK
Telephone: +44 (0)1223 420066, Fax: +44 (0)1223 420247
Email: booksales@rsc.org
Visit our website at www.rsc.org/books

Two-dimensional Inorganic Nanomaterials for Conductive Polymer Nanocomposites

Chaoying Wan
University of Warwick, UK
Email: Chaoying.wan@warwick.ac.uk

Xingyi Huang
Shanghai Jiao Tong University, China
Email: xyhuang@sjtu.edu.cn

and

Chris Bowen
University of Bath, UK
Email: c.r.bowen@bath.ac.uk

ROYAL SOCIETY
OF **CHEMISTRY**

Inorganic Materials Series No. 6

Print ISBN: 978-1-78801-843-2
PDF ISBN: 978-1-83916-259-6
EPUB ISBN: 978-1-83916-260-2
Print ISSN: 2472-3819
Electronic ISSN: 2472-3827

A catalogue record for this book is available from the British Library

The Royal Society of Chemistry is a charity, registered in England and Wales, Number 207890, and a company incorporated in England by Royal Charter (Registered No. RC000524), registered office: Burlington House, Piccadilly, London W1J 0BA, UK, Telephone: +44 (0) 20 7437 8656.

For further information see our website at www.rsc.org

Printed in the United Kingdom by CPI Group (UK) Ltd, Croydon, CR0 4YY, UK

Series Preface

Materials chemistry has grown hugely in the past few decades and continues to evolve rapidly. This is especially the case since the subject now finds itself central to many of the major challenges that face global society, such as those concerning energy and the environment; but applications of materials are also found in many everyday situations, with ongoing developments in technology around electronics and the miniaturisation of devices. Although the study of materials is a highly interdisciplinary topic, the role of innovative, fundamental chemistry continues to lie at the heart of the discovery of new materials, the understanding of their atomic-scale structure and the relationship of structure to a useful property.

With the *Inorganic Materials Series*, our aim is to provide a series of themed volumes, each reflecting the diversity of the topic as it continues to develop over the coming years, with many chosen to relate closely to emerging practical applications. The use of the title *Inorganic Materials* is by no means limiting, but instead emphasises the study of the chemistry of all elements in the Periodic Table and how they impact the development of new materials. The chapters in each book will have a pedagogical flavour with a target audience from first-year postgraduate student upwards, but with wide subject coverage ranging from continuous inorganic solids, to molecular and soft matter systems. We will use expert, active researchers as editors and contributors to provide an up-to-date perspective of their field. At the same time our aim is to provide an international perspective, so to reflect the diversity and interdisciplinarity of the now very broad area of the subject.

Duncan W. Bruce, York, UK
Dermot O'Hare, Oxford, UK
Richard I. Walton, Warwick, UK

Inorganic Materials Series No. 6
Two-dimensional Inorganic Nanomaterials for Conductive Polymer Nanocomposites
Edited by Chaoying Wan, Xingyi Huang and Chris Bowen

Published by the Royal Society of Chemistry, www.rsc.org

Preface

The development of functional, flexible and lightweight products is needed to meet the high demand for modern applications that range from microelectronics to energy storage devices. The majority of existing polymers are thermal and electrical insulators, which hinders their application in these areas. The conductivity of polymers has the potential to be significantly enhanced by the incorporation of conducting inorganic nanoparticles. The conductivity of polymer nanocomposites not only relies on the structure and function of the inorganic particles, but is also strongly determined by the morphology and dispersion of the nanoparticles, interfacial interactions and fabrication technologies used to create the composites.

A particularly new area for polymer nanocomposites is *two-dimensional (2D) inorganic nanomaterials*. These materials have found important applications for emerging energy technologies, due to their unique combination of electrochemical, electrical, thermal, optical and mechanical properties. 2D inorganic nanomaterials with ultrathin layered structures at single- or few-atom thicknesses have shown wide tunability of conductivity, large surface-to-volume ratio and good degree of functionalisation. This book covers a number of important 2D nanomaterials, from dielectric high-κ ceramics, thermally conductive boron nitride, electrical/thermal conductive graphene and MXenes, transition metal dichalcogenides, phosphorene, ionic exchangeable layered double hydroxides to hybrid fillers. The functionalisation and applications of 2D nanomaterials in polymer nanocomposites are discussed critically. In each chapter, an overview of the synthesis and surface modification methods of the nanoplatelets is presented and the current technical challenges in polymer nanocomposite manufacturing, such as nanoparticle dispersion, orientation and interfacial interactions, are addressed. The synergy of the inorganic nanoplatelets and polymer properties is

Inorganic Materials Series No. 6
Two-dimensional Inorganic Nanomaterials for Conductive Polymer Nanocomposites
Edited by Chaoying Wan, Xingyi Huang and Chris Bowen

Published by the Royal Society of Chemistry, www.rsc.org

discussed, with a focus on thermal conductivity, ionic conductivity and electrical properties.

The book will therefore highlight many recent developments in the synthesis, chemistry and applications of two-dimensional inorganic nanoplatelets in polymer nanocomposites. The conductive polymer nanocomposites have found a broad range of applications in the areas of batteries, supercapacitors, fuel cells, solar cells and the electronics industry.

We believe this book will provide a much-needed overview in this fast-developing area of contemporary inorganic materials science, and includes a historical context and overview of the fundamentals of the subject. This will provide advanced undergraduates, postgraduates and other researchers with a convenient introduction to this fascinating topic.

Chaoying Wan

Contents

Inorganic Materials Series No. 6
Two-dimensional Inorganic Nanomaterials for Conductive Polymer Nanocomposites
Edited by Chaoying Wan, Xingyi Huang and Chris Bowen

Published by the Royal Society of Chemistry, www.rsc.org

CHAPTER 1

2D High-κ Dielectric Ceramic Nanoplatelets for Polymer Nanocomposite Capacitors

HANG LUO,[a] SHENG CHEN,[b] RU GUO,[a] XUEFAN ZHOU[a] AND DOU ZHANG*[a]

[a] State Key Laboratory of Powder Metallurgy, Central South University, Changsha 410083, Hunan Province, China; [b] Key Laboratory of Polymeric Materials and Application Technology of Hunan Province, College of Chemistry, Xiangtan University, Xiangtan 411105, Hunan Province, China
*Email: dzhang@csu.edu.cn

1.1 Introduction

Polymer-based capacitor materials continue to attract interest in applications such as laser guns, railguns, radar systems, artificial skins, muscles, and flexible electronics due to their advantages of super-high power density.[1–3] Commercial dielectrics for capacitors are mainly based on biaxial tensile polypropylene (BOPP), and the energy density of BOPP is ~2 J cm^{-3} under a high breakdown strength of ~600 kV mm^{-1}. In this case, commercial devices based on BOPP have to be large and heavy.[4–6] Therefore, routes to enhance the energy density of capacitors is essential. The permittivity and breakdown strength are two key factors to determine the energy density of dielectrics. Ceramic/polymer composites are considered as good candidates for capacitors due to the advantages of combination of high permittivity from the ceramic phase and high breakdown strength of the polymer matrix.[6,7] High-permittivity

Inorganic Materials Series No. 6
Two-dimensional Inorganic Nanomaterials for Conductive Polymer Nanocomposites
Edited by Chaoying Wan, Xingyi Huang and Chris Bowen

Published by the Royal Society of Chemistry, www.rsc.org

ceramics, such as $Pb_{1-x}Zr_xTiO_3$ (PZT), $BaTiO_3$ (BT), TiO_2, and $Ba_{1-x}Sr_xTiO_3$ (BST), are often used to increase the permittivity of the polymer composites. The filler material, interface modification and interface polarisation within these composites, filler morphology, distribution, and loading are key factors in determining the properties of the ceramic/polymer composites.

In this chapter, we will provide an overview of the effects of filler morphology on the dielectric and energy storage properties of ceramic/polymer composites. The synthesis methods of two-dimensional (2D) nanoplatelets, the classification of the polymer-based composites, and the role of the intrinsic properties of different anisotropic morphologic nanofillers in the composites will be discussed. The design strategy of 2D nanocomposites for dielectric and high energy storage properties and the interfacial models and electric simulations for the composites with 2D fillers will be discussed in detail.

1.2 Synthesis Methods of 2D Ceramic Nanoplatelets

1.2.1 Hydrothermal/Solvothermal Synthesis

Hydrothermal and solvothermal methods are often used for synthesis of nanomaterials, which are generally defined as crystal synthesis or crystal growth, by chemical reactions of substances in a sealed heated solution under high-temperature and high-pressure conditions. For ceramic nanoparticle synthesis, hydrothermal/solvothermal methods reduce the time and energy consumption for processing, since high-temperature calcination, mixing, and milling steps are minimised. In addition, during crystallisation processes, crystallised ceramic powders are directly precipitated from solution. The rate and uniformity of nucleation, growth, and ageing can be highly regulated by adjusting the processing parameters, such as reaction temperature, duration, stirring rate, and types and concentrations of reactants, surfactants, and mineralisers, which results in excellent control of morphology and size, and reduced degree of agglomeration. The synthesis of 2D ceramic nanoplatelets by hydrothermal or solvothermal methods has attracted much attention recently due to the ability for precision control of the morphologies of a number of ceramic particles, such as TiO_2,[8] ZnO,[9] Fe_2O_3,[10] Al_2O_3,[11] MoS_2,[12] Bi_2Te_3,[13] $Bi_4Ti_3O_{12}$ (BIT),[14] and $BiFeO_3$.[15]

Xie *et al.* successfully synthesised ultrathin defect-free MoS_2 nanosheets using a hydrothermal method,[12] and He *et al.* utilised MoS_2 nanosheets synthesised by similar procedures to fabricate polymer nanocomposites with high dielectric performance.[16] In the hydrothermal method, Xie *et al.* designed a reaction with a high concentration of precursors. Hexa-ammonium heptamolybdate tetrahydrate ($(NH_4)_6Mo_7O_{24} \cdot 4H_2O$) and thiourea (Mo : S ratio = 1 : 2) were dissolved in deionised water under vigorous stirring to form a homogeneous solution. Then, the solution was heated and defect-free MoS_2 nanosheets were obtained by a quantitative reaction and oriented crystal growth. The thiourea was employed as the reductant to reduce Mo^{6+} to Mo^{4+} and was also effective to stabilise the ultrathin nanosheet morphology. In the work of

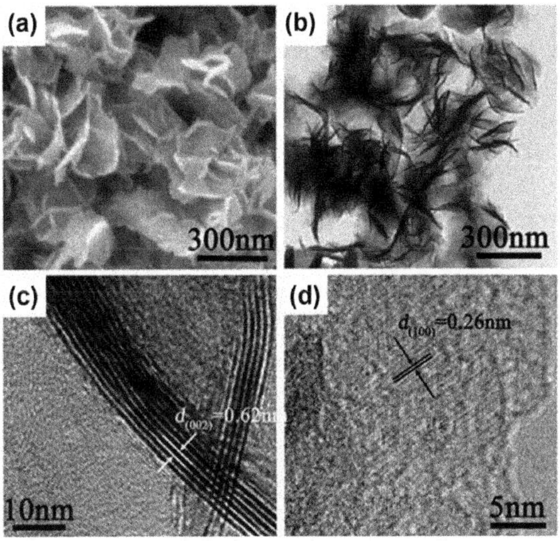

Figure 1.1 (a) SEM and (b) TEM images of MoS_2 nanosheets. (c, d) HRTEM images from the side and the top view.[16]
Reproduced from ref. 16 with permission from Elsevier, Copyright 2018.

He *et al.*, the lateral size of the MoS_2 nanosheets was in the range of ~100–200 nm, as shown in Figure 1.1(a). Figure 1.1(b) further illustrates the ultrathin nanosheet morphology of the MoS_2. Corresponding high-resolution transmission electron microscopy (HRTEM) images from the side and the top view are presented in Figure 1.1(c) and (d), respectively. The interplanar spacing was measured as 2.7 nm, which is consistent with the *d* spacing of (100) planes of hexagonal MoS_2. Furthermore, the thickness of the typical layer is approximately 6 ~ 8 nm, with an interlayer spacing of 0.62 nm.

Ultrathin anatase TiO_2 nanosheets with dominant {001} facets are fabricated by a one-step solvothermal reaction.[17] Tetrabutyl titanate ($Ti(OBu)_4$), hydrofluoric acid (HF), and absolute ethanol (C_2H_6O) were used as raw materials. Ethanol was employed as both a reaction solvent and alkoxy group source to tailor the growth of anatase TiO_2. Fluoride (F^-) was employed as a capping agent to stabilise {001} facets and enhance their exposure. The introduction of ethanol solvent effectively improved the content of surface-chemisorbed F^- on TiO_2. By this approach, anatase TiO_2 nanosheets were synthesised successfully, and exhibited a rectangular shape with a size of ~200 nm, a thickness of ~2.5 nm, and a side length of ~200 nm. Moreover, the nanosheets were dominated by {001} facets.

1.2.2 Molten Salt Synthesis

Molten salt synthesis is an effective method for preparation of high purity and nanoscale inorganic oxides with controllable compositions and morphologies. Inorganic molten salt serves as the reaction medium and the reactant oxides exhibit short diffusion distances and large mobility in the molten salts, which are beneficial for enhancing the reaction rate and reducing the reaction temperature. Moreover, the morphology of the final products can be controlled by adjusting the processing parameters, including the types and quantities of the used molten salts, heating temperatures and duration, and heating/cooling temperatures. In recent years, the molten salt synthesis method has been successfully used to prepare a variety of 2D ceramic nanoparticles, such as BIT,[18] $Na_{0.5}Bi_{4.5}Ti_4O_{15}$ (NBIT),[19] $Na_{0.5}Bi_{0.5}TiO_3$ (NBT),[19] BT,[20] $SrTiO_3$ (ST),[21] BST[22] and $NaNbO_3$ (NN)[23] platelets, that are applicable for the fabrication of ceramic/polymer composites.

As an example, by a two-step molten salt synthesis method, Jiang *et al.* synthesised 2D single-crystalline NBT-BT platelets with a perovskite structure, using the layer-structured NBIT plates as precursors,[24] which were used in sandwich-structured polymer nanocomposites.[25] Figure 1.2(a) and (b) shows

Figure 1.2 SEM images of (a) NBIT and (b) NBT-BT platelets. (c) HRTEM image of a partially converted NBIT platelet and schematic of the topochemical transformation from layer-structured NBIT to perovskite NBBT.[24] Reproduced from ref. 24 with permission from the Royal Society of Chemistry.

the prepared NBIT and NBT-BT platelets, demonstrating a high aspect ratio with an average size of 10 μm and thickness of 600 nm. First, NBIT platelets were synthesised in the first-step of the molten salt method. The reactant oxides Bi_2O_3, TiO_2, and Na_2CO_3 were fully dissolved within the molten salt NaCl, and the NBIT platelets were formed through a complete reaction between reactant oxides after heating at 1050 °C for 4 h. In the second -step of the molten salt method, the NBT-BT platelets were formed by topochemical transformation from NBIT precursors, as shown in Figure 1.2(c). In this process, NaCl molten salts were used as a reaction medium, and Na_2CO_3, $BaCO_3$, and TiO_2 were used as the reactant powders; the reaction was completed after heating at 950 °C for 4 h. It is known that NBIT exhibits an Aurivillius layer-type structure consisting of perovskite-like blocks interleaved with layers of $(Bi_2O_2)_n^{2n+}$, represented by a formula $(Bi_2O_2)_n^{2n+}(Na_{0.5}Bi_{2.5}Ti_4O_{15})_n^{2n-}$, where Bi^{3+} and Na^+ ions correspond to the perovskite A-site with 12-fold coordination, and Ti^{4+} is the perovskite B-site cation in six-fold coordination. During the topochemical transformation from NBIT to NBT-BT, the bismuthate layers were effectively replaced by perovskite layers, and Bi^{3+} ions throughout the structure were partially replaced by Na^+ and Ba^{2+} ions without extensive rearrangement of the structural framework. Therefore, the plate morphology was maintained after the second-step molten salt synthesis.

Similarly, NN platelets can also be prepared by a two-step molten salt synthesis method.[26] First, Bi_2O_3 and Nb_2O_5 were heated in a NaCl flux at 1000 °C for 2 h. The $Bi_{2.5}Na_{3.5}Nb_5O_{18}$ platelets were obtained and used as precursors for the second-step molten salt process. Then, the $Bi_{2.5}Na_{3.5}Nb_5O_{18}$ platelets were mixed with NaCl and reacted at 1000 °C for 2 h to obtain NN platelets. As shown in Figure 1.3(a) and (b), the $Bi_{2.5}Na_{3.5}Nb_5O_{18}$ and NN platelets both exhibited large aspect ratios. The perovskite-structured NN platelets with dimensions of ~2–5 μm and thicknesses of ~100–500 nm were prepared by topochemical conversion of layer-structured $Bi_{2.5}Na_{3.5}Nb_5O_{18}$ platelets as templates, as shown Figure 1.3(c) and (d).

1.2.3 Exfoliation of Layered Materials

Layered materials represent a diverse and largely untapped source of 2D systems with exotic electronic properties and high specific surface areas, such as transition metal oxides, transition metal dichalcogenides, transition metal carbides and nitrides, and other 2D compounds; for example, BN, Bi_2Te_3, and Bi_2Se_3. In the past decades, researchers have been exploring simple methods to exfoliate layered materials to obtain mono- or few-layer nanosheets, which have significant potential in sensing, catalysis, electronic, and energy storage applications.

In 2009, Zhi *et al.*[27] reported the chemical exfoliation method to prepare BN nanosheets in large quantities from the hexagonal BN (h-BN) particles. h-BN is used as an ultraviolet-light emitter, or as an electrically insulating but thermally conductive filler in composites. It is believed

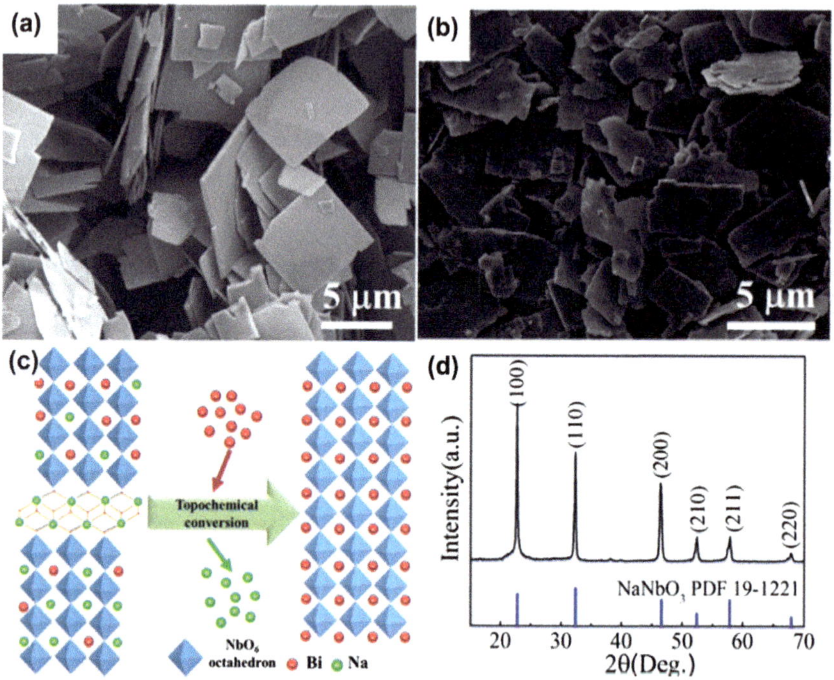

Figure 1.3 SEM images of (a) $Bi_{2.5}Na_{3.5}Nb_5O_{18}$ and (b) NN platelets. (c) Schematic of the topochemical conversion from $Bi_{2.5}Na_{3.5}Nb_5O_{18}$ precursors to NN platelets. (d) XRD patterns of NN platelets.[26]
Reproduced from ref. 26 with permission from Elsevier, Copyright 2017.

that h-BN nanosheets would be valuable to exploit many unique properties of a graphitic-like (002) plane, such as superb high thermal conductivity and mechanical strength. h-BN has a layered structure, where a sequence of atomic planes stack *via* van der Waals interactions to form a three-dimensional (3D) crystal. h-BN is stacked with the boron atom on the top of the nitrogen atom and *vice versa*. For the preparation of BN nanosheets, a strong polar solvent, *N*,*N*-dimethylformamide (DMF), was used to facilitate exfoliation due to the strong interactions between the polar DMF molecules and a BN surface. In experimental processes, h-BN powders were dispersed in DMF, and subjected to ultrasonication. Vigorous sonication was able to exfoliate BN nanosheets from the h-BN particles. Then, the solution was centrifuged in order to remove residual large-size h-BN particles. Due to a layered structure, most of the h-BN particles are thinner along one direction, with a thickness of micrometres. After exfoliation, the BN nanosheets were successfully prepared. The BN nanosheets exhibited micrometre dimensions, smaller than that of pristine particles, and a typical six-fold natural symmetry. Intriguingly, the thickness of the BN nanosheets was dependent on the centrifugation speed. At 5000 rpm, most of the BN nanosheets have less than 20 layers, with a thickness less than 7 nm. Wu *et al.*[28] successfully utilised a similar exfoliation method to prepare the BN nanosheet and fabricate polymer-based nanocomposites with large energy density. Figure 1.4(a) shows a schematic for the preparation of BN nanosheets. The morphology of BN nanosheets observed by scanning electron microscopy (SEM) exhibited the lateral size of ~500 nm to 1 μm, as shown in Figure 1.4(b). The TEM images in Figure 1.4(c) and (d) suggest that the thickness was approximately 2–10 nm. Figure 1.4(e) shows the statistical analysis on the thickness of the boron nitride nanosheets (BNNSs) over 106 pieces, indicating the successful exfoliation of h-BN through ultrasonication.

Coleman *et al.*[29] developed liquid exfoliation of layered transition metal dichalcogenides for producing 2D nanosheets. The transition metal

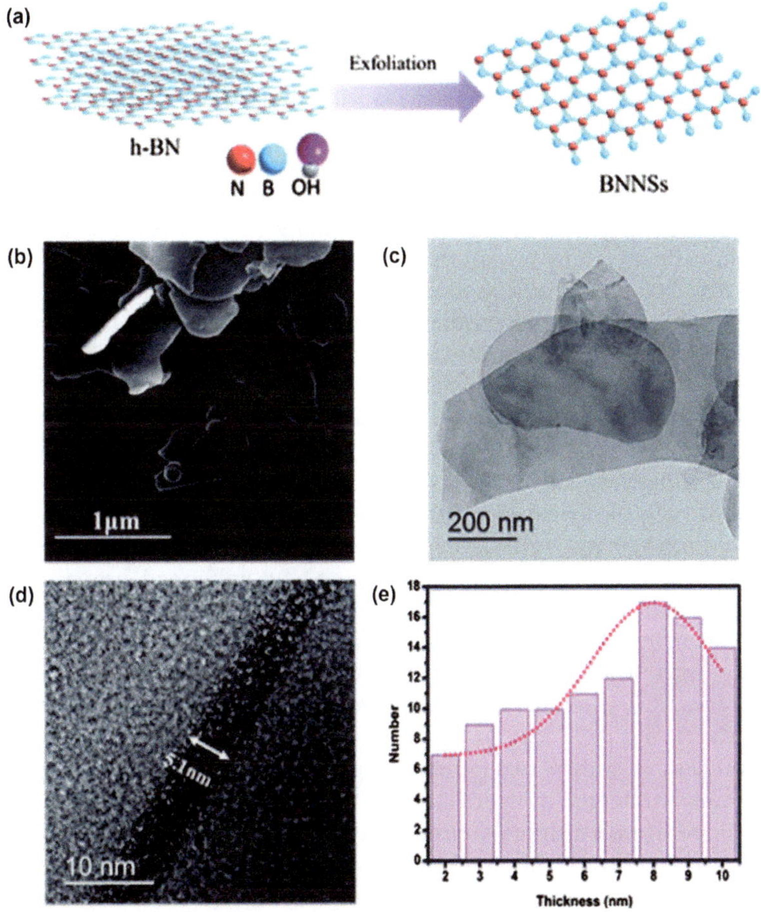

Figure 1.4 (a) Schematic for the preparation of BN nanosheets. (b) SEM and (c) TEM images of BN nanosheets, and (d) corresponding thickness. (e) Statistical analysis of the thickness of a set of BNNSs.
Reproduced from ref. 28 with permission from the Royal Society of Chemistry.

dichalcogenides consist of hexagonal layers of metal atoms (M) sandwiched between two layers of chalcogen atoms (X), with stoichiometry MX_2. Although the bonding within these trilayer sheets is covalent, adjacent sheets stack *via* van der Waals interactions to form a 3D crystal. The authors have demonstrated the exfoliation of bulk MX_2 crystal in common solvents to give mono- and few-layer nanosheets, including MoS_2, $MoSe_2$, WS_2, $MoTe_2$, $TaSe_2$, $NbSe_2$, and $NiTe_2$. In the experimental procedures, commercial MoS_2 and WS_2 powders were sonicated in a number of solvents with varying surface tensions. The resultant dispersions were centrifuged and the supernatant was decanted. The mass remaining in the supernatant was estimated by measuring the UV–visible–infrared (IR) absorption spectrum. It indicated that the amount of material retained (characterised by $A/l = \alpha C$, where A/l is the absorbance per length, α is the extinction coefficient, and C is the concentration) was maximised for solvents with a surface tension, γ, close to ~40 mJ m^{-2}. The analysis results showed that successful solvents were those with dispersive, polar, and H-bonding components of the cohesive energy density within certain well-defined ranges. In other words, the successful solvents are those that minimise the energy of exfoliation. Some solvents, such as *N*-methyl-pyrrolidone (NMP) and isopropanol (IPA), were indicated as promising for liquid exfoliation. The prepared $MoSe_2$ and WS_2 nanosheets show a lateral size of ~50–1000 nm and hexagonal symmetry. This exfoliation method is insensitive to air and water and can potentially be scaled up to produce large quantities of exfoliated material.

Layered ternary metal carbides and nitrides, referred to as the MAX phases ($M_{n+1}AX_n$, $n = 1$, 2 or 3), were also utilised to prepare nanosheets, where M is a traditional metal, A is a group IIIA or IVA element, and X is carbon or nitrogen atoms. MAX phases show unique combination of metallic and ceramic properties, such as excellent resistance to oxidation, heat, and corrosion, high electrical conductivity, high strength and elastic modulus, attributed to their inherent lamellar structure with alternately arranged MX and A layers. Usually, the 2D nanosheets show enhanced properties with respect to their corresponding bulk counterpart. However, the MAX phases possess relatively strong bonds between the MX and A layers, and are hard to exfoliate into ultrathin nanosheets using simple exfoliation processes. Naguib *et al.*[30] first reported exfoliation of the MAX phases: Ti_2AlC, Ta_4AlC_3, $(Ti_{0.5}, Nb_{0.5})_2AlC$, $(V_{0.5}, Cr_{0.5})_3AlC_2$, and Ti_3AlCN. A schematic of the exfoliation process is shown in Figure 1.5(a). In a typical process, the powders were simply immersed in hydrofluoric acid (HF) of varying concentrations at room temperature for times varying between 10 and 72 h. This procedure led to selective etching of the aluminium (Al) layers and their replacement by hydroxyl, OH, and fluorine (F) surface groups. By sonication, the 'A' group layer was removed from the MAX phases and 2D nanosheets were obtained, which were labelled as MXenes by the authors to denote the loss of the A element and emphasise their structural similarities with graphene. As a representative example, Figure 1.5(b) shows the SEM image of Ti_3AlCN after HF treatment, indicating the obvious exfoliation. Figure 1.5(c) shows the

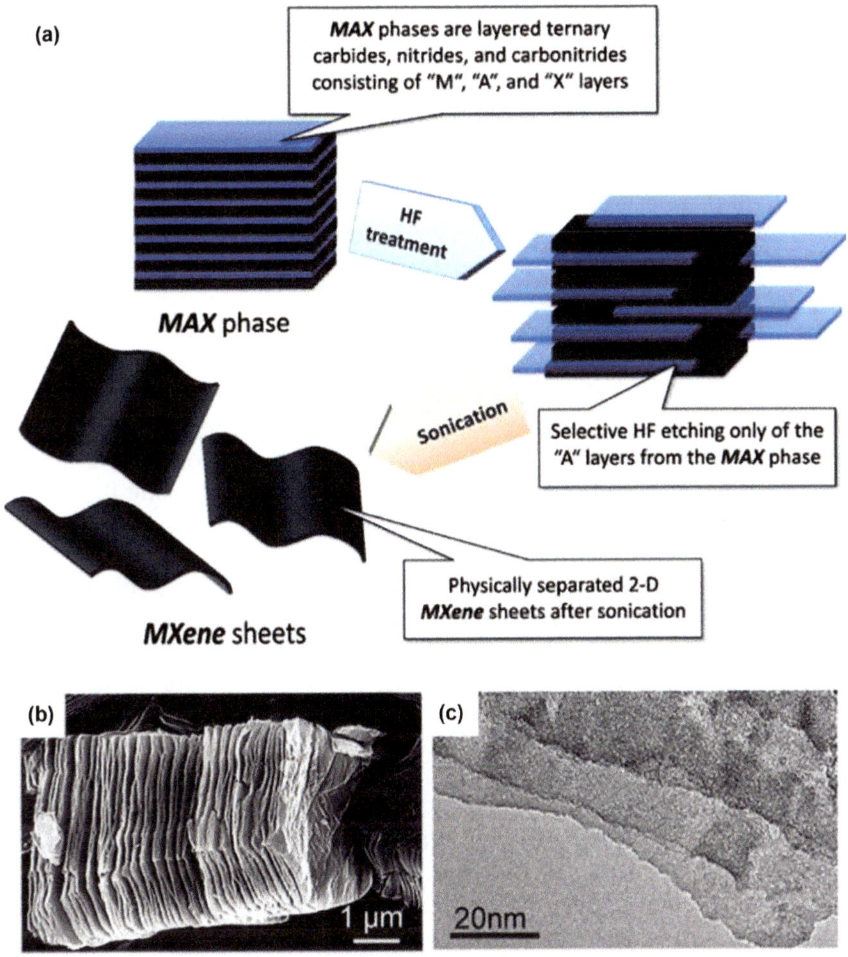

Figure 1.5 (a) Schematic of the exfoliation process of MAX phase. (b) SEM image of Ti_3AlCN after HF treatment. (c) TEM image of 2D Ti_3CN_x layers exfoliated from Ti_3AlCN.
Reproduced from ref. 30 with permission from American Chemical Society, Copyright 2012.

TEM image of 2D Ti_3CN_x layers exfoliated from Ti_3AlCN, which are transparent to the electron beam.

Zhang *et al.*[31] also successfully prepared ultrathin nanosheets of MAX phases with enhanced thermal and mechanical properties in polymeric compositions: $Ti_3Si_{0.75}Al_{0.25}C_2$. In this study, the authors developed a green, facile, and effective exfoliation method for the preparation of ultrathin nanosheets of MAX phases. Although MAX phases are chemically stable, the A layers are more reactive than the MX layers. By substituting a portion of the A atoms, the A layers were activated and a substitutional solid solution was obtained. The as-doped MAX phases could be exfoliated by breaking.

Using Ti_3SiC_2 MAX phase as an example, the doped phase of $Ti_3Si_{0.75}Al_{0.25}C_2$ (TSAC) was fabricated by a modified high-temperature self-propagation synthesis. The TSAC nanosheets were obtained by liquid exfoliation of TSAC powder in different solvents under ultrasound. In comparison, pure Ti_3SiC_2 powder cannot be exfoliated into nanosheets in these same solvents. Moreover, various solvents have been indicated to be effective for preparation of MAX phase nanosheets, such as ethanol, formamide, DMF, and NMP, indicating the wide applicability of the substitutional solid solution-based exfoliation method. The as-exfoliated TSAC nanosheets showed a size of ~100–200 nm and were indicated to be exfoliated along the *z*-axis and to possess single-crystalline nature.

There are also many transition metal oxides that can be produced in 2D states by chemical exfoliation methods, such as RuO_2,[32] MnO_2,[33] and MoO_3.[34] For instance, Coellho *et al.*[33] prepared MnO_2 nanosheets by a single-step liquid-phase exfoliation method that produces a mix of flat and flower-like nanosheets. In the experimental process, manganese oxide flower-like nanostructures (MOFN) were first synthesised. Manganese nitrate tetrahydrate ($Mn(NO_3)_2{\cdot}4H_2O$) was dissolved in deionised water with the addition of poly(ethylene glycol)-*block*-poly(propylene glycol)-*block*-poly(ethylene glycol) (PEG-PPG-PEG). The solution was heated up and maintained at a given temperature and was kept under vigorous stirring while $KMnO_4$ solution was added drop by drop. The brownish precipitate obtained was washed, vacuum filtrated, and dried. Then, the as-prepared MOFN was used to fabricate manganese oxide nanosheets (MON) by liquid-phase exfoliation. The MOFN was mixed with isopropyl alcohol and processed in an ultrasonic bath, followed by centrifugation and collection of the supernatant. The obtained dispersions consisted of MON and the quality can be assessed by UV–visible spectroscopy. Figure 1.6(a) and (b) present TEM images of the MOFN, showing a characteristic flower-like morphology. Figure 1.6(c) and (d) display the TEM and HRTEM images of the MON, showing typical dimensions of ~20–40 nm and favourable crystallinity.

1.3 Classification of Polymer-based Composites and the Role of Intrinsic Properties of Different Anisotropic Morphologic Nanofillers

In the study of polymer-based dielectric nanocomposites, the most investigated nanofillers are ferroelectric ceramics with the perovskite structure, such as $BaTiO_3$ (BT), $(Ba_{1-x}Sr_x)TiO_3$ (BST), and $Pb(Zr_{1-x}Ti_x)O_3$ (PZT). The relative permittivity of this kind of filler ranges from hundreds to thousands, which is much higher than that of the polymer matrix. Hence, it is a common strategy to incorporate nanofillers into the polymer matrix to raise the permittivity of the nanocomposite dielectrics. The nanofillers can be considered at a range of dimensions, namely, zero-dimensional (0D) nanofillers, which include spherical nanoparticles, nanocubes, and nanoparticles with

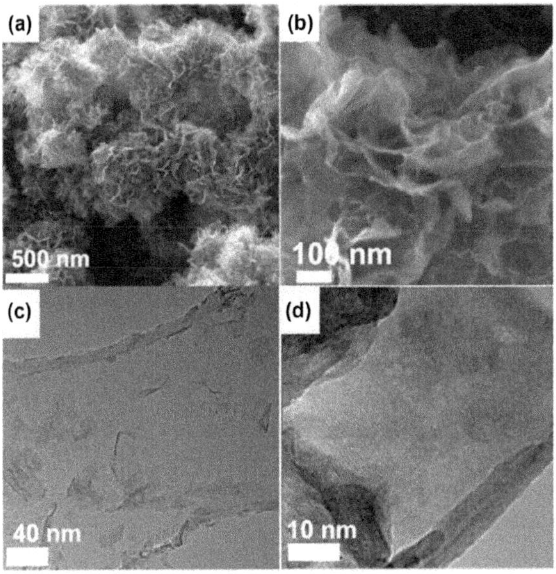

Figure 1.6 (a, b) TEM images of the MOFN. (c) TEM and (d) HRTEM images of MON. Reproduced from ref. 33 with permission from IOP Science, Copyright 2015.

irregular morphologies; one-dimensional (1D) nanofillers, which include nanowires, nanofibres, nanotubes, and nanoribbons; and two-dimensional (2D) nanofillers, which include nanoplates, nanosheets, and nanoclays.[35]

1.3.1 0D Nanoparticles

For composites filled with spherical nanoparticles, current research mainly focuses on the improvement of compatibility, interface bonding, and dispersion between nanoparticles and polymers through surface modification and coating of nanoparticles. Dang *et al.*[36] prepared a novel calcium copper titanite ($CaCu_3Ti_4O_{12}$; CCTO)/polyimide (PI) composite film. At a volume fraction of 40 vol% CCTO particles, the relative permittivity of the composite was 49 at 100 Hz, which was 14 times that of pure PI. The addition of nanoparticles to the polymer matrix could increase the permittivity of the system to some extent, but optimum dielectric properties are usually obtained with a high volume fraction of nanoparticles. Theoretical calculations show that when the loading of 0D nanoparticles reaches more than 39.7 vol%, holes and agglomerations are inevitably introduced into the composites, which act to deteriorate the dielectric properties and breakdown strength, as well as the mechanical properties of the composites.

Consequently, numerous studies are aimed at interfacial effects using the surface modification method to improve the dispersion and compatibility between the nanoparticles and the polymer matrix. Yu *et al.*[37] treated the surface of BT nanoparticles with polyvinylpyrrolidone (PVP), then compounded them with PVDF to prepare the composite film. At a filler volume fraction of 10 vol% BT, the breakdown strength of the film increased by 38% compared with the composite without PVP modification. Almadhoun *et al.*[38] introduced hydroxyl groups into the surface of BT to form h-BT bonds by means of the hydrogen peroxide water treatment. Hydrogen bonds were formed between the hydroxyl groups and fluorine of the P(VDF-TrFE) matrix, which enhanced the interface combination intensity. The h-BT/P(VDF-TrFE) composite achieved a lower leakage current, higher breakdown strength, and also showed better frequency and temperature stability.

Furthermore, the macromolecule coating of active functional groups can be directly grown on the surface of the filler by a chemical synthesis method, which can improve the effective dispersion of the filler and promote its bonding with the polymer matrix. Yang *et al.*[39] mixed BT nanoparticles with fluoroalkyl acrylate and coated BT particles with a layer of trifluoroethyl acrylate (PTFEA) approximately 5 nm thick by reversible addition-fragmentation chain-transfer polymerisation. Since both PTFEA- and PVDF-based polymer matrices contain fluorine, the dispersion of particles was obviously improved. In addition, the existence of an organic insulating layer limits the accumulation and migration of space charge, thus significantly reducing the dielectric loss of composites and increasing the energy storage density by 50%. Xie *et al.*[40] prepared BT/ poly(methyl methacrylate) (PMMA) composites by atom transfer radical *in situ* polymerisation (ATRP) on the surface of BT coated with PMMA of different thickness. The effects of different shell thickness on the dielectric properties of the composites were studied. The PMMA layer prepared by ATRP reaction has good binding with the BT nanoparticle filler and effectively prevents carrier migration and reduces dielectric loss. Subsequently, a branched aromatic polyamide (HBP) with a medium permittivity was placed between the PMMA and BT as a transition layer to form a core@double-shell structure. The existence of the HBP layer enhances the interfacial polarisation, restricts the dipole movement in PMMA side groups, and reduces the friction between molecules. The energy density of the dielectric nanocomposite showed a significant increase as a result of the further improved permittivity and weakened dielectric loss.

1.3.2 1D Nanowires, Nanorods, and Nanofibres

Compared with 0D nanofillers, 1D nanowires show the following advantages. First, the lower surface energy of nanowires caused by the larger aspect ratio and smaller specific surface area, which contributes to improving the dispersion of fillers in the polymer matrix. Second, because the nanowires have a large dipole moment, the dielectric property of the composites can be significantly increased even at low filler loading levels. At present, nanowire materials are mainly chosen as the fillers in dielectric composites, which include TiO_2, BT, BST, and PZT nanowires.

Many theoretical calculations and experiments have found that 1D fillers can improve the dielectric properties of composites more effectively than spherical or quasi-spherical fillers. Guo *et al.*[41] have compounded spherical BT, spherical TiO_2, and rod TiO_2 as fillers with the PP polymer matrix, respectively. At the same filler loading of 3 vol%, the relative permittivity of a rod TiO_2 composite was 4.1, which is higher than that of spherical TiO_2 composite and rod TiO_2 composite. The results showed that the enhancement effect of the rod-like TiO_2 was better than that of BT spherical particles with higher permittivity.

The length to diameter ratio of 1D fillers has an influence on the dielectric properties of the composites. Tang *et al.*[42] obtained BT nanowires with different length to diameter ratio by altering the reaction temperature of the hydrothermal method, then studied the effect of length to diameter ratio on the permittivity of BT/PVDF nanocomposites. At a BT nanowire volume fraction of 30 vol%, the relative permittivity of the composite with a filler length diameter ratio of 45.8 was found to be 44.3, which was 35.7% and 352% higher than that of the composite with the aspect ratio of 9.3 and the pure PVDF, respectively. Liu *et al.*[43] adopted high aspect ratio $SrTiO_3$ nanowires surface-modified by PVP to prepare $SrTiO_3$/PVDF composites. At a low filler loading of 2.5 vol%, the permittivity of the nanocomposite improved significantly without apparent decrease in breakdown strength. The reason for the enhanced permittivity is that the large aspect ratio makes it easier to reach the percolation threshold, increases the dipole motion, and reduces the interface area, thus inhibiting the agglomeration of nanofibres. As a result, the energy storage density of the $SrTiO_3$/PVDF nanocomposite is more than twice that of PVDF.

Owing to the existence of nanowire anisotropy, the orientation of the nanowires has a significant influence on the dielectric properties of composites. The construction of 1D filler alignment in nanocomposites is also an effective way to improve the dielectric and energy storage properties of dielectric capacitors. Sodano *et al.*[44] have aligned PZT nanowires (NWs) in the PVDF matrix and formed an array distribution of nanocomposites by a uniaxial strain assembly method. It was found that the relative permittivity in the nanocomposites aligned in the 3-direction (the orientation of PZT NWs parallel to the direction of the electric field) is much larger than that of the PVDF composites with random PZT at the same loading, *e.g.*, ≈34 for the aligned nanocomposite *versus* ≈25 for the random nanocomposite at 30 vol% PZT. It was also shown that the 3-direction-aligned PZT NW nanocomposites have a larger electric displacement than the random and 1-direction-aligned (the orientation of PZT NWs perpendicular to the direction of electric field) PZT NW nanocomposites, as a result of the higher dielectric permittivity of the samples. Moreover, the composite with 40 vol% aligned PZT in PVDF can achieve a maximal calculated energy density of $\approx 1.28\ \mathrm{J\,cm^{-3}}$ at a low electric field of $15\ \mathrm{MV\,m^{-1}}$. A similar investigation was conducted with aligned BT nanowires by Xie *et al.*[45] As shown in Figure 1.7, they fabricated the *X*–*Y*-aligned nanocomposite (nanowires aligned perpendicular to the applied electric field) with the help of shear stresses, and the *Z*-aligned nanocomposite (nanowires aligned in the direction of the

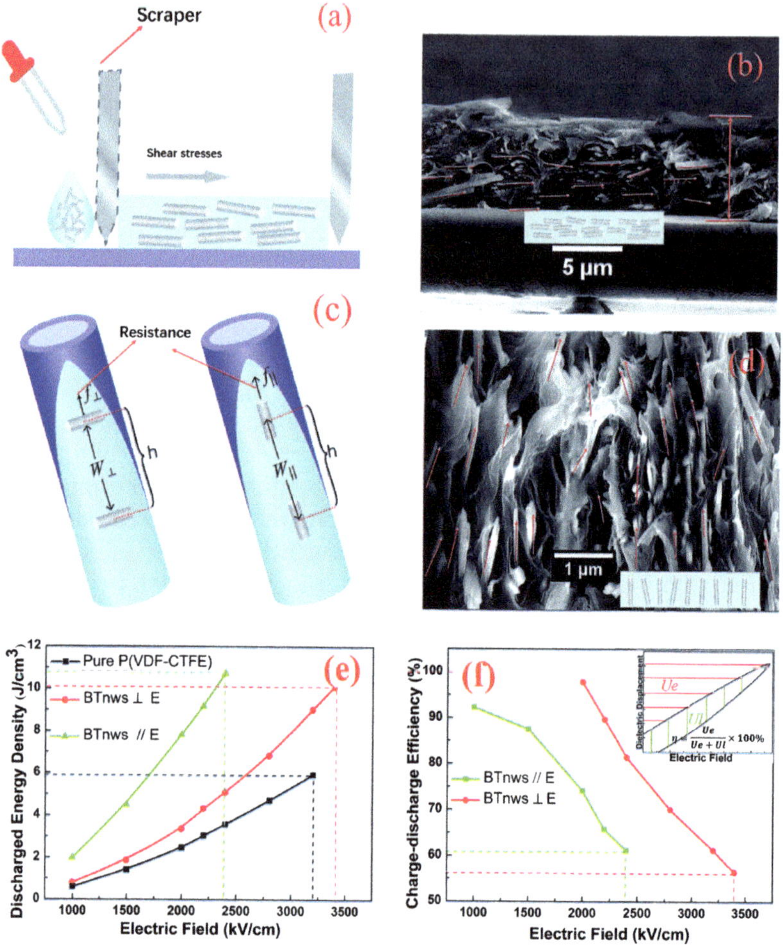

Figure 1.7 (a) Schematic of the aligning mechanism for BT nanowires (NWs) in a polymer matrix. (b) Representative cross-sectional SEM image of the X–Y-aligned nanocomposite film. (c) Schematic diagram of the moving mechanism for BT NWs on the inner wall of the elongated Teflon tube. (d) Representative cross-sectional SEM image of the Z-aligned nanocomposite film. (e) Variations of discharge energy density and (f) charge–discharge efficiency of pure P(VDF-CTFE), *X*–*Y*-aligned (BT NWs ⊥ E) and *Z*-aligned (BT NWs//E) nanocomposites.
Reproduced from ref. 45 with permission from the Royal Society of Chemistry.

applied electric field) under the guidance of the inner wall of a Teflon tube. The result showed that, compared with the pristine P(VDF-CTFE) matrix, the polarisation and breakdown strength of the *X*–*Y*-aligned nanocomposite were enhanced simultaneously, resulting in an improved energy storage performance. Furthermore, a superior performance was achieved through the Z-aligned nanocomposite, where the energy density and efficiency were improved to 10.8 J cm^{-3} and 61.4%, respectively, at a lower electric field of 2400 kV cm^{-1}.

1.3.3 2D Nanoplatelets

In recent years, 2D fillers with high aspect ratio in-plane, such as BNNSs,[28,46,47] MoS_2 nanosheets,[16,48] ZrO_2 nanosheets,[49] montmorillonite nanosheets (MMTs),[50,51] and graphene nanosheets (GNs),[52–55] have been extensively explored in ferroelectric polymer nanocomposites, to tailor the dielectric and energy storage properties. For example, conductive graphene has been introduced into a polymer enabling improvement of the dielectric permittivity due to the microcapacitor effect; while incorporation of insulating BNNSs or MMT nanoplatelets can enhance the breakdown strength, owing to their barrier effect. According to the characteristics of fillers and their action mechanisms in the composite, three types of insulation filler polymer systems and conductive filler polymer systems are classified and are summarised in the following sections.

1.3.3.1 Insulation Filler Polymer Systems

Theoretically, 2D insulation fillers, such as nanosheets and nanoribbons, can be regarded as a large number of barriers perpendicular to the direction of electric field that prevent the expansion of conductive channels and enhance carrier scattering. This plate-like structure is also conducive to heat dissipation and restraining the formation of thermal breakdown and electromechanical breakdown paths along the electric field direction, thus effectively enhancing the breakdown field strength of the composites. Among the 2D insulating fillers, BNNSs, MMTs, ZrO_2 nanosheets, BT nanosheets,[56] and $NaNbO_3$ platelets[57] have attracted the greatest interest from scientists and have been employed in nanoelectronics.

The material h-BN, referred to as 'white graphene' and the most stable crystalline form of boron nitride, is a layered van der Waals crystal, which is similar to graphite. It shows high mechanical strength and thermal conductivity resulting from a graphitic-like structure of strong B–N covalent sp^2 bonds in the (002) plane. In addition, insulating hexagonal boron nitride has a wide band gap (6 eV), relatively low relative permittivity (3–4), and high breakdown strength (800 kV mm^{-1}). This unique structure and physical performance of hexagonal boron nitride promote its applications as a dielectric layer in nanocomposites.[58–60]

Li *et al.*[59] obtained ultrathin BNNSs by a sonication–centrifugation method, and then prepared P(VDF-TrFE-CFE)/BNNS nanocomposite films *via* a simple solution-casting method, as shown in Figure 1.8. Due to the polar surface induced by B–N bonds, a proper fraction of bare BNNSs without any chemical modification can be uniformly dispersed in a matrix of polar polymer. Generally, inorganic fillers often have many defects in the

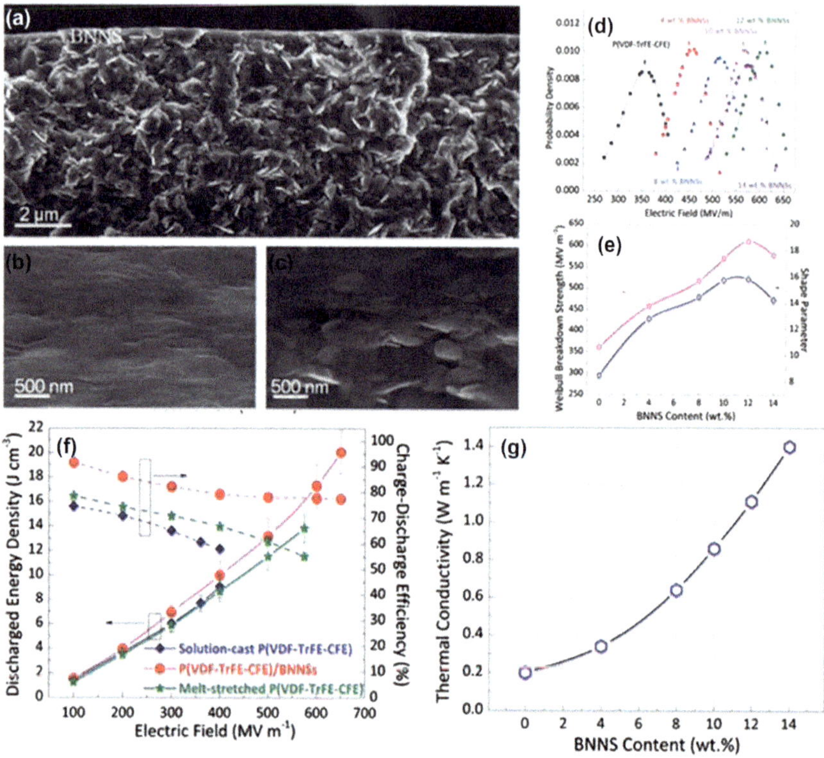

Figure 1.8 (a) Large-scale cross-section SEM image of the P(VDF-TrFE-CFE)/BNNS composite film with 12 wt% BNNSs. (b) Zoomed-in top view and (c) cross-section SEM images of the P(VDF-TrFE-CFE)/BNNS composite film with 12 wt% BNNSs. (d) Weibull plots for P(VDF-TrFE-CFE)/BNNS nanocomposites with various BNNS contents. (e) Weibull breakdown strength and shape parameter as functions of BNNS fraction. (f) Comparison of discharged energy density and charge–discharge efficiency of pristine P(VDF-TrFE-CFE) and P(VDF-TrFE-CFE)/BNNS nanocomposites with 12 wt% of BNNSs at different electric fields. (g) Thermal conductivity of P(VDF-TrFE-CFE)/BNNS nanocomposites with different filler contents.
Reproduced from ref. 59 with permission from the Royal Society of Chemistry.

polymer matrix due to their poor compatibility, which hinders improvement of the composite properties. Researchers usually use ligands with long hydrocarbon chains to activate the surface of the filler to improve the compatibility of the two phases for good dispersion and distribution. However, low permittivity of the modified surface results in a large contrast for the high-permittivity ferroelectric polymer and inorganic dopant phases, which becomes a vulnerable local region to localised electric fields in the nanocomposite, and, accordingly, degrades the overall dielectric strength. For BNNSs, the high aspect ratio and specific surface area not only simplifies the preparation process of the polymer nanocomposites, but also eliminates the steps of chemical modification. Thus, a dense network-like microstructure was formed in the polymer matrix *via* a simple solution- casting technique, which is conducive to increasing the breakdown field strength of the composites.

The results show that the breakdown strength of nanocomposites filled with 12 wt% of BNNSs reaches 610 MV m^{-1}, which is 70% higher than that of pure polymers, and the shape parameter β, standing for less scattering of E_b, also increases from 8.44 to 15.8, which means that the dielectric stability of the composites is significantly enhanced. It is reasoned that BNNSs within the polymer matrix create a robust scaffold to hamper the onset of electromechanical failure, and act as an efficient barrier against leakage current and space charge conduction. In addition, a discharge energy density of 20.3 J cm^{-3} was achieved at 650 MV m^{-1}, while the charge–discharge efficiency remains above 80% at 600 MV m^{-1}, which resulted from the suppressed conduction loss and high field resistivity of the presence of BNNSs under the electric field. It should be emphasised that with the introduction of BNNSs into the terpolymer matrix, the thermal conductivity is increased six-fold, from ~0.2 W m^{-1} K^{-1} for the terpolymer, to over 1.2 W m^{-1} K^{-1} for the nanocomposite filled with 14 wt% of BNNSs, which benefits the breakdown stability and lifetime of polymer capacitors.

In addition, the most common 2D insulation fillers for nanocomposites are MMT nanosheets, which are layered silicate sheets and are an environmentally friendly and naturally occurring smectite clay. This material becomes a good choice for designing polymer nanocomposites due to its low cost and rich intercalation chemistry, allowing improved compatibility with the polymer matrix, to enable higher polarisation and improved dielectric performance and breakdown strength.

Ghosh *et al.*[61] reported a flexible polymer nanocomposite (PCN) by incorporation of unmodified MMT nanoclay into the PVDF polymer matrix. The results showed that the uniform intercalation of PVDF chains into the layered silicates of unmodified MMT nanoclay induces a α–γ phase conversion in the PCN films, which was a consequence of interfacial interaction between the negatively charged surface of the MMT nanoplatelets and the electropositive CH_2 dipoles of PVDF, as shown in Figure 1.9. The obtained PVDF/MMT clay nanocomposite films stabilise the phase and increase the path tortuosity *via* strong intercalation of the PVDF matrix into inorganic layered silicates, without sacrificing the quality of the surface morphology. The PCN films exhibit superior dielectric properties (up to ε_r~28 and tan δ~0.032 at 1 kHz) compared with that of pure PVDF because the MMT clay acts as an efficient insulating barrier against the leakage current and space charge conduction. Moreover, the MMT nanoplatelets, arranged in a layer-by-layer structure within the PVDF matrix, formed tortuous paths between the electrodes that block the path of the applied electric field. As a result, a large increase in E_b of 873 kV mm^{-1} and U_e of 24.9 J cm^{-3} is achieved. Consequently, the PCN films possess more than 60% charge–discharge

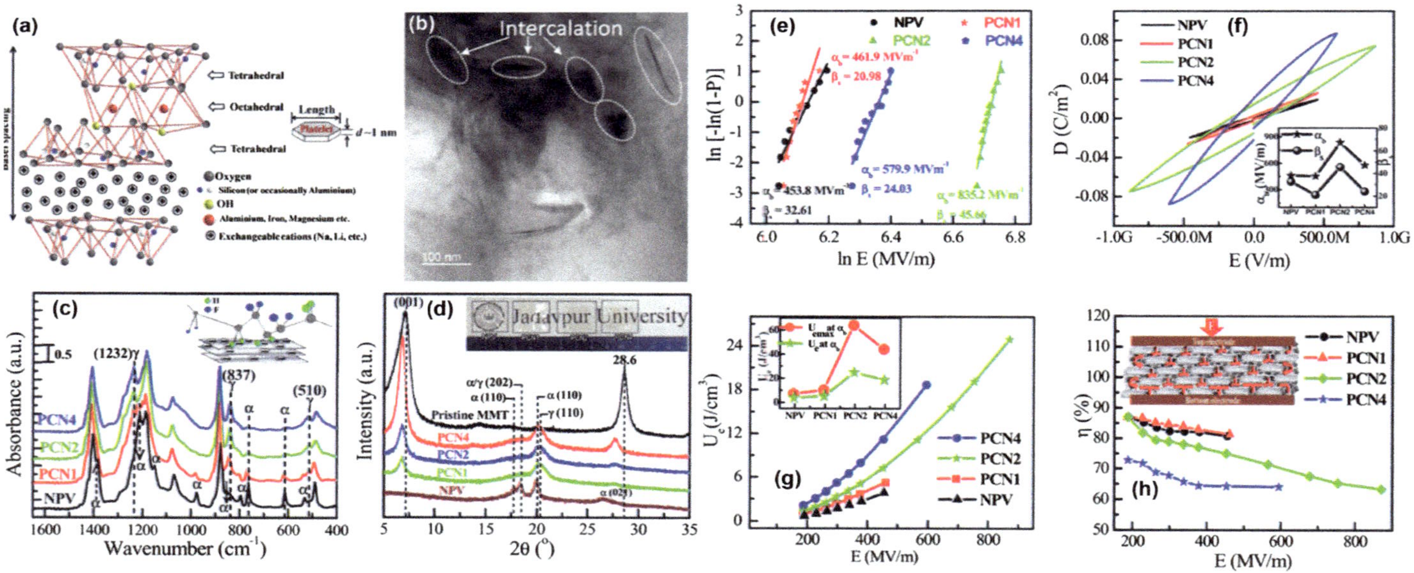

Figure 1.9 (a) Structure of MMT, 2 : 1 layered silicate showing two tetrahedral sheets of silicon oxide fused to an octahedral sheet of aluminium hydroxide and the platelet structure. (b) TEM image of the PCN1 film where intercalation is indicated by the marked (oval shaped) regions. (c) Fourier transform infrared spectra of the NPV and PCN films within the 1600–400 cm^{-1} frequency region (curves are offset for clarity). (d) Wide-angle XRD patterns of pristine MMT clay, NPV, and PCN films in the range $2\theta \sim 5$–35°. (e) Cumulative probability distribution as a function of applied electric field to statistically determine the breakdown strength (α_b) and scatter parameter (β_s) of NPV and PCN films. (f) Bipolar D–E loops of the NPV and PCN films under several applied fields at 100 Hz. The inset shows the variation of Weibull breakdown strength (α_b) and shape parameters (β_s) of the NPV and PCN films. (g) Measured discharged energy density (U_e) of the NPV and PCN films as a function of electric field. (h) Charge–discharge efficiency (η%) of the NPV and PCN films as a function of the applied electric field. In the inset of (h), the schematic of the device under high E represents the tortuous paths formed by MMT clay nanoplatelets. Reproduced from ref. 61 with permission from IOP Science.

efficiency even at higher electric field. This environmentally friendly MMT would not only simplify the fabrication of the PVDF/clay nanocomposites by eliminating the chemical modification step, but also enhance the breakdown strength and energy density of ferroelectric polymers as high energy density capacitors for green energy technologies.

1.3.3.2 Semiconductive Filler Polymer Systems

MoS_2 nanosheets, as an emerging class of transition metal dichalcogenides with an excellent on/off ratio of 10^8 and high carrier mobility of 200 $cm^2 V^{-1} s^{-1}$ at room temperature, have attracted considerable interest in recent years due to their outstanding thermal and mechanical properties, as well as attractive photocatalytic and electrical properties.[16] These characteristics enable them to be applied to transistors in lubricants, photoactive materials, and thin films. In particular, MoS_2 is also a semiconductor with an appreciable band gap of 1.8 eV, high specific surface area, and good chemical stability, and it has triggered increasing interest as a promising dielectric material for adjusting permittivity and electric field.

Cheng *et al.*[62] investigated the effects of MoS_2 nanosheets on the dielectric properties and breakdown strength of polymer composites. First, by using the liquid-phase exfoliation method, MoS_2 nanosheets were produced from their bulk powder in triethanolamine (TEOA) solution and then incorporated into poly(amic acid) (PAA) solution to thermally imidise and yield MoS_2/PI nanocomposite films. With increasing MoS_2 content, the relative permittivity of the nanocomposite films showed obvious improvement, which can be attributed to the relatively larger relative permittivity of the MoS_2, as well as the interfacial polarisation. Meanwhile, the dielectric loss of the composite maintained low values below 0.02 in the frequency range of 100 Hz to 1 MHz. For breakdown strength performance, a small quantity of MoS_2 nanosheets is beneficial to achieve an elevated breakdown strength, because of its efficient conduction barrier effect, which limits the charge migration and hinders the growing electric tree. However, when the MoS_2 content increases further to 1.5 vol%, concentration of the local electric field caused by aggregate defects in the composites leads to deterioration of the breakdown strength. Therefore, the MoS_2/PI nanocomposite film with 1 vol% MoS_2 reached a high energy density of 3.35 J cm^{-3} at 395 MV m^{-1} and maintained the energy efficiency at over 80%. In addition, self-passivated aluminium flakes (AFs) were introduced to MoS_2/PVDF ternary composites to obtain MoS_2/AFs/PVDF by He *et al.*[16] Figure 1.10(a) shows that this type of composite is comprised of a network of nanocapacitors. The dielectric spectra as temperature and frequency of pristine PVDF, MoS_2/PVDF, and MoS_2/AFs/PVDF composites are shown in Figure 1.10(b–d), respectively. The permittivity of pure PVDF matrix exhibits a slow increase with increasing temperature at low frequency. For the MoS_2/PVDF composite, the shape of the curves in the composites with MoS_2 nanosheets is similar to those of PVDF, but the permittivity exhibits a sharp enhancement and strong frequency dependence over the whole

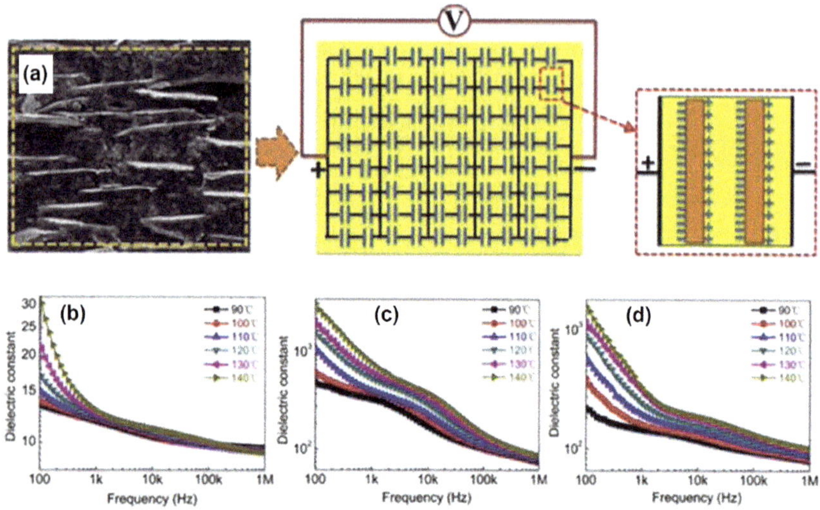

Figure 1.10 (a) Schematic of the formation of numerous nanocapacitors in the MoS_2/AFs/PVDF composites, and frequency dependence of the permittivity of (b) pristine PVDF, (c) MoS_2/PVDF, and (d) MoS_2/AFs/PVDF composites at 15 vol% MoS_2 loading in the temperature range 90–140 °C. Reproduced from ref. 16 with permission from Elsevier, Copyright 2018.

frequency range. However, the permittivity of the MoS_2/AFs/PVDF composite is obviously reduced with the addition of AFs, and the variation in the trend of permittivity of the three-phase composite is similar to that of the MoS_2/PVDF composite. It is speculated that the introduction of AFs successfully weakens the effect of the MoS_2-PVDF polarisation.

1.3.3.3 *Conductive Filler Polymer Systems*

Graphene is the most attractive 2D conductive filler in the study of percolation-type composites. Because of its excellent mechanical, electrical, and thermal conductivity, as well as large specific surface area (2630 $m^2 g^{-1}$), graphene is considered as an ideal filler for preparing high dielectric composites.[63,64] For example, graphene/polymer composites with a small volume fraction have excellent dielectric properties.

Due to its large specific surface area, numerous local microcapacitors can be constructed in the composite system to markedly improve the permittivity, while the percolation threshold can be reduced to 0.31 vol%, or even lower. For example, Tan *et al.*[65] prepared reduced graphene oxide (rGO) and dispersed it into a PVDF matrix to yield a rGO/PVDF dielectric composite. The experimentally measured percolation threshold of the composite was only 0.18 vol%. At the volume fraction of 0.177%, the relative permittivity and loss were 180 and 0.98 at 1 kHz, respectively. Conductive fillers such as graphene can effectively improve the permittivity and energy storage density of composites. However, during the preparation process it is necessary to control the number of fillers accurately. If too much filler is added, conductive paths are easily formed in the composites, resulting in larger leakage current and dielectric loss, thus reducing the breakdown strength of the composites. The reason for agglomeration of the filler is a strong van der Waals force between the graphene layers, and uniformly dispersing the conductive filler in the polymer matrix is a key requirement to improving the composite material. Therefore, the dispersibility can be improved by surface modification of graphene to increase the permittivity and reduce the dielectric loss.

Generally, chemical modification is used to improve the compatibility of the graphene sheet with the polymer matrix, such as introducing -OH, -COOH, or =O groups onto the surface of graphene. For example, Yang *et al.*[66] synthesised dopamine-coated rGO (PDA-rGO) and fluorinated functional group rGO (PF-PDA-rGO) to prepare dielectric composites. Compared with PDA-rGO obtained by traditional methods, the filler PF-PDA-rGO could be better dispersed in a ferroelectric polymer P(VDF-HFP) and has strong interfacial adhesion with the matrix, resulting in composites with a low percolation threshold and excellent flexibility. At a filler volume fraction of 10 vol%, the relative permittivity and loss of P(VDF-HFP)/PF-PDA-rGO composite are 107.9 and 0.07 at 1 kHz, respectively. Wen *et al.*[67] incorporated functionalised graphene in the double-bonded P(VDF-TrFE) copolymer. The covalent bonding effect between graphene and the matrix means that the filler is well dispersed and has a strong interfacial effect in the interface area. Thus, the relative permittivity of the composites reached 74 at 100 Hz, which is seven times higher than that of pure polymers. Tong *et al.*[68] produced a rGO/P(VDF-HFP) composite by the spin-coating method. Due to the good dispersibility and orientation distribution of graphene in the polymer matrix, at a rGO loading of only 0.7 vol% the relative permittivity of rGO/P(VDF-HFP) reached 54 and the dielectric loss was only 0.27. In addition, a three-phase co-filling system was employed to improve the properties of the composites. For example, Shen *et al.*[69] filled the PVDF with conductive rGO and ceramic BT to effectively improve the permittivity and breakdown strength of the composite. This co-filling system could overcome the shortage of leakage current caused by the conductivity of graphene itself, and at the same time reduce the amount of BT added in the matrix. Therefore, the combination of conductive material and ceramics as fillers and utilising the synergistic effect of the two phases is expected to produce high-performance dielectric materials.

1.4 The Design Strategy of 2D Nanocomposites for High Dielectric and Energy Storage Properties

1.4.1 Surface Functionalisation

Generally, poor dispersibility and aggregation of nanofillers in the polymer matrix tend to induce high dielectric loss and conductivity, resulting in a weakened breakdown strength and polarisation effect. Therefore, the interfacial compatibility between the polymer matrix and the nanofillers is crucial for high-performance composites. Interface design is achieved by introducing a series of interfacial layers to create a core–shell structure, including the use of inorganic, organic, and conductive outer shell layers on the nanofillers, for surface functionalisation.

Inorganic modifiers are often employed to enhance compatibility and dispersion of the filler in the polymer matrix. Commonly, several valve metal oxides, such as aluminium oxide (Al_2O_3), silica (SiO_2), and TiO_2, have been chosen as the shell layer coating on the nanofiller core–shell structured formation due to their high insulation and intermediate relative permittivity. For example, Chen *et al.*[70] have reported a novel nanocomposite film by incorporating the Bi_2Te_3@Al_2O_3 core–shell structure filler into a PVDF matrix, as shown in Figure 1.11. The schematic shows the accumulation of interfacial charges at the interfaces of Bi_2Te_3–Al_2O_3 and Al_2O_3–PVDF. As displayed, interfacial charges produced a Gouy–Chapman–Stern layer at the surface of the Bi_2Te_3, which is highly conductive and can be easily moved through by charges. The overlapped layers led to a large electrical leakage current and low breakdown field. It was the presence of the Al_2O_3 layer that formed the strong bonding with the polymer matrix, thus preventing the formation of a current channel of composites. As a result, the composite film loaded with 10 vol% 2D Bi_2Te_3@Al_2O_3 nanoplates exhibited a high relative permittivity of 140 and a relatively low dielectric loss of 0.05 at 1 kHz. Similarly, SiO_2 was subsequently employed as a filler shell to prepare composites, by Chen *et al.*[71] It was found that due to the highly insulating SiO_2 shell effectively reducing the concentration and mobility of the charge carriers, the alternating current conductivity of Bi_2Te_3@SiO_2/P(VDF-HFP) composite films is two orders of magnitude lower than that of Bi_2Te_3/P(VDF-HFP) composites.

Other novel nanocomposites are films comprising 2D core–shell $NaNbO_3$@Al_2O_3 platelets (2D NN@AO Ps) and poly(vinylidene fluoride-hexafluoropropylene) (P(VDF-HFP)), which were prepared by Pan *et al.*[72] The heterogeneous structure of the 2D NN@AO Ps was characterised by TEM, X-ray diffraction (XRD) patterns, and elemental mapping analyses, as shown in Figure 1.12, which suggest that a homogeneous and tight AO coating layer is obtained. The 2D NN@AO Ps/P(VDF-HFP) nanocomposite films have larger relative permittivity, ε_r, and lower tan δ compared to those of the 2D NN P/P(VDF-HFP) nanocomposite films. This could be explained by the following: (1) the highly insulating AO layer efficiently reduces the Maxwell–Wagner–Sillars (MWS) interfacial polarisation and space charge polarisation in the NN@AO P/P(VDF-HFP) nanocomposite films, thus improving the ε_r value; (2) the lower dielectric loss could be attributed to the reduction in the amorphous structural defects at the AO interface, resulting in enhanced microstructural uniformity of the nanocomposite film; and (3) the insulating AO layer can also effectively obstruct the current channels generated by the interfacial charge carriers. Finally, the NN@AO P/P(VDF-HFP) nanocomposite films yield the maximal energy density of 14.59 J cm^{-3} for an electric field of 440 kV mm^{-1}, as well as, simultaneously, an outstanding discharge efficiency of 70.1%.

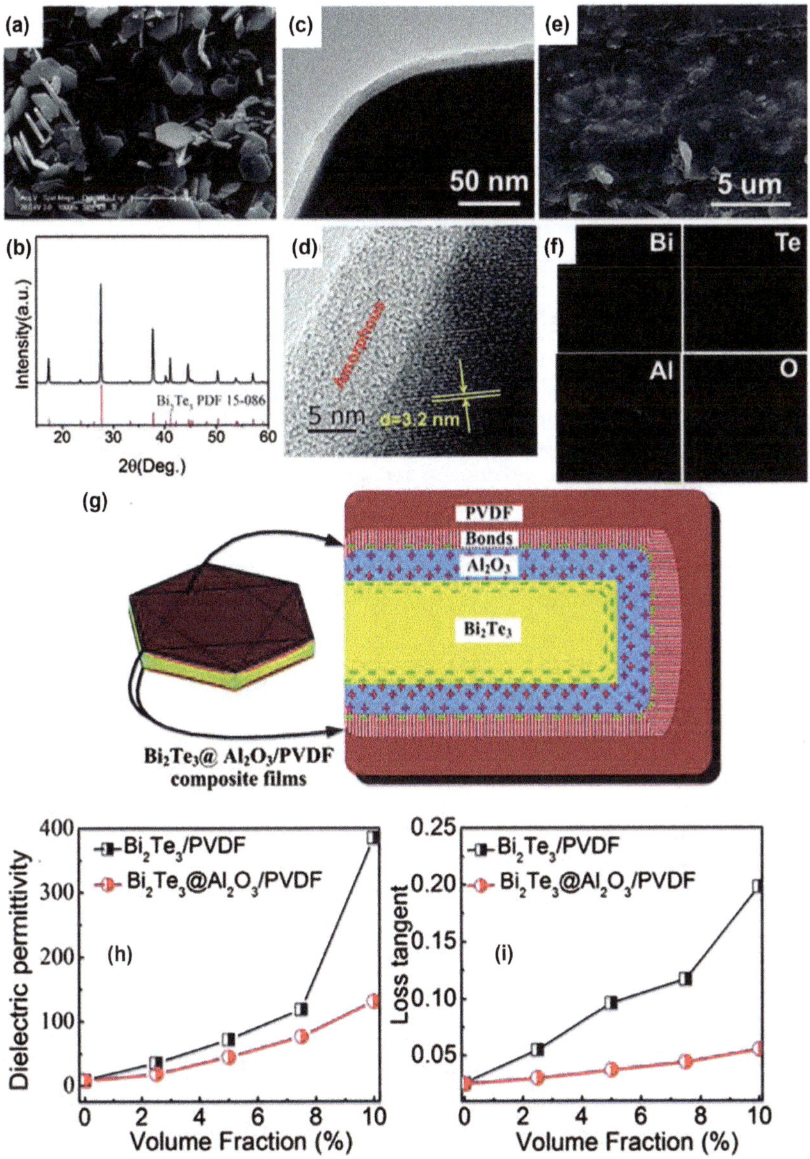

Figure 1.11 (a) Field emission SEM image and (b) XRD pattern of 2D Bi_2Te_3 nanoplates. (c) TEM image, (d) HRTEM image, and (f) element mapping of 2D Bi_2Te_3@Al_2O_3 nanoplates. (e) SEM image of a cross-section of 10 vol% Bi_2Te_3@Al_2O_3/PVDF composite film. (g) A schematic illustration of the accumulation of interfacial charges at the interfaces of Bi_2Te_3–Al_2O_3 and Al_2O_3–PVDF. (h) Permittivity and (i) dielectric loss of Bi2Te3@Al_2O_3/PVDF and Bi_2Te_3/PVDF composite films at 1 kHz.
Reproduced from ref. 70 with permission from the Royal Society of Chemistry.

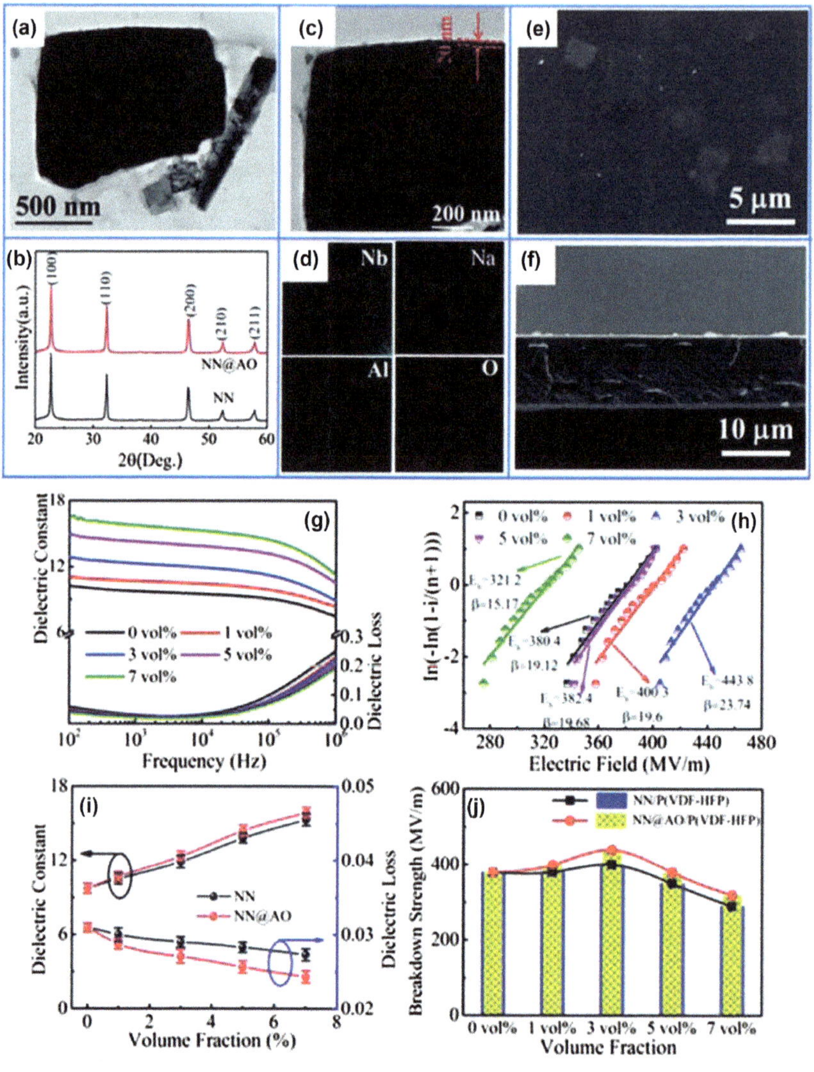

Figure 1.12 (a) TEM images, (c) magnified TEM images, and (d) elemental mappings of 2D NN@AO Ps. (b) XRD patterns of 2D NN and 2D NN@AO Ps. (e) Top view and (f) cross-sectional view of 3 vol% 2D NN@AO Ps/P(VDF-HFP) nanocomposite films. (g) Variations in the permittivity and loss with frequency for pure polymeric matrix and nanocomposite films with different contents of 2D NN@AO Ps. (h) Breakdown strength of nanocomposite films with 2D NN@AO Ps and 2D NN Ps. (i) Comparison of permittivity and loss at 10^3 Hz. (j) Weibull plots and determined characteristic breakdown strength of 2D NN@AO Ps/P(VDF-HFP) nanocomposite films.
Reproduced from ref. 72 with permission from the Royal Society of Chemistry.

Organic modifiers can be physically adsorbed onto the filler surface through electrostatic interactions, or by hydrogen bonding. A variety of modifiers have been used, which include dopamine, cetrimonium bromide, polyethyleneimine, poly(vinyl pyrrolidone), ethylene diamine, polyvinyl alcohol, paraffin, and so on. These organic compounds have been utilised to modify the surface of inorganic fillers and are of interest due to their simple treatment process, for example, by solution mixing. To reduce agglomeration and improve the poor dispersibility of filler in the polymer matrix, Wang *et al.*[51] choose cetrimonium bromide (CTAB) as the surfactant to modify the nanolaminate-shaped MMT clay. The long alkyl chain of CTAB could not only be intercalated into the interlayers of MMT to improve the dispersity of the nanolaminates, but also provide positive surface charge, which can tune the polarisation of negatively charged MMT nanolaminates. Thus, compared to the pristine PVDF film, the energy density was increased from 5.34 $J\,cm^{-3}$ to 5.91 $J\,cm^{-3}$ at 1 wt% MMT content and the maximum energy density could reach 10.2 $J\,cm^{-3}$. In addition, Wang *et al.*[73] treated MoS_2 nanosheets by the mussel-inspired co-modification of poly(dopamine) (PDA) and polyethyleneimine (PEI). Due to the formation of covalent bonds between PDA and PEI, a stable and uniform coating layer with rich polar groups ($-NH_2$ and –OH groups) was obtained. In addition, improvement of the interfacial compatibility between MoS_2 nanosheets and the poly(arylene ether nitrile) (PEN) matrix can be achieved by strong interfacial interactions, as shown in Figure 1.13. As a consequence, the MoS_2/PEN composites show excellent mechanical strength and thermal stability, high permittivity, low dielectric loss, and good frequency stability, and can be potentially used as high-performance dielectric materials.

Zhi *et al.*[74] developed a facile and efficient approach to simultaneously functionalise and tune the reduction state of graphene oxide (GO) with c-aminopropyl triethoxysilane (APTES), then incorporated APTES-functionalised GO sheets (GO-APTES) into nitrile butadiene rubber (NBR) by latex co-coagulation, to form GO-APTES/NBR composites, as in Figure 1.14. In the system of composites, APTES acted as a coupling agent, while its amine groups interact with the functional groups on the GO sheets to remove many surface groups of the GO sheets to increase the conductive areas between the insulating barriers. The results showed that the high aspect ratio of GO-APTES sheets leads to an increase in the permittivity of the composite, while the intrinsic barriers of GO limit the leakage current and avoid a high dielectric loss. The GO-APTES/NBR composites exhibit a relatively high permittivity of 30.8 and small loss factor of 0.04 at 1 kHz, as well as a good insulating property.

1.4.2 Orientation of the Fillers

Recently, many investigations have indicated that filler orientation and spatial distribution play an important role in determining and improving the energy storage capabilities of nanocomposites. Tomer *et al.*[75] prepared

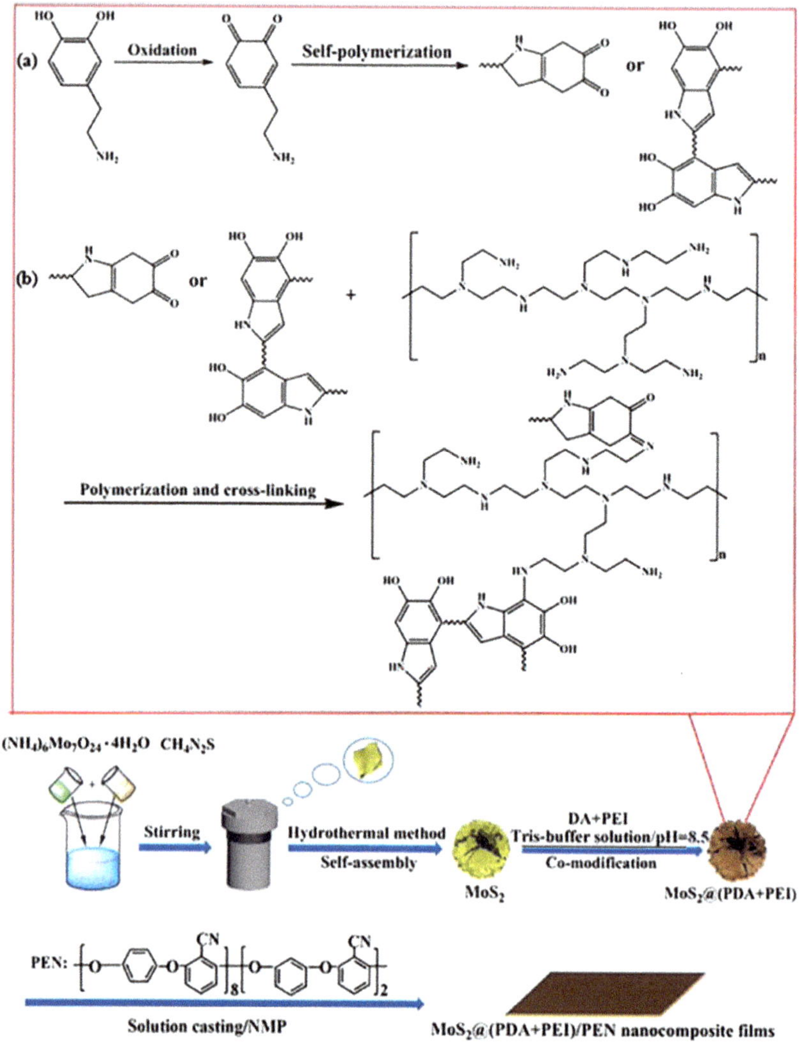

Figure 1.13 Preparation process of functionalised MoS_2/PEN nanocomposite films and possible chemical reaction between PDA and PEI.
Reproduced from ref. 73 with permission from Elsevier, Copyright 2019.

polyethylene/montmorillonite nanocomposites by an extrusion blow moulding and hot-pressing process method. The effect of the spatial arrangement of the high aspect ratio nanofillers on the dielectric and breakdown behaviour were investigated. The nanocomposite with randomly oriented fillers showed markedly higher losses (broader D–E loop) compared to the nanocomposite containing oriented fillers, which can be explained by the fact that oriented-filler composites are expected to provide more ordered trapping centres and more efficient scattering for the injected charge, thus

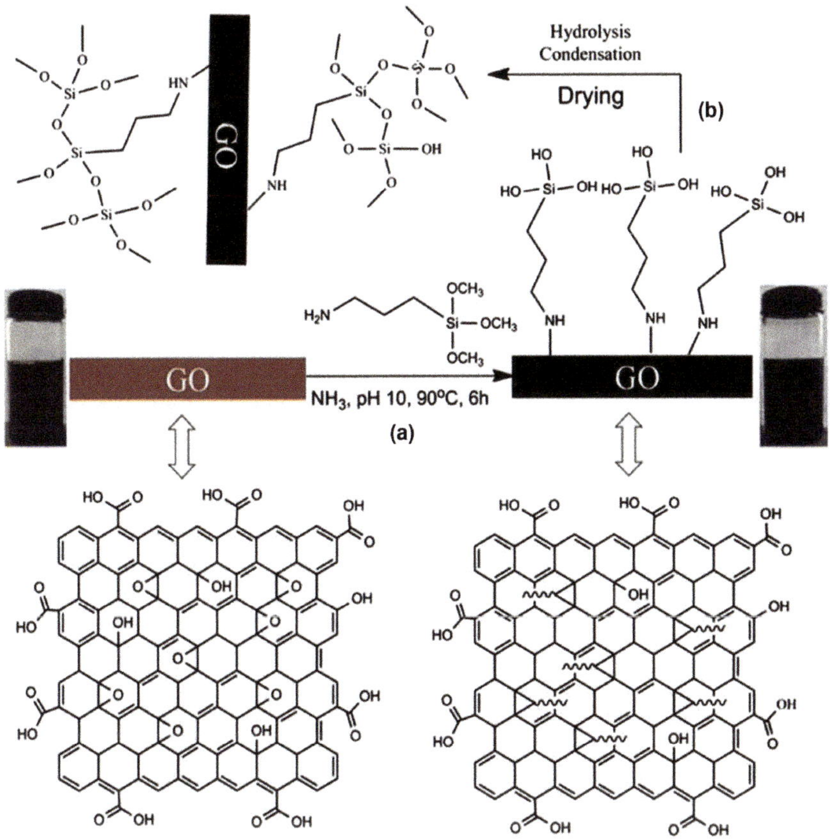

Figure 1.14 (a) Schematic of reaction of GO with APTES. (b) Hydrolysis condensation between different GO-APTES sheets to form a three-dimensional disordered structure.
Reproduced from ref. 74 with permission from Elsevier, Copyright 2015.

obstructing its ability to traverse the sample to the opposite electrode. Nevertheless, nanocomposites with no filler orientation had more losses at high electric fields due to the lower obstruction effect. In addition, Weibull statistics methods were used to analyse the electric breakdown strength of the composite, and composites with oriented fillers showed constantly larger electric breakdown strength than both the unfilled polymer and composites with random fillers. First, the array distribution of the filler can impede the progress of electric trees, since it needs a more tortuous path for electric trees to penetrate throughout the samples, causing the occurrence of breakdown. Second, oriented fillers can be used as structured scattering centres for conducting electrons, resulting in weakening of kinetic energies. More importantly, considering both the displacement and high electric field behaviour, nanocomposites with aligned fillers showed a marked improvement in recoverable energy density and the energy efficiency of the polymer.

Wang *et al.*[76] synthesised plate-like $(Ba_{0.6}Sr_{0.4})TiO_3$ (P-BST) *via* a topochemical microcrystal conversion, and fabricated P-BST/PVDF textured composites using a tape-casting and hot-pressing method, as shown in Figure 1.15. It was found that the particles with larger size are more sensitive to the shear force generated during tape-casting and to be textured along the casting direction. The textured arrangement enhanced the uniform dispersion of these particles in the matrix, while reducing the structural defects. In addition, BST microcapacitors are also easily connected in series and in parallel to form the capacitance in textured composites. As a result, P-BST/PVDF textured composites exhibited a higher relative permittivity of 62.2 and lower dielectric loss of 0.042 compared with the composite with a disordered array of particles. The maximum energy density of the composite, of 6.36 J cm^{-3}, was obtained at an electric field of 29 kV mm^{-1}.

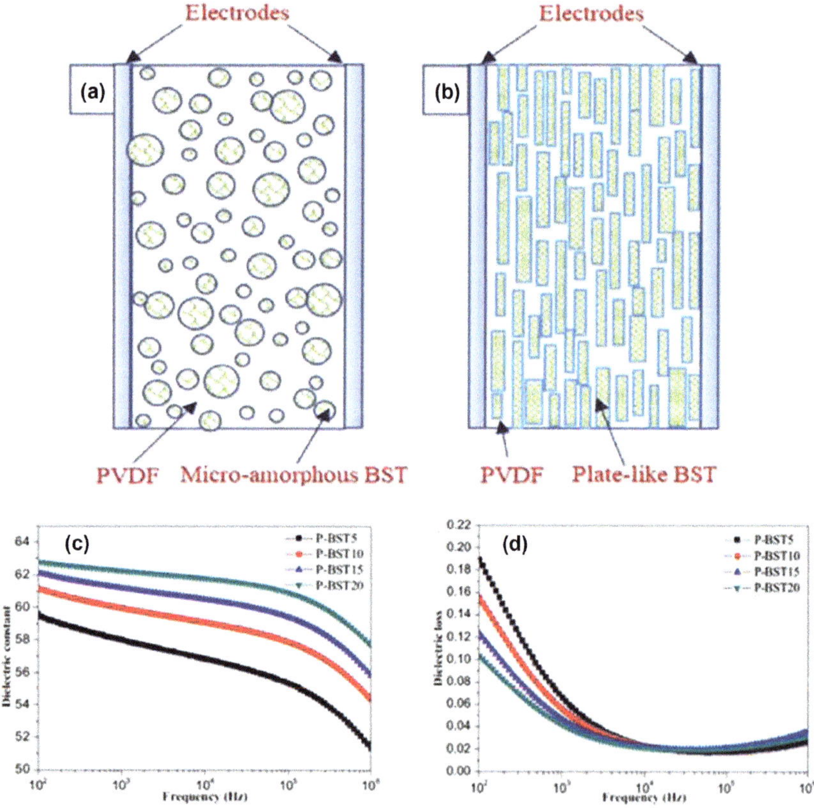

Figure 1.15 (a, b) Schematic showing particle layout of the (a) micro-irregular BST/PVDF and (b) P-BST/PVDF textured composites. (c, d) Frequency dependence of the dielectric properties measured for 40 vol% P-BST/PVDF textured composites with different P-BST particle size for (c) permittivity and (d) dielectric loss.
Reproduced from ref. 76 with permission from Elsevier, Copyright 2018.

1.4.3 Multilayer Structure Design

Sandwich or multilayer structure design is an effective strategy to synchronously enhance permittivity and electric breakdown strength so as to achieve high energy density of the polymer nanocomposites.[77] As shown in Figure 1.16, Pan *et al.*[57] synthesised 2D NN platelets by a two-step melt-salt process, and then designed a trilayered architecture composite with high-permittivity NN/PVDF as the two outer layers and high breakdown strength pristine PVDF as the middle layer. The results showed that the breakdown strength of the trilayered architecture composite films could be improved significantly compared with the single-layer composite films. For example, the 5-0-5 trilayered architecture composite films have the highest breakdown

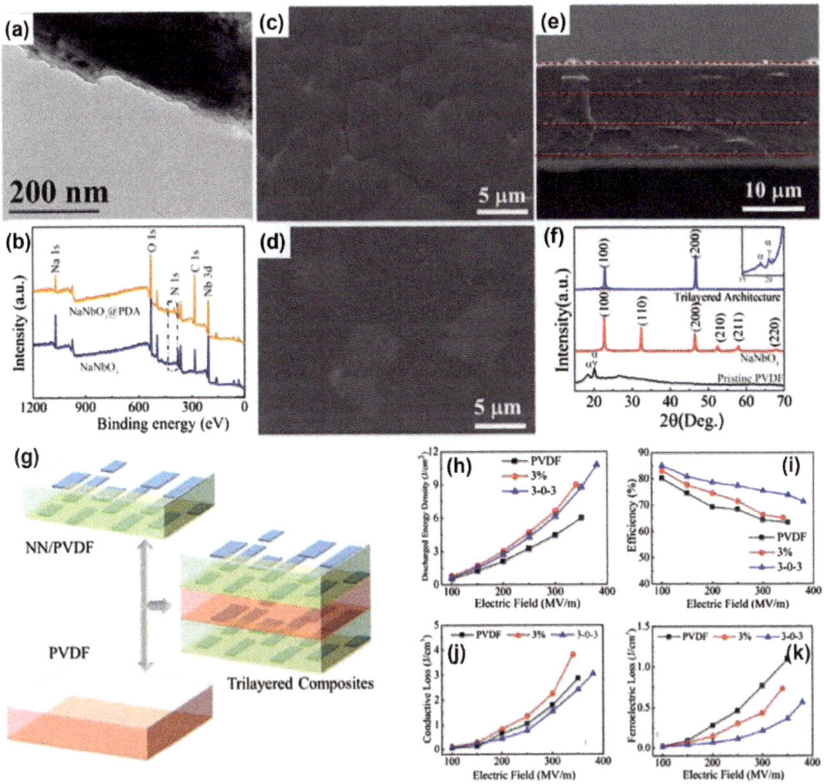

Figure 1.16 (a) TEM images of 2D NN@PDA platelets. (b) XPS patterns of 2D NN and 2D NN@PDA platelets and top view of (c) middle layer and (d) outer layer. (e) Cross-section of 3-0-3 trilayered architecture composite film. (f) XRD patterns of pristine PVDF, 2D NN platelets, and 3-0-3 trilayered architecture composite film. (g) Fabrication of trilayered architecture composite films. (h) Discharged energy density, (i) efficiency, (j) conductive loss, and (k) ferroelectric loss of pristine PVDF, 3 vol% NN/PVDF, and 3-0-3 composite film.
Reproduced from ref. 57 with permission from Elsevier, Copyright 2017.

strength of 400 MV m^{-1}, which is 14.3% higher than pristine PVDF (350 MV m^{-1}) and 25% higher than the corresponding single-layer composite film (300 MV m^{-1}). This can be explained by the following. First, the electric field redistribution caused by the incorporation of high-permittivity 2D NN platelets into the outer layers, and a weak electric field region is formed in the outer layers, so that the trilayered architecture composite films could tolerate a higher electric field than single-layer composite films at the same applied electric field. In addition, the strong interfacial barrier effect between the outer layer and middle layer also enhances breakdown strength in trilayered architecture composite films. Finally, due to the synergistically enhanced breakdown strength and electric displacement, the 5-0-5 composite film exhibited a higher discharge energy density of 13.5 J cm^{-3} and energy efficiency of 66.9% than the corresponding single-layer composite film.

As shown in Figure 1.17, 2D NBT-BT platelets coated with polyvinylpyrrolidone (PVP) were introduced as fillers for the preparation of NBT-BT@PVP/P(VDF-HFP) composites.[78] The multilayer films were designed with 1 vol% NBT-BT loadings as the central hard layers, and 30 vol% NBT-BT loadings as neighbouring soft layers. The five-layered (30-1-1-1-30) nanocomposites showed a maximum energy storage density of 14.95 J cm^{-3} at 258 kV mm^{-1}. The energy efficiency remained at 90% at 200 kV mm^{-1}. This excellent performance was attributed to the combination of the high relative permittivity and high dielectric strength of the multilayer structure. The former derives from the top and bottom soft layers (30 vol% NBT-BT loading), while the latter mainly comes from the central hard layers (1 vol% NBT-BT loading). The central hard layers act as electron traps that limit electron tunnelling through the film, resulting in improved breakdown strength.

Li *et al.*[79] developed a high-temperature dielectric polymer nanocomposite consisting of a sandwich structure, by sequentially casting three layers of c-BCB/(BNNSs), c-BCB/BT, and c-BCB/BNNSs, respectively, with a thickness ratio of 1 : 2 : 1 (c-BCB represents cross-linked divinyltetramethyldisiloxane-bis(benzocyclobutene)). Compared with a conventional single-layer configuration, this nanocomposite is capable of integrating the complementary properties of spatially organised multicomponents in a synergistic fashion to raise permittivity, and subsequently improve discharge energy densities, while retaining low loss and high charge–discharge efficiency at elevated temperatures. For example, an energy density of 1.1 J cm^{-3} with energy efficiency of 93% have been realised in a sandwich-structured polymer nanocomposite at 150 °C and 200 MV m^{-1}.

1.5 Interfacial Models and Electric Simulations

The dielectric properties of polymer-based nanocomposites with inorganic nanofillers are dependent not only on the intrinsic properties of polymers and nanofillers, but also on the properties of the interfacial regions. Moreover, the volume fraction of the interface becomes more significant with a decrease in the

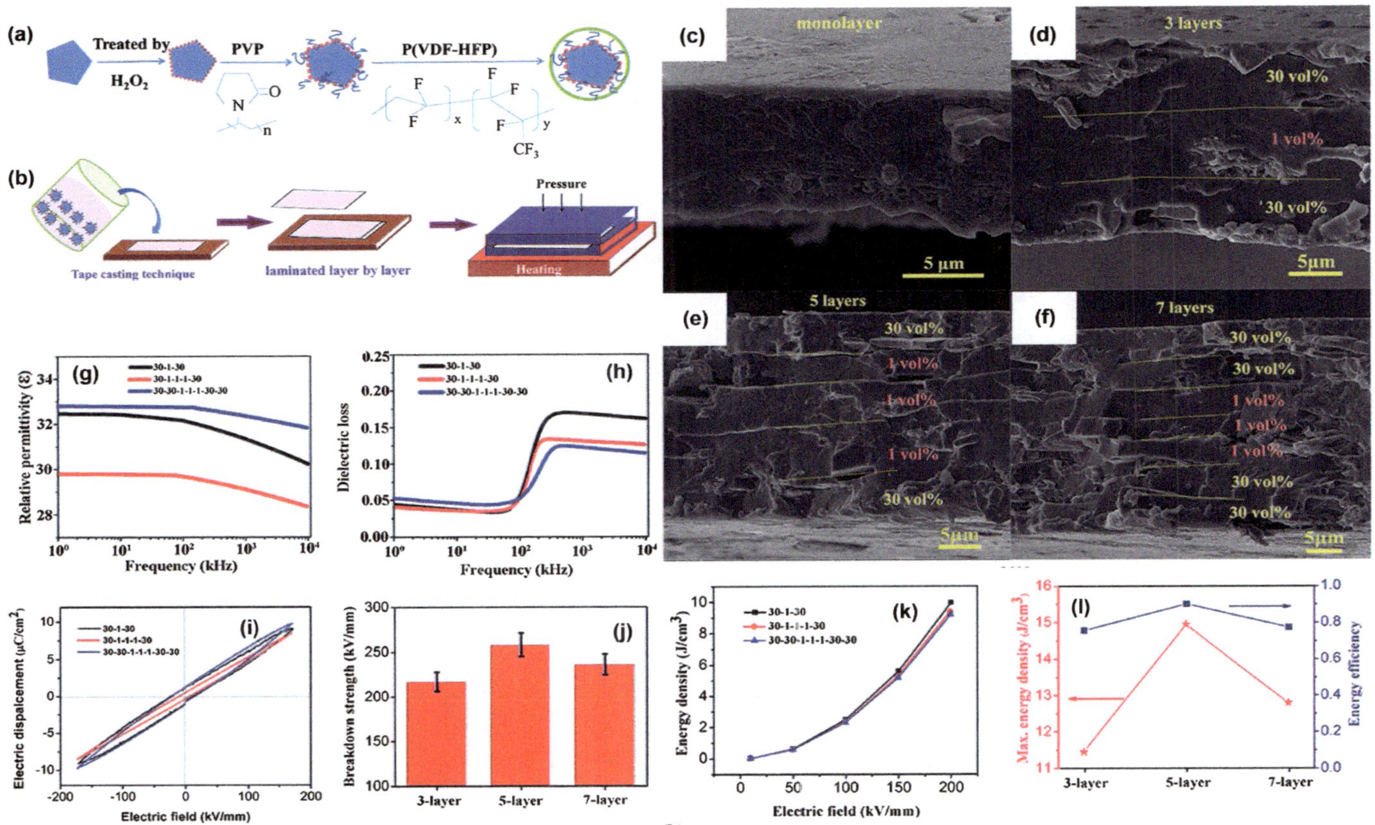

Figure 1.17 (a) Schematic of the modification process of the NBT-BT platelets and (b) the fabrication process of PVP@ NBT-BT/P(VDF-HFP) composites and the sandwich-structured multilayer films. (c–f) Cross-sectional SEM images of an NBT-BT@PVP/P(VDF-HFP) monolayer composite. Frequency dependence of (g) relative permittivity and (h) dielectric loss. (i) *D–E* loops and (j) the breakdown strength for multilayer NBT-BT@PVP/P(VDF-HFP) films. The dependence of the energy densities on (k) varying electric field and (l) variation of maximum energy density and energy efficiency at 200 kV mm^{-1} with layering.
Reproduced from ref. 78 with permission from the Royal Society of Chemistry.

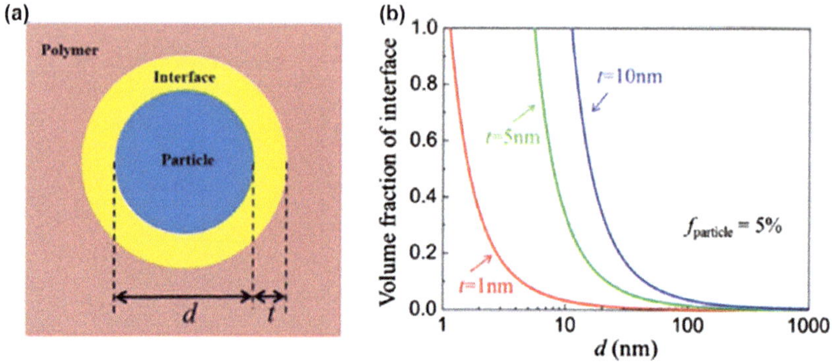

Figure 1.18 (a) Schematic of the ceramic/polymer interface structure in a nanocomposite. (b) Variation in volume fraction of interface in the nanocomposites with the diameter of nanoparticles and interface thickness. Reproduced from ref. 35 with permission from the Royal Society of Chemistry.

filler size. For example, the volume fraction of the interfacial region ($f_{\text{interface}}$) of a spherical nanocomposite filler can be calculated by eqn (1.1):

$$f_{\text{interface}} = f_{\text{particle}}\left[\left(1+\frac{2t}{d}\right)^{3}-1\right] \tag{1.1}$$

where t is the thickness of the interfacial region, d is the diameter of the nanoparticle, and f_{particle} is the volume fraction of the filler.[80] As presented in Figure 1.18(b), when f_{particle} is fixed at 5% and t is fixed at 5 nm, the interface volume fraction ($f_{\text{interface}}$) is less than 10% for nanoparticles with $d = 10$ nm, but the $f_{\text{interface}}$ dramatically increases to over 50% for nanoparticles with $d < 10$ nm. In these regions, the chemical and physical properties are different to the nanofillers and polymer matrix, such as polymer chain mobility, chain confirmation, crystallinity, and coulombic potential. Therefore, the polymer-based nanocomposites contain three phases: continuous phase of basic polymer, dispersed phase of inorganic filler, and interfacial phase (see Figure 1.18(a)). To understand the effect of interfaces in polymer nanocomposites, several models and theories concerning the interfacial regions have been hypothesised.[81,82] In this section, the effects of interfaces on the dielectric properties are discussed.

1.5.1 Tanaka's Multicore Model

A multilayered core model resulting from the 'interaction zones' was proposed by Tanaka and is displayed in Figure 1.19(a).[83] This model is constructed on the basis of both chemical and electrical analysis, but it is still a working hypothesis. As can be seen from Figure 1.19(a), the interfacial layer consists of three layers, including a bonded layer, a bound layer, and a loose layer. The first layer corresponds to the bonded layer. Usually, inorganic fillers are

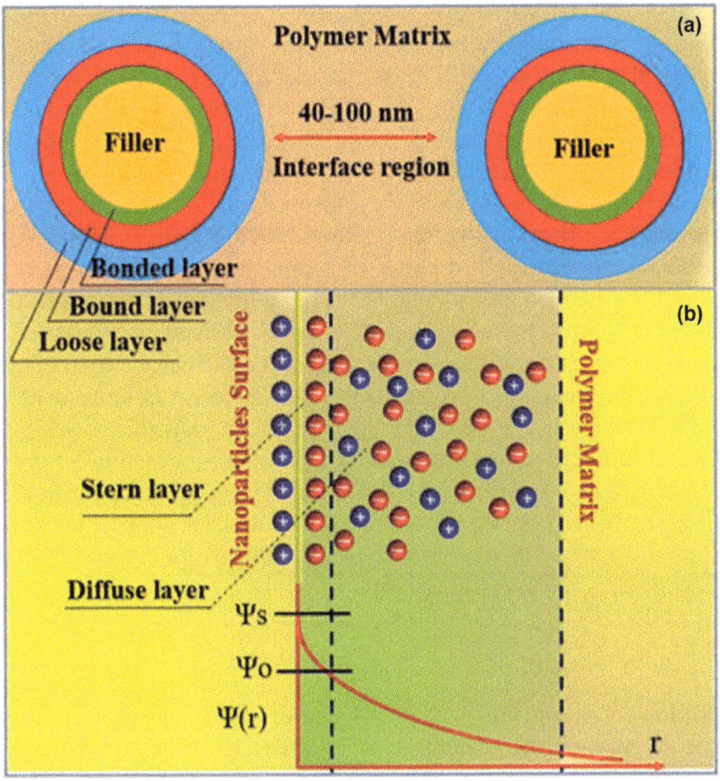

Figure 1.19 (a) Multicore model for polymer nanocomposites. (b) Charge distribution of a diffuse electrical double layer.
Reproduced from ref. 35 with permission from the Royal Society of Chemistry.

difficult to fully disperse in the organic polymer matrices due to surface tension between the two phases. In order to reduce the difference in surface tension, a silane coupling agent is widely used to modify the surface of the nanoparticles, resulting in a bonded layer being formed *via* hydrogen bonding between the polymer and silane coupling. The total thickness of the bonded layer is approximately 1 nm. The second layer is a bound layer in which a layer of polymer chains is strongly bound and/or interacts with the first layer and the surface of the inorganic particle. The thickness of this layer ranges from 2 to 9 nm, depending on the strength of the polymer–particle interaction. The third layer, denoted as a loose layer, is a region that loosely couples and interacts with the bound layer. Compared with the bonded layer and bound layer, the thickness of the loose layer can extend to several tens of nanometres. In this loose layer, the polymer produces different chain conformation, chain mobility, free volume, or crystallinity compared with the polymer matrix.

Although the multilayered core model is a working hypothesis that is not fully verified, there are many dielectric and electrical insulation phenomena that can be interpreted with it, such as permittivity, dielectric loss, conduction, space charge, thermally stimulated current, electroluminescence, photoluminescence, dielectric breakdown strength, electrical treeing, partial discharge resistance, tracking, glass transition temperature, and so on. For example, Tanaka explained the PD (partial discharge) resistance of polyamide-layered silicate nanocomposites according to the multicore model.[83] Tanaka considered that the third layer was the most susceptible to PD compared with the other two, resulting in PD degradation starting from the loose layer. As soon as the PD erodes the third layer, the second and the first would resist the extension of the PD, leading to that PD activity invading the weaker regions, such as the polymer region or the third layer of an adjacent nanoparticle. Hence, a zig-zag path was formed around the nanoparticles due to PD activity. In this way, the polymer nanocomposites exhibited an enhanced PD resistance. With regards to the permittivity of polymer nanocomposites, Tanaka considered that the first and second layers in the multicore model played the critical role of reduced permittivity.[84] The third layer played a positive role in increasing permittivity because it contained dipoles and ion carriers and larger free volume. Moreover, when the third layer trapped ions, the dielectric loss would become large at low frequency.

1.5.2 Lewis' Model

The addition of nanofillers in the polymer matrix produces numerous interfaces in nanometric dimensions. In order to understand these interfacial electrical and dielectric behaviours, Lewis proposed the diffuse electrical double layer model in the polymer nanocomposites, and the model is presented in Figure 1.19(b).[85,86] It can be seen that the surface of the nanofillers will be charged due to the difference in chemical potential or Fermi levels of the polymer matrix and the nanoparticles. In addition, the matrix develops counter charges near the surface of the nanoparticles, resulting in suppression of charge accumulation on the nanoparticles. For simplicity, in the Lewis model, the nanofillers possess a positively charged surface and the surface is considered to be planar; see Figure 1.19(b). For counter charges, there are two mechanisms to consider. One is that they refer to the induced polarisation of the polymer matrix, including the electronic polarisation and orientation of permanent dipoles, which usually occurs when a polymer contains polar components. Another is related to migration and redistribution of mobile charges on the polymer side due to coulombic attraction, leading to the formation of an electrical double layer consisting of a Stern layer and a Gouy–Chapman diffused layer (see Figure 1.19(b)). Due to the positively charged surface, a positive potential ψ_s is exhibited around the nanoparticles, which involves immobile charged impurities, trapped carriers, mobile electrons, and holes in the nanoparticle. Close to the positive layer, the negatively charged counter charges or dipoles are tightly absorbed to the particle surface to form the so-called Stern or Helmholtz double layer, which contains small molecules, special absorbed ions, and solvated ions, and cannot move freely. Outside this layer is the Helmholtz plane (OHP) with an electrical potential of ψ_o. The thickness of the Helmholtz double layer is approximately 1 nm.

The Gouy–Chapman diffused layer is formed outside and around the Stern layer by the distribution of negative and positive ions. The thickness of this layer is affected by the ion concentration in the polymer, as well as the potential difference between the polymer matrix and the nanoparticle surface. If the polymer matrix is highly conductive, a number of charge carriers concentrate at the polymer matrix region, leading to the diffuse layer shrinking into the OHP. However, if the polymer matrix is insulating, the diffuse layer may extend 10 nm or more. More importantly, this diffused layer potentially works as an 'interaction zone' to determine the dispersion of nanoparticles. The dielectric and electrical properties of the nanocomposite relate to the contained mobile charges, especially near the percolation threshold of the nanofillers (see Figure 19(b)). The distribution of charges in this layer is related to the electrical potential $\psi(r)$ across the interface region. The value of $\psi(r)$ is changed by the distance (r) from the particle surface and can be described with a combined Poisson–Boltzmann equation, and the $\psi(r)$ function is shown as follows:[85]

$$\nabla^2\psi(r) = -e\varepsilon^{-1}\sum_i z_i n_i(\infty)e^{-z_i e\psi(r)/kT} \tag{1.2}$$

where ε is the permittivity of the medium, k is the Boltzmann constant, and z_i and $n_i(\infty)$ are the ion valency and concentration of ion species i in the bulk matrix, respectively. When the potential is small, the $\psi(r)$ can be reduced further to the simple Debye–Hückel form and is shown below:

$$\psi(r) = \psi_0 e^{-kr} \tag{1.3}$$

$$k = \left(\frac{2e^2}{\varepsilon kT}\sum_i z_i^2 n_i(\infty)\right)^{1/2} \tag{1.4}$$

in which, κ is the Debye–Hückel parameter and has units of m^{-1}. The inverse of κ is called the Debye length and defines the extent of the exponential decay of the double layer. This equation indicates the potential variation in the diffused part of the double layer starting from the Stern layer.

The charge density (ρ_i) in the double layer is expressed as follows:[3]

$$\rho_i(r) = \rho_i(\infty)e^{-zie\psi(r)/kT} \tag{1.5}$$

At the nanoparticle surface $(r = 0)$, the total charge density is

$$\sum_i \rho_i(0) = \sum_i \rho_i(\infty) + \frac{\sigma^2}{2\varepsilon kT} \tag{1.6}$$

and is associated with the surface conductivity (σ) of the nanoparticles. The equation implies that ρ_i in the double layers can be increased be improving σ,

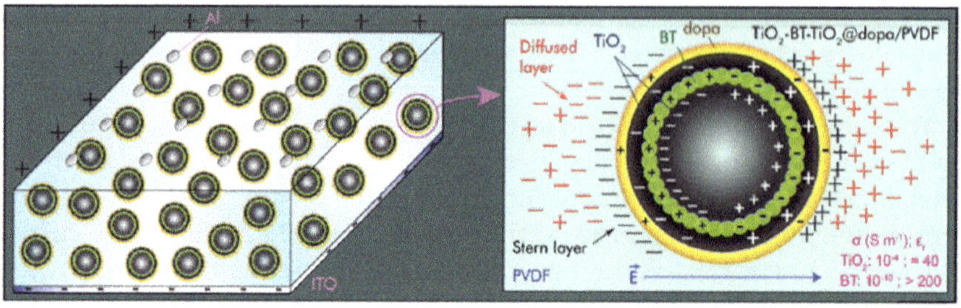

Figure 1.20 Schematic of the TiO_2–BT–TiO_2@dopa/PVDF device (left) and enlarged view of the nanomaterial interface (right) under the application of an electric field.
Reproduced from ref. 87 with permission from American Chemical Society, Copyright 2019.

which is useful to induce polarisation at opposite ends of the nanoparticles under an applied field E because the charges at the interfaces are efficiently transferred.

Based on Lewis' model, Gupta *et al.* applied polydopamine (dopa)-functionalised TiO_2–BT–TiO_2 (TiO_2–BT–TiO_2@dopa) core@double-shell nanoparticles in PVDF for high energy density capacitor applications.[87] The composites showed a higher relative dielectric permittivity, breakdown strength, and energy density, which is accounted for by the polarisation at the interfaces inside the nanoparticles, as well as space charge accumulation at the nanoparticle interfaces. Due to the Fermi level difference between the n-type semiconductor TiO_2 and BT (>0.5 eV), the accumulation of positive and negative space charges is focused at the surface of TiO_2 and BT (see Figure 1.20). In addition, in order to maintain charge neutrality, the surface charge densities (σ') in different layers of TiO_2–BT–TiO_2 nanoparticles follow the trend of $\sigma'(TiO_2)$ core > σ'(BT) middle layer > $\sigma'(TiO_2)$. The dopamine layer, in turn, responds by developing negative charges due to polar interaction with the positive TiO_2 layer. The interfacial charges present in the polymer matrix possessed a Gouy–Chapman–Stern layer at the interface of the TiO_2–BT–TiO_2 nanoparticles, resulting in an enhanced interfacial polarisation. Consequently, the PVDF-based nanocomposites displayed a high relative dielectric permittivity. In another case of core–shell BT@TiO_2 nanoparticle-filled P(VDF-HFP) composites, a large relative dielectric permittivity (>110) was obtained, which was more than three times higher than that of the P(VDF-HFP)/BT nanocomposites.[88] The improvement in the dielectric property was associated with the Maxwell–Wagner interfacial polarisation as a result of space charge accumulation and formation of Gouy–Chapman–Stern layers at the highly interactive interfaces among the multiple dielectric materials (see Figure 1.21). Pan *et al.* studied the dielectric behaviour of the Bi_2Te_3/PVDF and Bi_2Te_3@Al_2O_3/PVDF nanocomposites.[89] Compared with the Bi_2Te_3/PVDF nanocomposite, the Bi_2Te_3@Al_2O_3/PVDF composite film showed larger

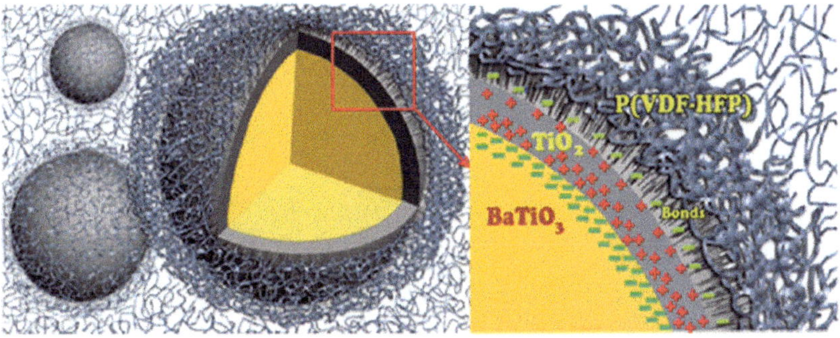

Figure 1.21 Schematic for the microstructure of the BT@TiO_2 core–shell nanoparticles in the P(VDF-HFP) polymer matrix. Enlarged view of the area marked by the red square (right), showing space charge in BT, TiO_2, and P(VDF-HFP). The charges in this illustration are bonded charges, and the charges of opposite polarity in the layers are not shown. Reproduced from ref. 88 with permission from the PCCP Owner Societies.

breakdown strength and lower dielectric loss. This is believed to be due to the reduced space charge polarisation by thc highly insulating Al_2O_3 shell layer, and thus decreased thickness of Gouy–Chapman–Stern layer. In the Bi_2Te_3@Al_2O_3/PVDF composite systems, interfacial charges concentrated at the interfaces of Bi_2Te_3–Al_2O_3 and Al_2O_3–PVDF. These interfacial charges would form a Gouy–Chapman–Stern layer at the surface of Bi_2Te_3. When the highly conductive Gouy–Chapman–Stern layers of nanoparticles overlap, free charge movement may happen over a distance, leading to a large electrical leakage current and low breakdown field. The decrease in thickness of the Gouy–Chapman–Stern layer was favourable to prevent the formation of a current channel. Moreover, the strong bonding between the Al_2O_3 shells and the polymer matrix results in uniform dispersion of the core–shell nanoparticles, further restraining the formation of a current channel. Lei *et al.* reported the dielectric behaviour of a sandwich-structured PVDF-based composite, which has upper and lower layers filled with $0.5Ba(Zr_{0.2}Ti_{0.8})O_3$–$0.5(Ba_{0.7}Ca_{0.3})TiO_3$ nanofibres and a middle layer filled with hybrid particles of hexagonal BNNSs coated by ferroferric oxide (Fe_3O_4@BNNSs).[77] The upper and lower layers can possess good ferroelectric hysteresis, and the middle layer contributes to high breakdown strength. In particular, the hybrid Fe_3O_4@BNNSs also enhanced the electric displacement and suppressed the remnant displacement of sandwiched nanocomposites. With the help of Lewis' model, the authors considered that the increase in the dielectric properties of the PVDF-based composites originated from the interfacial polarisation, which may arise from the diffusion layer of the semiconductor–insulator structure at the interfaces between the insulating BNNSs and the semiconducting Fe_3O_4 (see Figure 1.22). Overall, some of the interfacial issues in the polymer-based nanocomposites can be explained by means of Lewis' model.

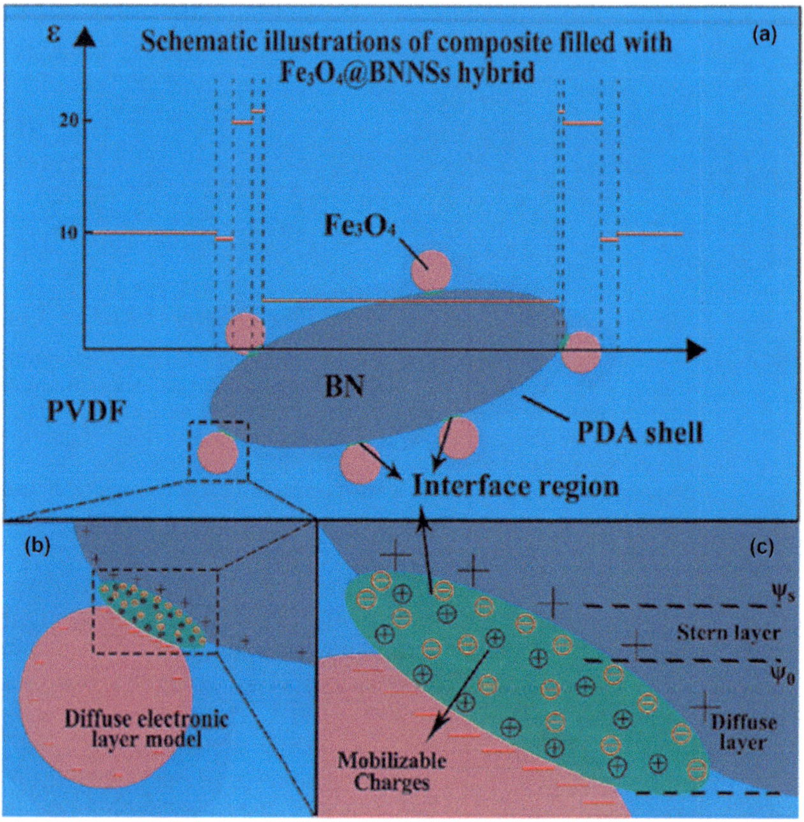

Figure 1.22 Schematic of composites filled with Fe_3O_4@BNNSs hybrid. (a, b) Distribution of dielectric permittivity was superimposed in different regions of the composite. Abrupt increase of dielectric permittivity indicated the enhanced interfacial polarisation in the interfacial regions. Interfacial region in (c) is magnified to show the diffuse electronic layer of the semiconductor–insulator structure.
Reproduced from ref. 77 with permission from The American Chemical Society, Copyright 2018.

1.5.3 Finite Element Simulation and Phase-field Simulation

1.5.3.1 Finite Element Simulation

To understand the polarisation mechanism and breakdown behaviour of polymer-based nanocomposites, finite element simulation offers a simple and effective way to calculate the distribution of permittivity, polarisation, and electric field.[26,90,91] In the case of P(VDF-HFP)/TiO_2 nanocomposites containing different shapes of TiO_2 particles,[8] it is found that 5 wt% 2D nanosheets produce a super-high discharged energy density of 13.0 J cm^{-3} at 570 MV m^{-1}. Finite element simulation was conducted to investigate the electric field distribution in nanocomposites, and the results are displayed in Figure 1.23.

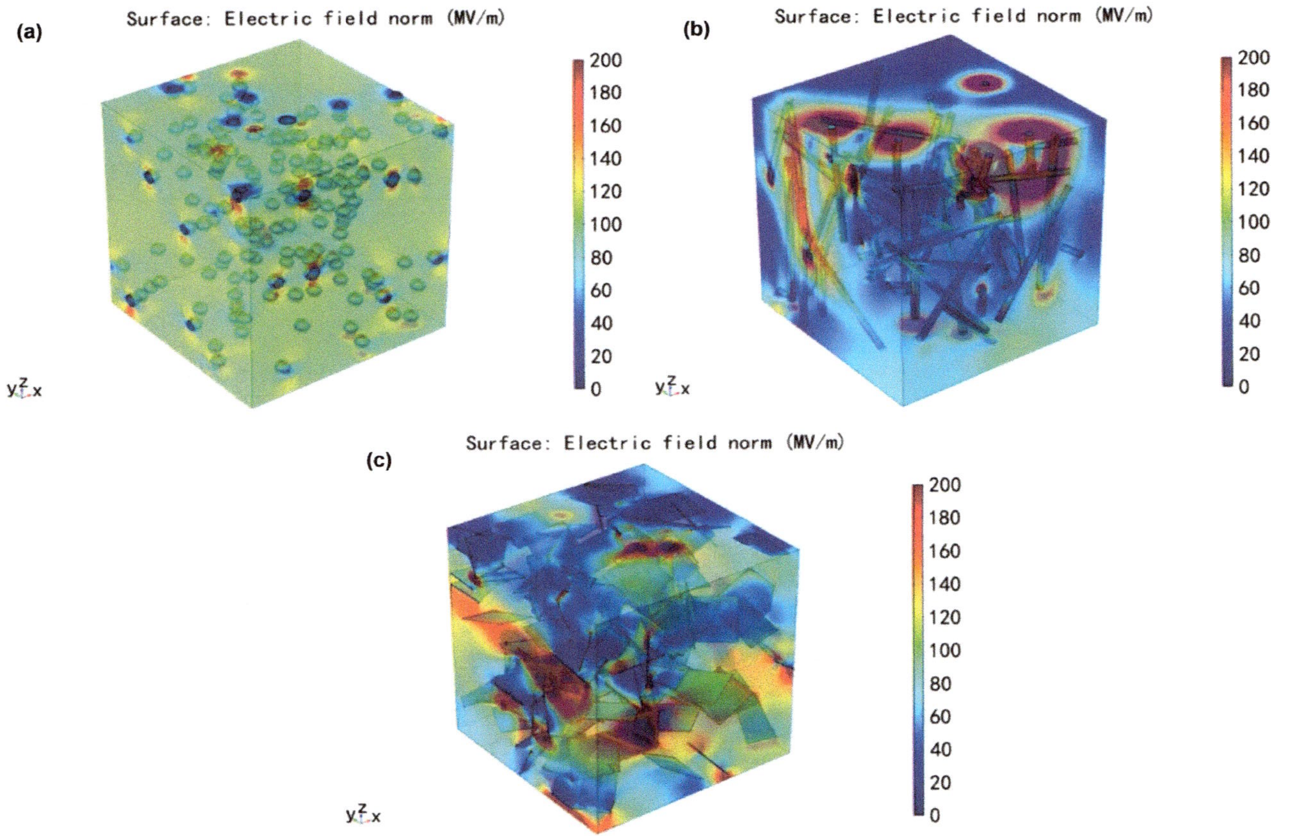

Figure 1.23 Schematic of electric field distribution of nanocomposites with 5 wt % of (a) 0D, (b) 1D, and (c) 2D TiO_2 nanofiller. Reproduced from ref. 8 with permission from American Chemical Society, Copyright 2018.

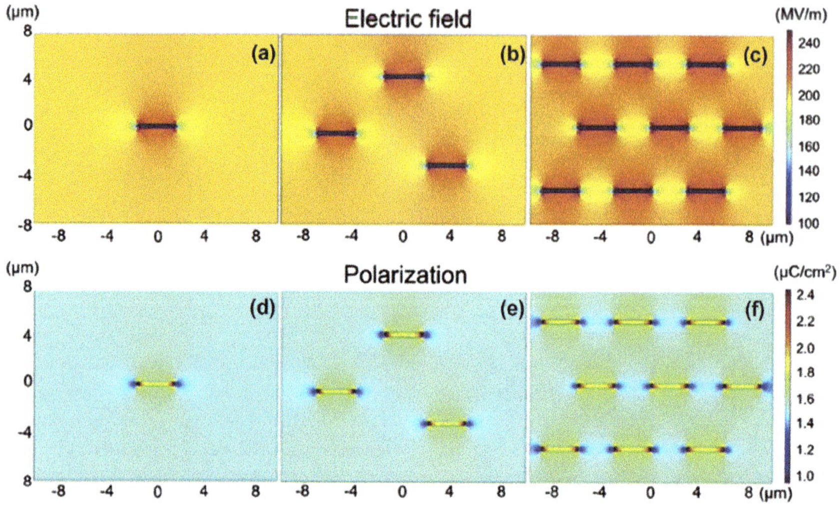

Figure 1.24 Simulation of local electric field of composite with 2D platelets of BT content at (a) 1 wt%, (b) 3 wt%, and (c) 9 wt%. Local polarisation distribution of composite with BT content at (d) 1 wt%, (e) 3 wt%, and (f) 9 wt%.
Reproduced from ref. 92 with permission from Elsevier, Copyright 2019.

Compared with the 1D nanofiller, nanocomposites with 2D nanofiller exhibit a more uniform electric field, resulting in a higher discharged energy density due to the suppressed charge injection from electrode. To study the effects of 2D nanoplatelet content on the distribution of the local electric field and local polarisation in 2D BT/PVDF nanocomposites, finite element simulation using COMSOL Multiphysics was conducted, and the results are shown in Figure 1.24.[92] As displayed in Figure 1.24(a–c), the surface of platelets produced higher local electric field, suggesting that the nanoplatelets would amplify the local electric field strength. With increments in the nanofiller content, the local electric field became more concentrated on the surface of the nanoplatelets, resulting in a sharp decrease in the breakdown strength of the nanocomposites. Although a higher content of nanoplatelets leads to higher local electric field, the amplified local electric field tends to improve the induced polarisation. A higher polarisation achieved at a larger filler content at the same external electric field is presented in Figure 1.24(d–f). Utilising the finite element simulation, Fu *et al.* found that the distribution of electric field and energy density for PVDF composites containing 2D $Ni(OH)_2$ platelets was more uniform than for spheres and 3D flower-type $Ni(OH)_2$ (see Figure 1.25). As a result, the platelet 2D $Ni(OH)_2$/PVDF composite possessed the highest relative permittivity (16.3), breakdown strength (421 kV mm^{-1}), and energy density (17.3 J cm^{-3}).[93] Through finite element simulation, Pan *et al.* confirmed that P(VDF-HFP)-based

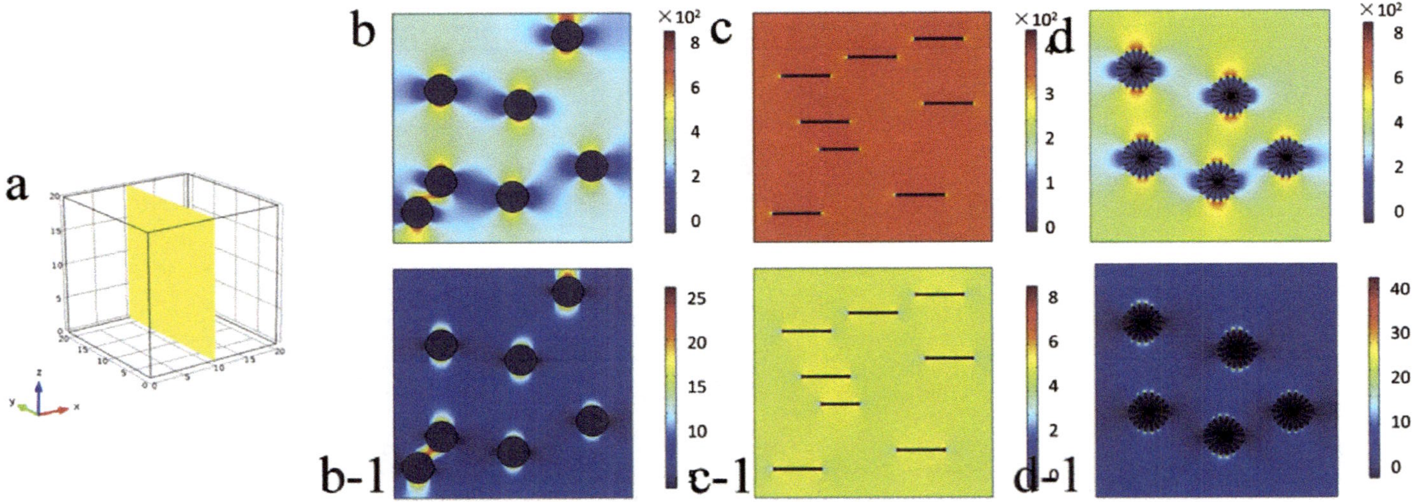

Figure 1.25 (a) A perpendicular cross-section model of $Ni(OH)_2$/PVDF composite film. (b) Electric field density and (b-1) energy density of sphere-$Ni(OH)_2$/PVDF composite block. (c) Electric field density and (c-1) energy density of platelet-$Ni(OH)_2$/PVDF composite block. (d) Electric field density and (d-1) energy density of 3D-Ni $(OH)_2$/PVDF composite block. Reproduced from ref. 93 with permission from Elsevier, Copyright 2019.

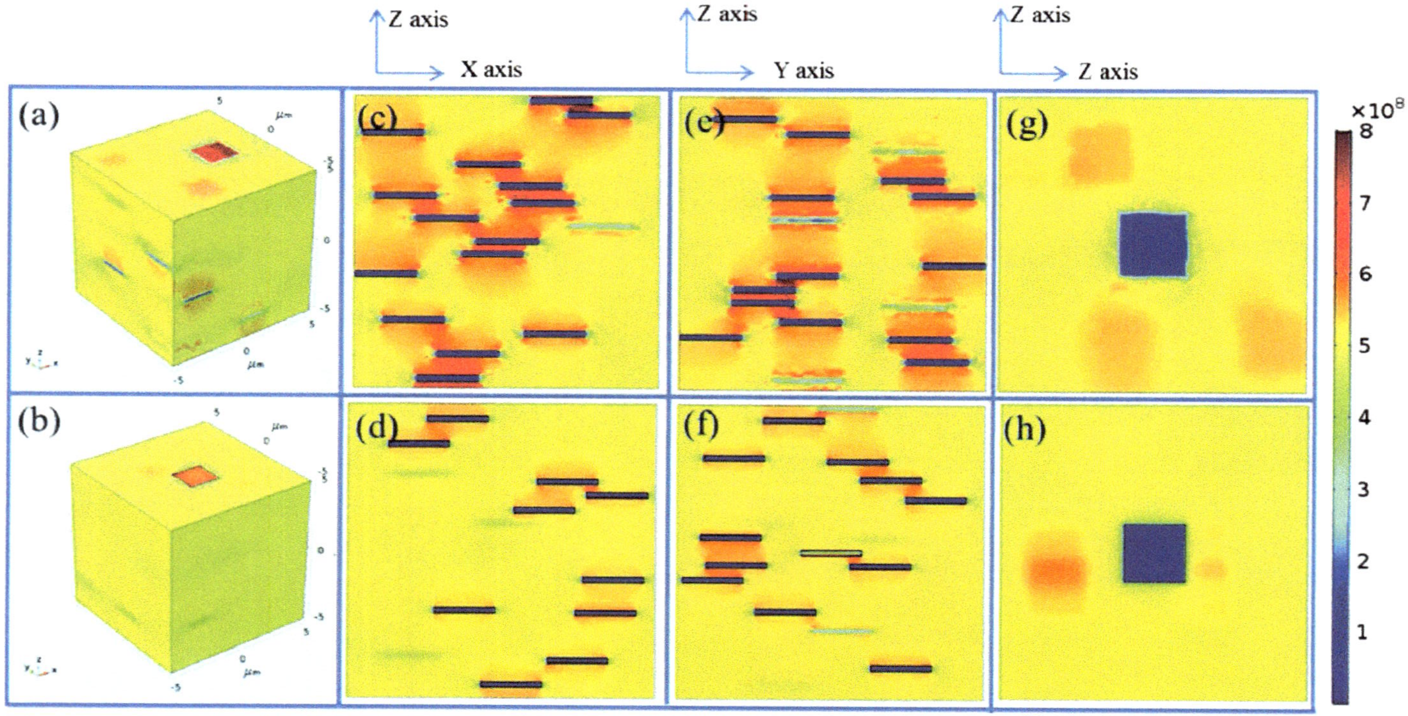

Figure 1.26 Finite element simulations of nanocomposite films with 2D $NaNbO_3$ platelets (a) and 2D $NaNbO_3@Al_2O_3$ platelets (b). Cross-sectional images (*Z–X*, *Z–Y*, and *Z–Z* axes) of the nanocomposite films with 2D $NaNbO_3$ platelets (c, e, and g) and $NaNbO_3@Al_2O_3$ platelets (d, f, and h).
Reproduced from ref. 72 with permission from the Royal Society of Chemistry.

nanocomposite films with 2D $NaNbO_3@Al_2O_3$ platelets could endure a higher applied electric field than nanocomposite films with 2D $NaNbO_3$ platelets, as shown in Figure 1.26.[72] To better understand the electrical breakdown mechanism of nanocomposite films with 3 vol% 2D NN Ps and 2D NN@AO Ps, finite element simulations were performed. To determine a more exhaustive electric field distribution of nanocomposite films with 2D NN Ps and 2D NN@AO Ps, the cross-sectional images (*Z–X*, *Z–Y*, and *Z–Z* axes) are shown in Figure 1.26(c–h). Highly distorted regions in the electric field are observed between the adjacent nanofiller regions along the direction perpendicular to the electric field. The local electric field distortion of 2D NN Ps/P(VDF-HFP) nanocomposite films is greater than that of the NN@AO Ps/P(VDF-HFP) nanocomposite films. Therefore, the finite element simulation confirms that the breakdown field of such nanocomposite films could be improved by adding 2D NN@AO Ps. Furthermore, the highly insulating Al_2O_3 shell could alleviate the electric field concentration because the relative permittivity of Al_2O_3 (~11) was similar to that of the polymer matrix (~9.76), leading to the formation of a buffer layer between the high-ε_r $NaNbO_3$ platelets and the polymer matrix.

1.5.3.2 Phase-field Simulation

The electrostatic breakdown propagation of dielectric materials is complex; however, the use of a phase-field model is a good approach to understand the breakdown behaviour arising from electrostatic stimuli.[94] The model considers that the complete breakdown path of the composite is similar to the process of crack propagation, although it incorporates electric energy, gradient energy, and phase separation energy. The calculation process was described in detail by Shen's group as follows.[95] In the phase-field model, a continuous phase-field variable $\eta(r, t)$ is introduced to record the electro-thermal damage distribution in the nanocomposite, in which $\eta(r, t) = 1$ is the breakdown phase, $\eta(r, t) = 0$ is the non-breakdown phase, and the diffuse transitional region is the interface area. In the dielectric inhomogeneous system, the free energy, which is related to the phase separation, the interface, the temperature, and the electric field, can be expressed by

$$F = \int_V \left[f_{\text{sep}}(\eta(r)) + \frac{1}{2}\gamma|\nabla\eta(r)|^2 + f_{\text{elec}}(r) \right] \mathrm{d}V \tag{1.7}$$

A modified Allen–Cahn equation was used to display the breakdown phase evolution,

$$\frac{\partial\eta(r,t)}{\partial t} = -L_0 H(f_{\text{elec}} - f_{\text{critical}}) \left[\frac{\partial f_{\text{sep}}(\eta)}{\partial\eta(r,t)} \right] - \gamma\nabla^2\eta(r,t) + \frac{\partial f_{\text{elec}}(r)}{\partial\eta(r,t)} \tag{1.8}$$

where L_0 represents the kinetic coefficient referred to the interface mobility, $H(f_{\text{elec}} - f_{\text{critical}})$ represents the Heaviside unit step function ($H(f_{\text{elec}} < f_{\text{critical}}) = 0$

and $H(f_{elec} > f_{critical}) = 1$), and $f_{critical}$ represents a position-dependent material constant referring to the maximal energy density of each component in the nanocomposite. In addition, the Heaviside function was introduced into the Allen–Cahn equation in order to assure the growth of breakdown phase only if the electric energy of a local point is greater than its maximal energy endurance. The spectral iterative perturbation method is used to gain the electric field distribution during the microstructure evolution process.

Through a continuum phase-field model, the microstructural effects on the effective relative permittivity, breakdown strength, and the energy density of polymer nanocomposites can be systematically studied, including the shape, orientation, and volume fraction of the nanofiller. In addition, in the case of a sandwich/multilayer-structured nanocomposite, phase-field simulation has also been used effectively to analyse the spatial distribution of the electric field in the nanocomposites. For example, Yang *et al.* studied the breakdown phase evolution of five types of 3D nanocomposites containing vertical nanofibres (S1), vertical nanosheets (S2), random nanoparticles (S3), parallel nanofibres (S4), and parallel nanosheets (S5).[96] Based on the results of simulation, the nanocomposite with vertical nanofibres (S1) is the easiest to break down, and the nanocomposite with parallel nanosheets (S5) is the hardest to break down. Moreover, according to the high-throughput calculation, Yang *et al.* designed the sandwich microstructure based on PVDF–BT nanocomposite in which the upper and lower layers were filled with parallel nanosheets and the middle layer was filled with vertical nanofibres. The results showed that the energy density of the nanocomposite was 2.44 times larger than that of pure PVDF polymer. Recently, Zeng and Nan *et al.* fabricated a series of single-layer and multilayer-structured P(VDF-HFP)-based nanocomposites with parallel BNNSs as the nanofiller.[97] The detailed structures of the nanocomposites and the simulated spatial distribution of the electric field are presented in Figure 1.27. The results of

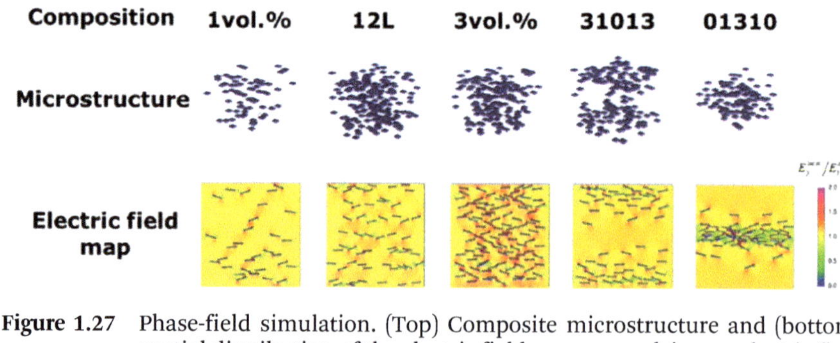

Figure 1.27 Phase-field simulation. (Top) Composite microstructure and (bottom) spatial distribution of the electric field map on applying an electric field of polymer nanocomposites.
Reproduced from ref. 97 with permission from AIP Publishing, Copyright 2018.

phase-field simulation implied that the 12L (12-layer) structured nanocomposites produced the most homogeneous spatial distribution of electric field compared with other structured nanocomposites, which benefits suppression of dielectric loss. The multilayer-structured nanocomposites containing a multi-interfacial region can effectively block the growth of electrical trees, and thus improve the breakdown strength and energy density. These results indicate that developing phase-field models and performing high-throughput calculations help the design of polymer nanocomposites with high storage properties and potential architectures.

More importantly, a comprehensive electrical–thermal–mechanical breakdown phase-field model was developed to study the breakdown process of polymer-based dielectrics by Yang *et al.*[98] Taking the P(VDF-HFP) as an example, the electrical, thermal, and mechanical effects on the breakdown strength were studied *via* parameterising the permittivity, electrical conductivity, and Young's modulus. The results showed that when incorporating electric, Joule heat, and strain energies, the phase-field model could predict breakdown strengths ($E_b^{\text{elec+Joule+strain}}$) of P(VDF-HFP) that were consistent with experimental measurements in the specified temperature range (295–363 K) (see Figure 1.28(a)). In addition, the breakdown strength of P(VDF-HFP)-based nanocomposites with 5 vol% nanofillers calculated from high-throughput phase-field simulations with different permittivity, electrical conductivity, and Young's modulus are shown in Figure 1.28(b). As shown, the breakdown strength is more related to the electrical conductivity than to the permittivity and Young's modulus.

Overall, the development of phase-field models and high-throughput calculations helps the design of polymer nanocomposites for high energy storage properties.

1.6 Conclusion

Polymer-based composites with 2D fillers, such as nanosheets and nanoplatelets are promising for high energy density capacitor applications. The synthesis methods of 2D nanosheets and nanoplatelets are introduced, including the hydrothermal/solvothermal method, molten salt method, and exfoliation method. The thickness of the synthesised 2D nanosheets and nanoplatelets ranges from nano- to micrometres. The 2D ceramic platelets tend to align in the perpendicular direction to the surface of the composite film with tape-casting treatment. Therefore, the polymer-based composites with 2D fillers can produce outstanding properties, such as high breakdown strength, low dielectric loss, and high energy density. Finite element simulation and phase-field simulation are useful methods to determine the polarisation and electric field distribution in the composites, and provide guidance for material design. The incorporation of 2D nanosheets and nanoplatelets into polymer matrices is an effective route to achieve high energy density capacitors.

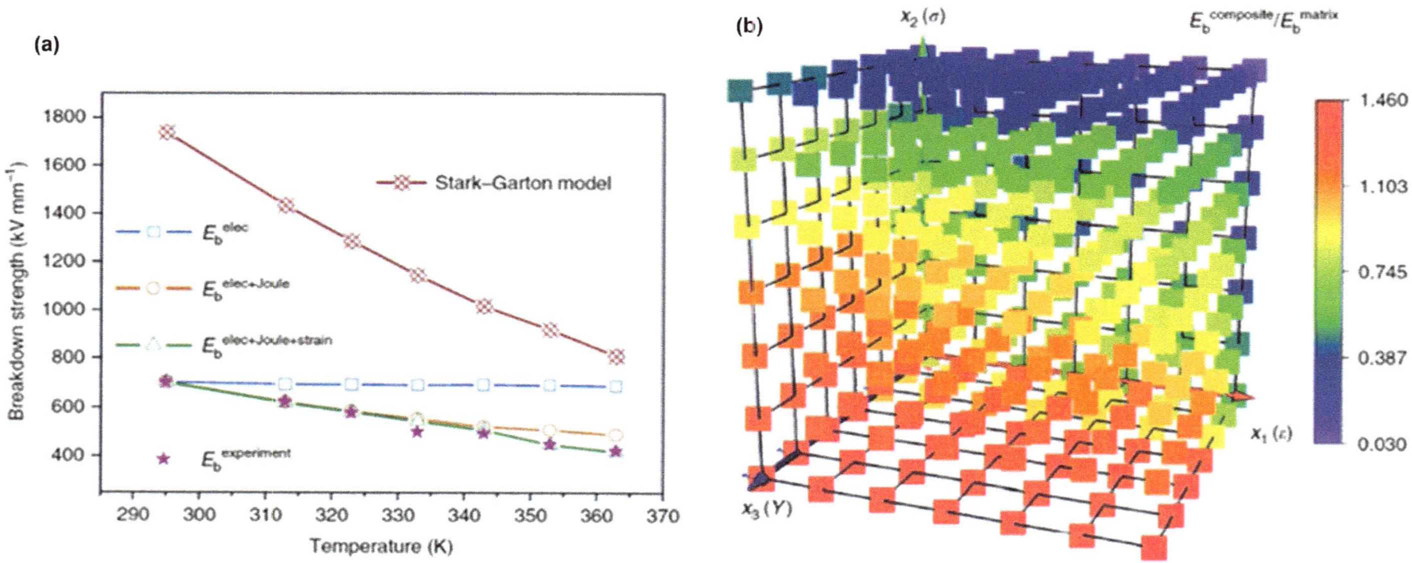

Figure 1.28 (a) Breakdown strengths of P(VDF-HFP) as a function of temperature obtained from experimental measurements and predictions from the Stark–Garton model and the phase-field model. E_b^{elec}, $E_b^{elec+Joule}$, and $E_b^{elec+Joule+strain}$ denote the breakdown strengths predicted by the phase-field model considering only electric energy, the combination of electric and Joule heat energy, and the combination of electric, Joule heat, and the strain energy, respectively. (b) High-throughput simulation results of the breakdown strength for P(VDF-HFP) nanocomposites filled by 5 vol% nanofillers with different permittivity, electrical conductivity, and Young's modulus. Here, $x_1(\varepsilon)$, $x_2(\sigma)$, and $x_3(Y)$ represent $\varepsilon_{filler}/\varepsilon_{matrix}$, $\sigma_{filler}/\sigma_{matrix}$, and Y_{filler}/Y_{matrix}, respectively.
Reproduced from ref. 98 with permission from Springer Nature, Copyright 2019.

Acknowledgements

The authors gratefully acknowledge support from National Natural Science Foundation of China (51672311), Hunan Natural Science Foundation (2019JJ40349), Science and Technology Project of Hunan Province, China (2016WK2022), and State Key Laboratory of Powder Metallurgy, Central South University, Changsha, China.

References

1. Q. Li, L. Chen, M. R. Gadinski, S. Zhang, G. Zhang, H. U. Li, E. Iagodkine, A. Haque, L. Q. Chen, T. N. Jackson and Q. Wang, *Nature*, 2015, **523**, 576.
2. X. Huang, B. Sun, Y. Zhu, S. Li and P. Jiang, *Prog. Mater. Sci.*, 2019, **100**, 187.
3. Z. Pan, L. Yao, J. Zhai, X. Yao and H. Chen, *Adv. Mater.*, 2018, **30**, 1705662.
4. Y. Wang, L. Wang, Q. Yuan, J. Chen, Y. Niu, X. Xu, Y. Cheng, B. Yao, Q. Wang and H. Wang, *Nano Energy*, 2018, **44**, 364.
5. Z. Yao, Z. Song, H. Hao, Z. Yu, M. Cao, S. Zhang, M. T. Lanagan and H. Liu, *Adv. Mater.*, 2017, **29**, 1601727.
6. Z. M. Dang, J. K. Yuan, S. H. Yao and R. J. Liao, *Adv. Mater.*, 2013, **25**, 6334.
7. F. Liu, Q. Li, J. Cui, Z. Li, G. Yang, Y. Liu, L. Dong, C. Xiong, H. Wang and Q. Wang, *Adv. Funct. Mater.*, 2017, **27**, 1606292.
8. Y. Zhu, H. Yao, P. Jiang, J. Wu, X. Zhu and X. Huang, *J. Phys. Chem. C*, 2018, **122**, 18282.
9. S. Boudjadar, S. Achour, N. Boukhenoufa and L. Guerbous, *Int. J. Nanosci.*, 2011, **09**, 585.
10. F. X. Ma, X. Y. Sun, K. He, J. T. Jiang, L. Zhen and C. Y. Xu, *J. Magn. Magn. Mater.*, 2014, **361**, 161.
11. W. L. Suchanek, J. M. Garcés, P. F. Fulvio and M. Jaroniec, *Chem. Mater.*, 2010, **22**, 6564.
12. J. Xie, H. Zhang, S. Li, R. Wang, X. Sun, M. Zhou, J. Zhou, X. W. Lou and Y. Xie, *Adv. Mater.*, 2013, **25**, 5807.
13. J. Fu, S. Song, X. Zhang, F. Cao, L. Zhou, X. Li and H. Zhang, *CrystEngComm*, 2012, **14**, 2159.
14. G. Xu, Y. Yang, H. Bai, J. Wang, H. Tian, R. Zhao, X. Wei, X. Yang and G. Han, *CrystEngComm*, 2016, **18**, 2268.
15. S. Li, R. Nechache, I. A. V. Davalos, G. Goupil, L. Nikolova, M. Nicklaus, J. Laverdiere, A. Ruediger, F. Rosei and D. Damjanovic, *J. Am. Ceram. Soc.*, 2013, **96**, 3155.
16. D. He, Y. Wang, L. Zhang, S. Song and Y. Deng, *Compos. Sci. Technol.*, 2018, **159**, 162.
17. M. Li, Y. Chen, W. Li, X. Li, H. Tian, X. Wei, Z. Ren and G. Han, *Small*, 2017, **13**, 1604115.
18. Z. Zhao, X. Li, H. Ji and M. Deng, *Integr. Ferroelectr.*, 2014, **154**, 154.

19. A. Hussain, J. U. Rahman, F. Ahmed, J. S. Kim, M. H. Kim, T. K. Song and W. J. Kim, *J. Eur. Ceram. Soc.*, 2015, **35**, 919.
20. M. M. Kržmanc, B. Jančar, H. Uršič, M. Tramšek and D. Suvorov, *Cryst. Growth Des.*, 2017, **17**, 3210.
21. M. E. Ebrahimi, M. Allahverdi and A. Safari, *J. Am. Ceram. Soc.*, 2005, **88**, 2129.
22. L. Wang, F. Gao, J. Xu, K. Zhang, J. Kong, M. Reece and H. Yan, *Compos. Sci. Technol.*, 2018, **158**, 112.
23. K. Ishii, S. Tashiro, A. Kikuchihara, T. Kimura and D. Damjanovic, *J. Am. Ceram. Soc.*, 2013, **96**, 2382.
24. C. Jiang, D. Zhang, K. Zhou, H. Yan, H. Zhang and I. Abrahams, *J. Mater. Chem. A*, 2017, **5**, 15780.
25. C. Jiang, D. Zhang, K. Zhou, X. Zhou, H. Luo and I. Abrahams, *J. Mater. Chem. A*, 2016, **4**, 18050.
26. Z. Pan, B. Liu, J. Zhai, L. Yao, K. Yang and B. Shen, *Nano Energy*, 2017, **40**, 587.
27. C. Zhi, Y. Bando, C. Tang, H. Kuwahara and D. Golberg, *Adv. Mater.*, 2009, **21**, 2889.
28. L. Wu, W. Kai, D. Liu, H. Rui and F. Qiang, *J. Mater. Chem. A*, 2018, **6**, 7573.
29. J. N. Coleman, M. Lotya, A. O'Neill, S. D. Bergin, P. J. King, U. Khan, K. Young, A. Gaucher, S. De, R. J. Smith, I. V. Shvets, S. K. Arora, G. Stanton, H. Y. Kim, K. Lee, G. T. Kim, G. S. Duesberg, T. Hallam, J. J. Boland, J. J. Wang, J. F. Donegan, J. C. Grunlan, G. Moriarty, A. Shmeliov, R. J. Nicholls, J. M. Perkins, E. M. Grieveson, K. Theuwissen, D. W. McComb, P. D. Nellist and V. Nicolosi, *Science*, 2011, **331**, 568.
30. M. Naguib, O. Mashtalir, J. Carle, V. Presser, J. Lu, L. Hultman, Y. Gogotsi and M. W. Barsoum, *ACS Nano*, 2012, **6**, 1322.
31. X. Zhang, J. Xu, H. Wang, J. Zhang, H. Yan, B. Pan, J. Zhou and Y. Xie, *Angew. Chem.*, 2013, **52**, 4361.
32. S. Lee, X. Jin, I. Y. Kim, T.-H. Gu, J.-W. Choi, S. Nahm and S. J. Hwang, *J. Phys. Chem. C*, 2016, **120**, 11786.
33. J. Coelho, B. Mendoza-Sánchez, H. Pettersson, A. Pokle, E. K. McGuire, E. Long, L. McKeon, A. P. Bell and V. Nicolosi, *2D Mater.*, 2015, **2**, 025005.
34. D. Hanlon, C. Backes, T. M. Higgins, M. Hughes, A. O'Neill, P. King, N. McEvoy, G. S. Duesberg, B. Mendoza Sanchez, H. Pettersson, V. Nicolosi and J. N. Coleman, *Chem. Mater.*, 2014, **26**, 1751.
35. H. Luo, X. Zhou, C. Ellingford, Y. Zhang, S. Chen, K. Zhou, D. Zhang, C. R. Bowen and C. Wan, *Chem. Soc. Rev.*, 2019, **48**, 4424.
36. Z. M. Dang, T. Zhou, S. H. Yao, J. K. Yuan, J. W. Zha, H. T. Song, J. Y. Li, Q. Chen, W. T. Yang and J. Bai, *Adv. Mater.*, 2009, **21**, 2077.
37. Y. Ke, Y. Niu, Y. Zhou, Y. Bai, W. Hong and C. Randall, *J. Am. Ceram. Soc.*, 2013, **96**, 2519.

38. M. N. Almadhoun, U. S. Bhansali and H. N. Alshareef, *J. Mater. Chem. A*, 2012, **22**, 11196.
39. Y. Ke, X. Huang, Y. Huang, L. Xie and P. Jiang, *Chem. Mater.*, 2013, **25**, 2327.
40. L. Y. Xie, X. Y. Huang, C. Wu and P. K. Jiang, *J. Mater. Chem. A*, 2011, **21**, 5897.
41. N. Guo, S. A. Dibenedetto, P. Tewari, M. T. Lanagan, M. A. Ratner and T. J. Marks, *Chem. Mater.*, 2015, **22**, 1567.
42. H. X. Tang, Z. Zhou and H. A. Sodano, *ACS Appl. Mater. Interfaces*, 2014, **6**, 5450.
43. S. Liu and J. Zhai, *J. Mater. Chem. A*, 2014, **3**, 1511.
44. H. Tang, Y. Lin and H. A. Sodano, *Adv. Energy Mater.*, 2012, **2**, 393.
45. B. Xie, H. B. Zhang, Q. Zhang, J. D. Zang, C. Yang, Q. P. Wang, M. Y. Li and S. L. Jiang, *J. Mater. Chem. A*, 2017, **5**, 6070.
46. Y. Wu, Z. Wang, X. Shen, X. Liu, N. M. Han, Q. Zheng, Y. W. Mai and J. K. Kim, *ACS Appl. Mater. Interfaces*, 2018, **10**, 26641.
47. X. H. Peng, X. L. Liu, P. Qu and B. Yang, *J. Mater. Sci.: Mater. Electron.*, 2018, **29**, 16799.
48. Q. C. Jia, X. Y. Huang, G. Y. Wang, J. C. Diao and P. K. Jiang, *J. Phys. Chem. C*, 2016, **120**, 10206.
49. S. Yang, J. Du, Z. Xin, H. Xin and C. W. Nan, *Mater. Express*, 2016, **6**, 277.
50. S. P. Fillery, H. Koerner, L. Drummy, E. Dunkerley, M. F. Durstock, D. F. Schmidt and R. A. Vaia, *ACS Appl. Mater. Interfaces*, 2012, **4**, 1388.
51. Y. Wang, J. Li and Y. Deng, *RSC Adv.*, 2015, **5**, 85884.
52. D. R. Wang, Y. R. Bao, J. W. Zha, J. Zhao, Z. M. Dang and G. H. Hu, *ACS Appl. Mater. Interfaces*, 2012, **4**, 6273.
53. D. R. Wang, T. Zhou, J. W. Zha, J. Zhao, C. Y. Shi and Z. M. Dang, *J. Mater. Chem. A*, 2013, **1**, 6162.
54. K. Han, Q. Li, Z. Y. Chen, M. R. Gadinski, L. J. Dong, C. X. Xiong and Q. Wang, *J. Mater. Chem. C*, 2013, **1**, 7034.
55. D. R. Wang, X. M. Zhang, J. W. Zha, J. Zhao, Z. M. Dang and G. H. Hu, *Polymer*, 2013, **54**, 1916.
56. X. Lv, L. Hang, C. Sheng, X. Han, M. Chao, X. Zhou, W. Liu, W. Zhong, K. Zhou and Z. Dou, *Mech. Syst. Signal Process.*, 2018, **108**, 48.
57. Z. Pan, B. Liu, J. Zhai, L. Yao, Y. Ke and S. Bo, *Nano Energy*, 2017, **40**, 587.
58. L. Qi, H. Kuo, G. Matthew Robert, Z. Guangzu and W. Qing, *Adv. Mater.*, 2014, **26**, 6244.
59. Q. Li, G. Z. Zhang, F. H. Liu, K. Han, M. R. Gadinski, C. X. Xiong and Q. Wang, *Energy Environ. Sci.*, 2015, **8**, 922.
60. L. X. Xing, Q. Li, G. Z. Zhang, X. S. Zhang, F. H. Liu, L. Liu, Y. D. Huang and Q. Wang, *Adv. Funct. Mater.*, 2016, **26**, 3524.
61. S. K. Ghosh, W. Rahman, T. R. Middya, S. Sen and D. Mandal, *Nanotechnology*, 2016, **27**, 215401.
62. D. K. Cheng, H. Wang, B. Liu, S. Wang, Y. Li, Y. S. Xia and C. X. Xiong, *J. Appl. Polym. Sci.*, 2019, **136**, 7.

63. C. Andrews, Y. Lin and H. A. Sodano, *Smart Mater. Struct.*, 2010, **19**, 14.
64. H. Tang, G. J. Ehlert, Y. Lin and H. A. Sodano, *Nano Lett.*, 2012, **12**, 84.
65. V. K. Thakur, E. J. Tan, M. F. Lin and P. S. Lee, *J. Mater. Chem. A*, 2011, **21**, 3751.
66. M. Zhu, X. Huang, K. Yang, X. Zhai, J. Zhang, J. He and P. Jiang, *ACS Appl. Mater. Interfaces*, 2014, **6**, 19644.
67. F. Wen, Z. Xu, S. Tan, W. Xia, X. Wei and Z. Zhang, *ACS Appl. Mater. Interfaces*, 2013, **5**, 9411.
68. W. S. Tong, Y. H. Zhang, L. Yu, X. L. Luan, Q. An, Q. Zhang, F. Z. Lv, P. K. Chu, B. Shen and Z. L. Zhang, *J. Phys. Chem. C*, 2014, **118**, 10567.
69. S. Yang, Y. Guan, Y. Hu, Y. Lei and C. W. Nan, *Appl. Phys. Lett.*, 2013, **103**, 1939.
70. J. Chen, X. Wang, X. Yu, L. Yao, Z. Duan, F. Yun, Y. Jiang, Y. Zhou and Z. Pan, *J. Mater. Chem. C*, 2018, **6**, 271.
71. J. Chen, X. Wang, X. Yu, F. Yun, Z. Duan, Y. Jiang, F. Yang and Y. Zhou, *Appl. Surf. Sci.*, 2018, **447**, 704.
72. Z. Pan, Q. Ding, L. Yao, S. Huang, S. Xing, J. Liu, J. Chen and J. Zhai, *Nanoscale*, 2019, **11**, 10546.
73. Y. J. Wang, Y. Kai, L. F. Tong, Y. You, Y. M. Huang and X. B. Liu, *Appl. Surf. Sci.*, 2019, **481**, 1239.
74. Z. Xin, Y. Mao, Z. Yu, S. Wen, L. Yan, L. Zhang, T. W. Chan and L. Li, *Composites, Part A*, 2015, **76**, 194.
75. V. Tomer, G. Polizos, C. A. Randall and E. Manias, *J. Appl. Phys.*, 2011, **109**, 5.
76. L. Wang, F. Gao, J. Xu, K. Zhang, J. Kong, M. Reece and H. Yan, *Compos. Sci. Technol.*, 2018, **158**, 112.
77. Y. Zhang, T. Zhang, L. Liu, Q. Chi, C. Zhang, Q. Chen, Y. Cui, X. Wang and Q. Lei, *J. Phys. Chem. C*, 2018, **122**, 1500.
78. C. Jiang, D. Zhang, K. Zhou, X. Zhou and I. Abrahams, *J. Mater. Chem. A*, 2016, **4**, 18050.
79. Q. Li, F. Liu, T. Yang, M. R. Gadinski, G. Zhang, L. Q. Chen and Q. Wang, *Proc. Natl. Acad. Sci. U. S. A.*, 2016, **113**, 9995.
80. X. Zhang, B. W. Li, L. Dong, H. Liu, W. Chen, Y. Shen and C. W. Nan, *Adv. Mater. Inter*, 2018, **5**, 1800096.
81. D. Pitsa and M. G. Danikas, *NANO*, 2011, **06**, 497.
82. M. Danikas, A. Bairaktari, R. Sarathi and A. B. Bin Abd Ghani, *Eng. Technol. Appl. Sci. Res.*, 2014, **4**, 636.
83. T. Tanaka, M. Kozako, N. Fuse and Y. Ohki, *IEEE Trans. Dielectr. Electr. Insul.*, 2005, **12**, 669.
84. T. Tanaka, *2006 IEEE 8th International Conference on Properties & applications of Dielectric Materials*, 2006.
85. T. J. Lewis, *IEEE Trans. Dielectr. Electr. Insul.*, 2004, **11**, 739.
86. Prateek, V. K. Thakur and R. K. Gupta, *Chem. Rev.*, 2016, **116**, 4260.
87. Prateek, R. Bhunia, S. Siddiqui, A. Garg and R. K. Gupta, *ACS Appl. Mater. Interfaces*, 2019, **11**, 14329.

88. M. Rahimabady, M. S. Mirshekarloo, K. Yao and L. Lu, *Phys. Chem. Chem. Phys.*, 2013, **15**, 16242.
89. J. Chen, X. Wang, X. Yu, L. Yao, Z. Duan, Y. Fan, Y. Jiang, Y. Zhou and Z. Pan, *J. Mater. Chem. C*, 2018, **6**, 271.
90. Z. Pan, L. Yao, G. Ge, B. Shen and J. Zhai, *J. Mater. Chem. A*, 2018, **6**, 14614.
91. L. Yao, Z. Pan, S. Liu, J. Zhai and H. H. D. Chen, *ACS Appl. Mater. Interfaces*, 2016, **8**, 26343.
92. F. Wen, H. Lou, J. Ye, W. Bai, L. Wang, L. Li, W. Wu, Z. Xu, G. Wang, Z. Zhang and L. Zhang, *Compos. Sci. Technol.*, 2019, **182**, 107759.
93. W. Ji, H. Deng, C. Sun and Q. Fu, *Compos. Sci. Technol.*, 2019, **172**, 117.
94. Z. H. Shen, J. J. Wang, J. Y. Jiang, Y. H. Lin, C. W. Nan, L. Q. Chen and Y. Shen, *Adv. Energy Mater.*, 2018, **8**, 1800509.
95. J. Jiang, Z. Shen, J. Qian, Z. Dan, M. Guo, Y. Lin, C.-W. Nan, L. Chen and Y. Shen, *Energy Storage Mater.*, 2019, **18**, 213.
96. Z. H. Shen, J. J. Wang, Y. Lin, C. W. Nan, L. Q. Chen and Y. Shen, *Adv. Mater.*, 2018, **30**, 1704380.
97. Y. Zeng, Z. H. Shen, Y. Shen, Y. H. Lin and C. W. Nan, *Appl. Phys. Lett.*, 2018, **112**, 4.
98. Z. H. Shen, J. J. Wang, J. Y. Jiang, S. X. Huang, Y. H. Lin, C. W. Nan, L. Q. Chen and Y. Shen, *Nat. Commun.*, 2019, **10**, 1843.

CHAPTER 2

Surface Engineering of Boron Nitride Nanoplatelets for Thermal Conductivity Enhancement of Polymers

YU CHEN,[a] PINGKAI JIANG,[a] JIE KONG[b] AND XINGYI HUANG*[a]

[a] Shanghai Key Laboratory of Electrical Insulation and Thermal Ageing, State Key Laboratory of Metal Matrix Composites, Shanghai Jiao Tong University, Shanghai 200240, China; [b] MOE Key Laboratory of Materials Physics and Chemistry in Extraordinary Conditions, Shaanxi Key Laboratory of Macromolecular Science and Technology, School of Science, Northwestern Polytechnical University, Xi'an 710072, P.R. China
*Email: xyhuang@sjtu.edu.cn

2.1 Introduction

Today, polymer materials play an irreplaceable role in the microelectronics, electrical equipment and aerospace sectors. With increasing power density and continuous miniaturisation of electronic devices and electric equipment, excessive accumulation of heat can result in reduction of the operational efficiency and potential explosion of modern devices, which makes thermal management of these electronic devices an urgent and important issue.[1] However, the low thermal conductivity of polymer materials ranging from 0.1 to 0.5 $W m^{-1} K^{-1}$ severely limits the development of the modern electronics industries.[2] Therefore, improving the thermal conductivity of

Inorganic Materials Series No. 6
Two-dimensional Inorganic Nanomaterials for Conductive Polymer Nanocomposites
Edited by Chaoying Wan, Xingyi Huang and Chris Bowen

Published by the Royal Society of Chemistry, www.rsc.org

polymer materials without losing their insulation properties has become key to solving this problem. The addition of highly thermally conductive inorganic fillers into polymers has been a widely used and effective method for enhancement of the thermal conductivity of polymers.[2,3]

Different types of fillers have been explored for improving the thermal conductivity of polymers, such as metal materials,[4,5] carbon materials[6–8] and ceramic materials.[9–11] Boron nitride nanoplatelets (BNNP) have attracted growing attention due to their ultra-high thermal conductivity, excellent insulating properties, outstanding chemical stability and mechanical properties.[12–17] To date, intensive research activity has been carried out on the synthesis and exfoliation of BNNP,[18,19] surface modification,[20,21] agglomerates and orientation of BNNP[22–24] in polymer composites.

2.2 Boron Nitride Nanoplatelets

2.2.1 Properties of BN

Boron nitride (BN) is a crystalline compound, consisting of equal numbers of boron (B) and nitrogen (N) atoms. BN was previously considered as only being synthesised artificially, mainly from boric acid (H_3BO_3) or boron trioxide (B_2O_3). However, recently, it has also been found in nature.[25] There are four main crystalline forms of boron nitride (Figure 2.1): hexagonal boron nitride

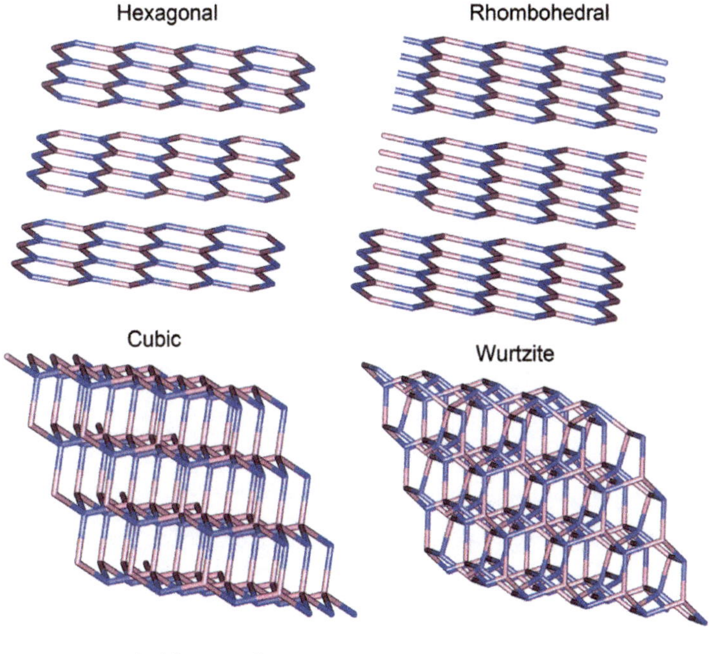

Figure 2.1 Boron nitride crystal structures.
Reproduced from ref. 26 with permission from the Royal Society of Chemistry.

(h-BN) with an analogous structure to graphite, cubic boron nitride (c-BN) with an analogous structure to diamond, wurtzite boron nitride (w-BN) with an analogous structure to lonsdaleite, and rhombohedral boron nitride (r-BN). Of these, h-BN and c-BN are the most common forms.[26] Similarly to graphite, h-BN has a typical layered structure built from sp^2-hybridised hexagonal rings. Each layer is an infinitely extended hexagonal honeycomb structure composed of alternately arranged B and N atoms. Hexagonal boron nitride is also known as 'white graphite' because of its white colour.

Unlike the non-polar C–C bonds formed by the same atom in graphite, the B–N bonds in h-BN have characteristics of both a covalent bond and an ionic bond, due to the different electronegativity of the B and N atoms (boron = 2.04, nitrogen = 3.04).[14] The bond length and bond energy of the B–N bond in h-BN are 1.44 Å and 4 eV, respectively, while the bond length and energy of the C–C bond in graphite are 1.42 Å and 3.7 eV, respectively.[27] Meanwhile, the partial ionic bond characteristic of the B–N bond means that h-BN has a wide band gap (5.5–6.0 eV), endowing h-BN with excellent electronic insulation properties.[28] The resistivity of h-BN is 10^{17} Ω cm at room temperature and 10^4 Ω cm at 200 °C, which is significantly higher than other ceramic materials.[17]

The interlaminar structure of h-BN is different from that of graphite. Besides, van der Waals forces, so-called 'lip–lip' interactions, exist in the adjacent BN layers, which makes it more difficult to exfoliate compared with graphene. The interlayer spacing of h-BN is 0.333 nm, which is slightly smaller than that of graphite (0.335 nm). Moreover, B and N atoms in h-BN are stacked according to AA′ stacking, while C atoms in graphite are stacked with AA′ stacking (Figure 2.2). The neighbouring layers of h-BN are parallel and strictly aligned along the *c*-axis with every B atom eclipsed by the

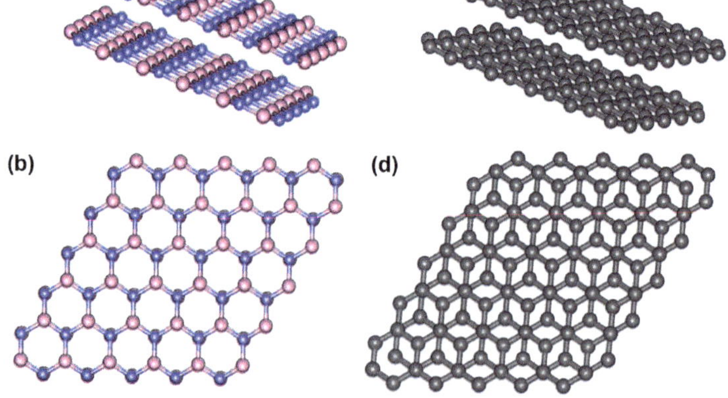

Figure 2.2 Structural models of (a, b) h-BN sheets and (c, d) graphite sheets displaying the difference in their stacking sequence.
Reproduced from ref. 26 with permission from the Royal Society of Chemistry.

N atoms in the upper and lower layers, and similarly for every N atom.[26] This stable and strong structure ensures that h-BN has excellent thermal and chemical stability.

Hexagonal-BN possesses high thermal conductivity and high temperature stability. The thermal conductivity of bulk h-BN has been reported to be 390 $W m^{-1} K^{-1}$ at room temperature.[29] In vacuum conditions, the thermal decomposition temperature of h-BN is about 2700 °C and the melting point is 3000 °C (inert gas atmosphere), while sublimation occurs slightly below 3000 °C (nitrogen atmosphere). In an oxidising atmosphere, h-BN maintains oxidation resistance and chemical stability for temperatures over 800 °C, while the oxidation temperature can reach above 2000 °C if in an inactive reducing atmosphere.[30] The super-excellent chemical stability of h-BN is another point that makes it attractive for many applications.[31] Hexagonal-BN has good corrosion resistance, is unreactive with ordinary inorganic acids, bases and oxidants at room temperature, and is chemically inert to almost all molten metals.

In addition, h-BN has a low relative permittivity of ~5, low dielectric loss and high breakdown voltage of 800 $MV m^{-1}$,[32] showing the potential to be applied in the fields of electronics and energy storage.[33] Due to the weak forces between the near planes, the friction coefficient of h-BN is as low as 0.16 even at high temperature. Therefore h-BN has also been extensively used in high-temperature solid lubricants.[34]

2.2.2 Synthesis and Exfoliation of BNNP

Since the discovery of graphene in 2004,[35] various types of two-dimensional nanomaterials have been emerging. Hexagonal-BN is isostructural with graphite, while BNNP possesses a crystal structure similar to graphene. Compared to other BN nanostructures, the (002) crystal plane of BNNP is exposed to a greater extent, further exploiting the superiority of its in-plane phonon transport. The intrinsic thermal conductivity of BNNP is significantly higher than that of bulk h-BN. In addition, BNNP provides unique properties over bulk h-BN, such as mechanical strength,[13] high optical transparency,[36] adjustable photoluminescence,[37] ultraviolet cathode luminescence[38] and atom-thick electron tunnelling barrier.[39]

The methods for producing BNNP can be mainly divided into top-down and bottom-up, corresponding to exfoliation from h-BN or synthesis from B and N precursors. Both types have their own advantages and shortcomings. As mentioned above, the lip-lip interaction between the adjacent h-BN layers makes exfoliation of BNNP more difficult than that of graphene. The exfoliation method is subject to strong dependence on the raw h-BN, and it is not easy to precisely control the lateral sizes and layers of BNNP.[40] The bottom-up methods, represented by chemical vapour deposition (CVD), face problems of high technical requirements, harsh synthesis conditions and low production efficiency.[41] To date, it is still a challenge to produce high-quality BNNP on a large scale at normal temperature and pressure.

2.2.2.1 *Top-down Methods*

2.2.2.1.1 Mechanical Exfoliation. The scotch-tape method and the micromechanical cleavage method, which were originally used to obtain graphene,[35] are also applicable to exfoliate BNNP. Pacile *et al.* prepared two-dimensional boron nitride nanoplatelets by the Scotch tape method.[16] The specific experimental procedure is as follows. The hexagonal boron nitride powder with a lateral size of 10 μm was adhered to the SiO_2 substrate with a thickness of 300 nm using scotch tape, and then boron nitride was forcibly separated into small fragments and, finally, two-dimensional boron nitride nanoplatelets of only a few atomic layers were attained. Based on the scotch tape method, Alem *et al.* exfoliated the treated h-BN into BNNP of few layers and a single layer by an ion etching technique, improving the success rate and controllability.[42] Monolayer BNNP can be attained by these two methods, but the yield is exceeding low, and only suitable for scientific research.

The exfoliation efficiency of BNNP was improved by the ball milling method.[43] The ball milling method utilises the collision of the hard ball with boron nitride to wrinkle or even peel off the boron nitride sheets (Figure 2.3). Xia *et al.* studied the effect of the duration of ball milling on the exfoliation of BNNP.[44] As the milling time was prolonged, the average diameter of the BNNP decreased significantly, with the crystal structure also affected. Li *et al.* reported a milder ball milling method, in which benzyl benzoate was added and acts as the lubricant, greatly improving the crystallinity of the resulting BNNP.[40]

In the fluid exfoliation method, BNNP can be obtained by taking advantage of the sheer force of the high-speed flowing fluid, which promotes slippage between boron nitride sheets.[45] Chen *et al.* used the self-made eddy

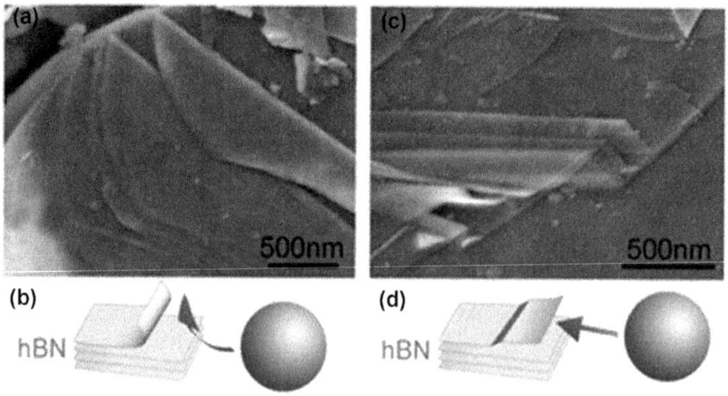

Figure 2.3 Scanning electron microscopy (SEM) images and corresponding diagrams illustrating two observed exfoliating mechanisms under the shear force created by milling balls. (a, b) Cleavage from the edge of an h-BN particle. (c, d) Thin sheets peeling off the top surface of an h-BN particle. Reproduced from ref. 40 with permission from the Royal Society of Chemistry.

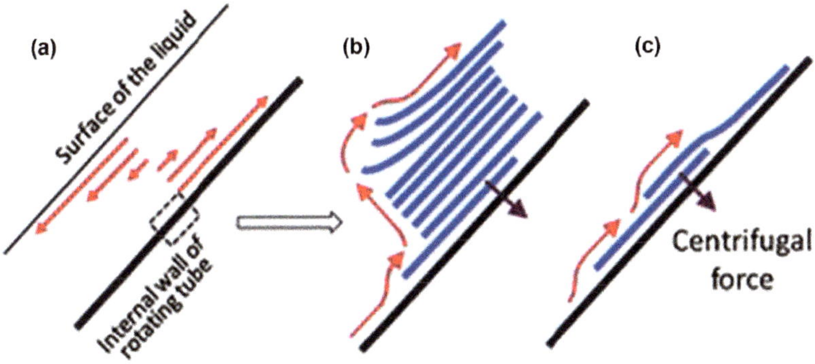

Figure 2.4 Cartoons of (a) the microfluidic flow velocity indicated by red arrows for a section of the rotating tube, (b) the exfoliation process and (c) slippage on the inner surface of the tube.
Reproduced from ref. 46 with permission from the Royal Society of Chemistry.

current from centrifugal rotating equipment to exfoliate BNNP.[46] A small amount of h-BN powder was dispersed in *N*-methyl pyrrolidone (NMP) in a glass tube placed at an angle of 45°. Under the high-speed rotation of 8000 rpm, a 5% monolayer of BNNP was attained (Figure 2.4).

2.2.2.1.2 Sonication-assisted Liquid-phase Exfoliation. Sonication-assisted liquid-phase exfoliation has been widely used in the preparation of graphene.[47–49] The mechanism of this method is that ultrasonic cavitation breaks the force between the layers of h-BN dispersed in the solvent to realise the exfoliation.[50] Solvents preventing the formation of boron nitride play a decisive role in this process.[51] The choice of solvents can refer to Hansen solubility parameter theory (HSP),[52] which has been proved by many experiments:

$$R_a = [4(\delta_{D,solv} - \delta_{D,solu})^2 + (\delta_{P,solv} - \delta_{P,solu})^2 + (\delta_{H,solv} - \delta_{H,solu})^2]^{0.5} \quad (2.1)$$

where R_a is the Hansen dissolution parameter distance; δ_D, δ_P, δ_H are the dispersive, polar and hydrogen bonding solubility parameters; and the labels solv and solu represent solvent and solute, respectively. The smaller the value of R_a, the higher the potential solubility.

Coleman *et al.* systematically studied the relationship between solvent surface energy and two-dimensional layered nanomaterials according to Hansen solubility parameter theory.[18] It is believed that NMP and isopropanol (IPA) have a good effect on the exfoliation of BNNP due to their similar surface energy to boron nitride. It is possible to get a smaller R_a value by mixing multiple solvents, thus improving the dispersing and exfoliating performance of the solvents. Zhou *et al.* took a mixed solution of ethanol and water with an ethanol volume fraction of 55% as solvent, and achieved a concentration of 0.75 mg mL^{-1} BNNP with a thickness of 3–4 nm.[53] Figure 2.5 shows their results.

Some solvents that deviate from Hansen solubility parameter theory can also realise the exfoliation of BNNP. Lin *et al.* used water as the solvent to

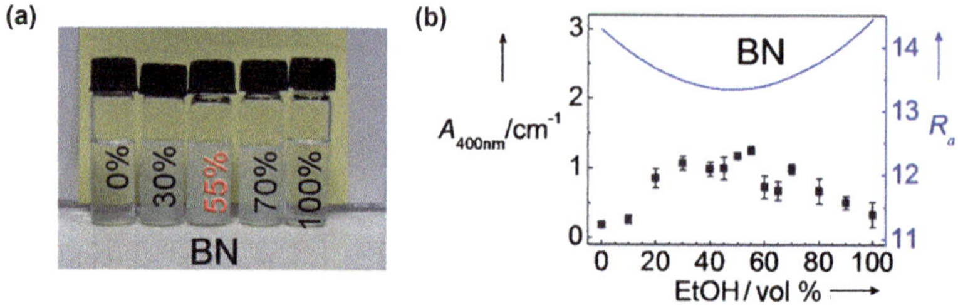

Figure 2.5 Photographs of (a) BN dispersions in various ethanol/water mixtures, which have been stored under ambient conditions for a week. (b) The absorbance of the BN suspensions in ethanol/water mixtures, with different composition shown as dots and the calculated R_a values as solid lines. Reproduced from ref. 53 with permission from John Wiley & Sons, Copyright 2011 WILEY-VCH Verlag GmbH & Co. KGaA, Weinheim.

prepare a 2 mg mL^{-1} aqueous solution of boron nitride and sonicated for 10 h, obtaining pure water-dispersed BNNP.[50] Wang *et al.* dispersed boron nitride powder in methane sulfonic acid (MSA) and the yield of BNNP reached as high as 0.3 mg mL^{-1} after the ultrasonic treatment of 8 h.[54]

2.2.2.1.3 Chemical Exfoliation. Some molecules can be attached to the surface of boron nitride. Under certain reaction conditions, the functional groups carried by the molecules interact with boron nitride, making the layer of boron nitride curl, wrinkle and finally peel off. Since the mechanism involves chemical reactions, chemical exfoliation is often accompanied by functionalisation of BNNP.

Li *et al.* added h-BN powder to a mixture of NaOH and KOH, and heated it to 180 °C for 2 h in a poly(tetrafluoroethylene)-lined stainless steel autoclave, finally attaining boron nitride nanosheets with curly edges.[55] The cations and anions of the hydroxide are inserted into the pleats of the boron nitride surface driven by the chemical potential, and induce spontaneous ringing of the boron nitride atomic layer while driving more ion insertion (Figure 2.6).

Du *et al.* mixed h-BN powder and NH_4F at 180 °C for 24 h and then attained fluoride boron nitride (F-BNNP) of less than three layers, realising the exfoliation and fluorination of boron nitride in one step.[56] The exfoliating process was as follows. The F ions bind to the B atoms on the BN surface, and the force between F and B buckles boron nitride sheets on the surface, followed by the insertion of ammonium ions and exfoliation.

Wang *et al.* proposed a hydrothermal exfoliation method where bulk h-BN powder was dispersed in IPA, heated at 180 °C for 12 h under strong stirring (500 rpm), with lithium chloride (LiCl) as the additive.[57] On the one hand, the lithium ion (Li^+) provided by LiCl is inserted between layers of h-BN in the process, increasing the distance between the layers; on the other hand, Li^+ turns into Li^0, which reacts with the solvent to generate hydrogen gas during the reaction, further propping up the space between layers and finally improving the exfoliation efficiency.

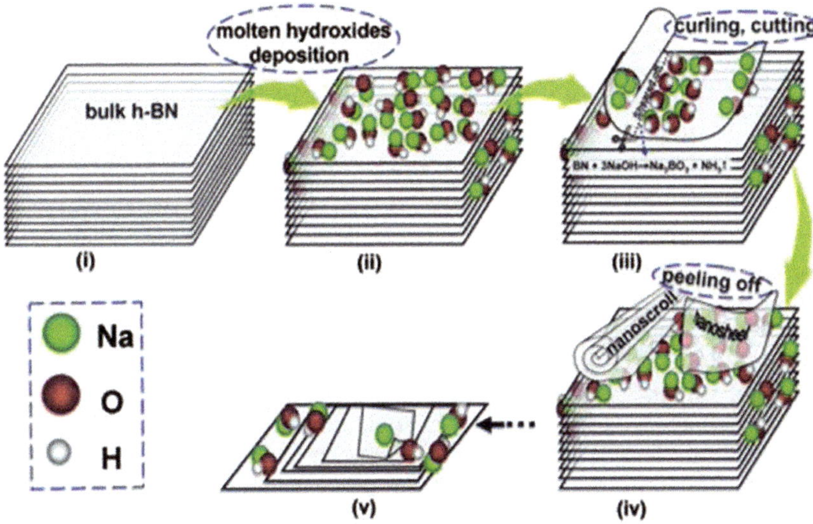

Figure 2.6 Illustration of the exfoliation mechanism. Schematic representation of the exfoliating process following the sequence deposition of hydroxides on h-BN, peripheral self-curling of sheets, insertion of hydroxides, cutting by the reaction of the reactants, and exfoliation of sheets. Reproduced from ref. 55 with permission from John Wiley and Sons, Copyright 2013 WILEY-VCH Verlag GmbH & Co. KGaA, Weinheim.

2.2.2.2 Bottom-up Methods

2.2.2.2.1 Chemical Vapour Deposition. The preparation of hexagonal boron nitride nanoplatelets by CVD can be traced back at least to the 1960s. In 1968, Rand *et al.* used ethyl borane (B_2H_6) and ammonia gas (NH_3) as precursors to synthesise BNNP on Si, Ta, Mo, Ge and molten quartz substrates, at a temperature range of 600–1080 °C.[58] Since then, CVD technology has attracted continuous attention for providing a relatively controllable method for attaining large scale and high quality boron nitride films. CVD technology has made remarkable progress in the selection of precursors, substrate templates, catalysts and reaction conditions.

Auwarter *et al.* prepared high quality boron nitride films with controllable layers by using the gaseous source trichloroborane on a nickel substrate.[59] The single film was triangular and regular in shape with a clean and flat surface. The growth of boron nitride films is mainly by virtue of dehydrogenation between molecules at high temperature. This study specifically studied the details of the experimental process, laying a foundation for the direct preparation of boron nitride films by dehydrogenative coupling between molecules.

In the CVD method, the B source and the N source for synthesising boron nitride can be derived from different chemicals (*e.g.* BCl_3/NH_3,[60] B_2H_6/NH_3[61,62]) or a single chemical (*e.g.* NH_3–BH_3[63]). Although relatively less stable, the latter is often preferred due to its low toxicity and boron: nitrogen atom ratio (1 : 1).[64] Song *et al.* first demonstrated that CVD can obtain boron nitride films of large size (up to several square centimetres) and several atomic layer thickness at normal pressure.[63] Under the protection of an Ar/H_2 atmosphere, ammonia borane (NH_3–BH_3) as the precursor was deposited at 1000 °C and then deposited on the Cu template substrate.

The morphology of the substrate has an effect on the nucleation and growth of the boron nitride film. Due to the influence of gas flow, most of the gas molecules are first adsorbed in the rough or impurity sites of the substrate during the growth process. Lee *et al.* studied the effect of copper foil surface morphology on the growth of boron nitride film.[65] A smoother copper foil surface was obtained by electrochemical polishing, with the resulting attainment of impurity-free and atomically flat boron nitride films.

2.2.2.2.2 Chemical Reaction. BNNP can be synthesised directly through chemical reaction. Rao *et al.* heated boric acid (H_3BO_3) and urea ($CO(NH_2)_2$) to 900 °C in a nitrogen atmosphere to synthesise BNNP of several atomic layers thickness.[66] In addition, there is a relationship between the concentration of reactants and the thickness of the prepared BNNP, where increasing the proportion of urea is beneficial to reduce the thickness of BNNP. The chemical reactions in this process are as follows:

$$2H_3BO_3 \rightarrow B_2O_3 + 3H_2O$$

$$NH_2CONH_2 \rightarrow NH_3 + HNCO$$

$$B_2O_3 + 2NH_3 \rightarrow 2BN + 3H_2O$$

In addition to the development of new direct synthesis methods, such as the combustion-annealing technique[67] and the substitution reaction,[68] the combination of chemical synthesis and ultrasonic treatment has also drawn increasing attention. For example, few-layered BNNP with lateral dimensions of about 100 μm are produced by chemical blowing reactions of NH_3BH_3 with the assistance of ultrasonication.[69,70]

2.2.3 BNNP for Enhancing Thermal Conductivity

2.2.3.1 Thermal Conductivity of BNNP

There are three main forms of heat transfer, namely thermal radiation, thermal convection and thermal conduction. For solid materials, heat transfer mainly takes the form of heat conduction.[71] Solid thermal conduction refers to the transfer of vibration energy from one particle to the neighbouring particle without changing the mass centre. Solid heat conduction is usually measured by thermal conductivity (κ). It is defined as the amount of heat transmitted through the unit horizontal sectional area per unit of time when the temperature gradient is 1 °C m^{-1}, and the unit of thermal conductivity (κ) is W m^{-1} K^{-1}. Thermal conductivity can be determined from:[3]

$$\kappa = \alpha \cdot C_p \cdot \rho \tag{2.2}$$

Table 2.1 Main properties of BNNP.

Characterisation	Properties	References
Colour	Transparent/white	74
Bonding	Covalent/ionic	74
Electronic structure	5.0–6.0 eV band gap	74
Raman	1365–1373 cm^{-1}	63, 194, 195
Infrared	1370–1378 cm^{-1} in plane B–N stretching; 800–830 cm^{-1} out of plane B–N–B bending	195–198
Ultraviolet	Sharp peak at 203 nm; transparent in the range of 250–900 nm	63
X-ray diffraction	Peak at $2\theta = 26$; lattice parameters: $a = 0.250$ nm, $c = 0.666$ nm; interlayer spacing: 0.333 nm	194, 196–198
Elastic modulus	220–510 $N\,m^{-1}$ (thickness about 1–2 nm)	74
Fracture strength	70–217 GPa	63, 199, 200
Specific surface areas	2600 $m^2\,g^{-1}$	141
Thermal conductivity	300–2000 $W\,m^{-1}\,K^{-1}$ (theoretical), 300–360 $W\,m^{-1}\,K^{-1}$ (experimental)	73, 74
Thermal stability	Over 600 °C	74

where, κ is thermal conductivity, C_p is the specific heat capacity at constant pressure, ρ is density and α is thermal diffusivity. The thermal conductivity (κ) reflects the ability of the material to conduct heat, the higher the value, the stronger the ability of the material to conduct heat. Thermal conductivity is an intrinsic property of the material itself. For isotropic materials, the thermal diffusivity is the same in all directions and the thermal conductivity is the same, while for anisotropic materials, the thermal conductivity also depends on the direction of the heat flux.[72]

Generally, the thermal conductivity of a substance can be attained by theoretical calculations and experimental tests. The room temperature thermal conductivity of BNNP is as high as 300–2000 $W\,m^{-1}\,K^{-1}$ by theoretical calculation and 300–360 $W\,m^{-1}\,K^{-1}$ by experiment.[73,74] Except for thermal conductivity, the other main properties of BNNP are shown in Table 2.1.

2.2.3.2 *Theoretical Basis for Enhancing the Thermal Conductivity of Polymers*

There are three types of thermal carriers inside solids: electrons, photons and phonons. For solid materials with electrical conductivity, due to the presence of a large number of moving electrons, the electrons carry a large amount of energy in the process of electron migration, so their thermal conductivity is usually very high.[75,76] Materials where photons play a major role in heat conduction must have good transmission properties, which can only be special glass or single crystals at a specific temperature. For most polymers and boron nitride, the main mechanism of heat conduction is through phonons because of their degree of electric insulation.[75,77,78]

A phonon is a quantum concept, not a particle that actually exists, but a collective vibration of a series of atoms coupled to each other that are excited.[79–82] When it is in contact with the heat source or at a high temperature, the particles on the surface of the material start to vibrate, the energy is given to the adjacent particles and, finally through particle vibration, throughout the material.[83–87] For structurally intact inorganic materials, there is little phonon scattering in the crystal plane due to their nearly complete crystal structure, and the mean free path of phonons is large, so they have high intrinsic thermal conductivity.[88–91] However, most polymers have low crystallinity, that is, incomplete crystallisation. There is a large amount of phonon scattering at the interface of the crystal and amorphous region, leading to a smaller mean free path of phonons.[92–95] As a consequence, the thermal conductivities of polymers are usually extremely low (0.1–0.5 $W m^{-1} K^{-1}$), severely limiting their application for thermal management in various fields.

From the point of view of directly improving the intrinsic thermal conductivity of polymers, intrinsic high thermal conductivity polymer molecules need to have a conjugate structure or perfect crystal-oriented structure. The preparation process is complicated so that large-scale production is still not practical. Therefore, it is a mainstream method to produce high thermal conductivity polymer-based composites by adding filler with high thermal conductivity into the polymer matrix.[2] The thermal conductivities of inorganic materials, carbon materials and metals are much higher than those of polymers. However, considering the good electrical conductivity of carbon-based materials and metallic materials, they are not suitable to be widely used as thermally conductive filler in the preparation of polymer composites with insulating requirements. Thus BNNP, with good dielectric properties and insulation properties, is widely used as the thermally conductive filler in the preparation of highly thermally conductive insulating polymer composites.[96–101]

2.2.3.3 *Factors that Influence Enhancing the Thermal Conductivity of Polymers*

2.2.3.3.1 The Filler/Matrix Interface. When fillers are dispersed in polymer matrix, the polymer/filler interface cannot be avoided. During the conduction of heat in composite materials, the incomplete contact between polymer and filler at the interface leads to scattering of the crystal surface and obvious interfacial scattering, decreasing the mean free path of phonons and resulting in thermal resistance.[102,103] Improving the compatibility of the filler in the matrix is an important means of thermal conductivity enhancement. The functionalisation of nanofillers is a method that has been proven to be effective in a number of experiments.[104–112] The type of interaction at the filler/matrix interface also has an important influence on the thermal conductivity of the composites. Yu *et al.* modified BNNP by a non-covalent bond and a covalent bond, respectively, designing three interfacial forces of different strengths. The results indicate that the particle dispersion, compatibility and thermal conductivity of the three composite materials adhere to the following rules: the covalent bond bonding > surface physical entanglement > directly dispersed. Under the same conditions, the stronger the interaction, the more the thermal conductivity of the polymer material is improved.[113]

2.2.3.3.2 The Distribution of the Filler. In the preparation of composites, most fillers added to the matrix tend to aggregate, especially in the curing process of thermosetting polymer matrix composites. This problem can also be improved by functionalisation of nanofillers.[105,114] Furthermore, thermal conductivity can be anisotropic due to its directionality. In recent years, some researchers have improved the thermal conductivity of polymers by aligning the non-spherical filler in one direction. Controlling the distribution of filler can be achieved in various ways: processing methods, such as extrusion[115] and injection moulding;[116] external field action, such as magnetic field,[117,118] and electric field;[10,23] and structural design, such as ice-templated[119] assembly strategy and bidirectional freezing techniques.[120]

2.2.3.3.3 The Crystal Structure Integrity of the Filler. Moreover, the thermal conductivity of polymer composites is also affected by the crystal structure integrity of the filler, the geometry of the filler and the processing process. For composite materials, the interface, defects, impurities or any structural discontinuities in crystal structure will cause strong phonon scattering and greatly reduce thermal conductivity.[121–123] The geometry of the filler, such as dimension, length, lateral dimension, sheet thickness and aspect ratio, has an influence on the thermal conductivity of the composite.[124–126] At the same amount of addition, the high aspect ratio filler system is superior to the spherical filler system. The corresponding thermal conductivity of polymers or their composites prepared by different processing is different for their different structures, defects and levels of impurities.[123,127,128]

2.3 Surface Engineering of BNNP

For a long time, surface engineering of boron nitride has been a subject of great attention from researchers. Surface engineering of BNNP is not only an efficient way to control the filler/matrix interface reasonably and to improve the compatibility of the filler in the matrix, but also an effective way to give boron nitride new properties that can be predicted and customised. In recent years, researchers have constantly tried to perform surface functionalisation of BNNP for many new features and applications to be implemented this way. Although the chemical stability and reaction inertness of BNNP make this work challenging, there are some technologies that have been successfully developed, with both non-covalent bond and covalent bond functionalisation applied.

2.3.1 Non-covalent Bond Functionalisation

Boron nitride is oleophobic and hydrophobic, and easy to aggregate and deposit in common solvents, which is disadvantageous for the subsequent preparation and processing of polymer composites. The advantage of non-covalent functionalisation compared to covalent bond functionalisation is that it does not destroy the structure of the BNNP and does not alter the two-dimensional properties of the nanomaterial.[129] Therefore, non-covalent force modification is often used to improve the dispersion of BNNP.

2.3.1.1 *Lewis Acid–Base Interaction*

There is a mutually attractive force in Lewis acid and base pairs. Lin *et al.* used a Lewis acid–base interaction mechanism between lipophilic and hydrophilic amine molecules and B atoms on the surface of h-BN to simultaneously exfoliate h-BN and functionalise BNNP, which can be stably dispersed in water and most organic solvents.[130] In the above mechanism, the terminal amino groups of alkylamine are the electron-rich groups acting as the Lewis base, while the B atoms in BNNP are the electron-deficient groups acting as the Lewis acid.

Lin *et al.* found that the Lewis acid–base interaction between the terminal NH_2 of octadecylamine and the electron-deficient B sites on the surface of boron nitride is effective for both exfoliation and non-covalent functionalisation of boron nitride, after increasing the treatment temperature and time.[130] Yu *et al.* functionalised BNNP with octadecylamine, making octadecylamine adsorb onto the surface of BNNP.[113] The long-chain octadecylamine molecule on the surface of the modified nanosheets will undergo molecular interfacial entanglement with the epoxy molecule during the curing process, which improves the dispersibility of boron nitride to some extent.

2.3.1.2 *π–π Stacking Interaction*

A frequently used method is to functionalise graphene by a π–π stacking interaction, which is also applied to surface engineering of BNNP due to their structural similarity. As early as 2008, Han *et al.* found that poly-(*m*-phenylenevinylene-*co*-2,5-dictoxy-*p*-phenylenevinylene) can modify BNNP through π–π interaction.[131] However, such copolymers are particularly expensive and the solvents used are relatively rare, so they are subject to considerable limitations in application. In 2013, Liu *et al.* used the 'Janus' modifier block copolymer of polystyrene and polymethyl methacrylate to modify BNNP, in which the polystyrene chain has a π–π stacking interaction with BNNP (Figure 2.7).[132] Subsequently, in 2015, Ma *et al.* reported a method to modify BNNP by an organic polymer, using polythiophene or polyvinylpyrrolidone.[133]

Due to their aromatic structure, dopamine and polydopamine (PDA) can be used for the surface modification of boron nitride, through π–π stacking interactions and van der Waals forces, for improving the dispersion of boron nitride in the matrix and reducing the interface thermal resistance.[134,135] Shen *et al.* added h-BN powder to a proportionally mixed Tris buffer and ethanol solution with stirring for 6 h.[24] The phenolic hydroxyl and amine groups of PDA gave the polydopamine-modified boron nitride good hydrophilicity.

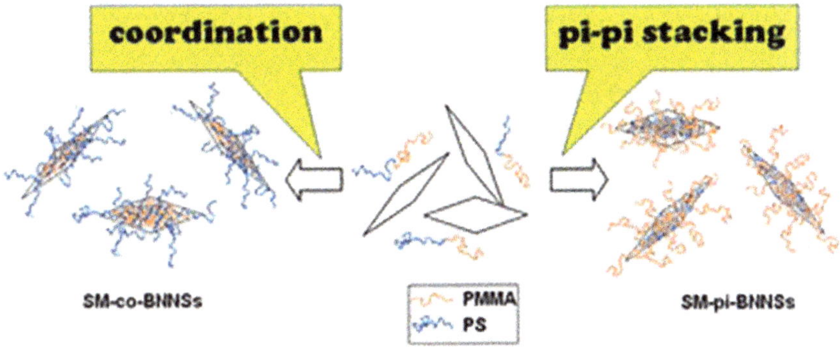

Figure 2.7 Schematic of the 'Janus' modifier for modifying BNNP.
Reproduced from ref. 132 with permission from the Royal Society of Chemistry.

Figure 2.8 Photographs showing the differences in the dispersibility of microplatelets in water for (a) h-BN (left) and h-BN@PDA (right) platelets just added into water, (b) after 10 min of ultrasonication and (c) after intense shaking and another 10 min of ultrasonication.
Reproduced from ref. 24 with permission from American Chemical Society, Copyright 2015.

The attained powder can be easily dispersed in water (Figure 2.8) and more uniformly dispersed in the polyvinyl alcohol matrix, which makes the thermal conductivity of this polymer composite as high as 5.4 $W\,m^{-1}\,K^{-1}$ at 10 vol% filler loading. Yu *et al.* simplified the exfoliation and polydopamine function of boron nitride into a one-step reaction by ball milling.[107] The resulting BNNP@PDA was further mixed with cellulose nanocrystal (CNC), fabricating a material with thermal conductivity up to 40 $W\,m^{-1}\,K^{-1}$ with 94 wt% BNNP.

In contrast to being applied directly, modified boron nitride can also be used for more complex chemical treatment. Su *et al.* first adsorbed dopamine molecules onto boron nitride by π–π stacking interactions, followed by chemical treatment to attain polydopamine-modified boron nitride, and then mixed with multilayer graphene, added in an alicyclic epoxy resin, and finally obtained a polymer composite with electric insulativity (electrical conductivity $<1.5\times10^{-10}$ $S\,cm^{-1}$) and high thermal conductivity (through-plane ~1.27 $W\,m^{-1}\,K^{-1}$, in-plane ~1.31 $W\,m^{-1}\,K^{-1}$).[110]

2.3.1.3 Hydrogen Bond Interaction

Hydrogen bonding is an important non-covalent interaction force in BNNP modification for surface engineering of BNNP. Zeng *et al.* used ball milling technology to grind sodium cholate with h-BN, not only exfoliating h-BN, but also recombining sodium cholate and BNNP through van der Waals forces. Using a hydrogen bond interaction between the modified BNNP and the polyvinyl alcohol matrix, a 'nacre-like' structure with outstanding mechanical properties and high thermal conductivity was obtained.[136]

2.3.1.4 Decoration with Inorganic Particles

Functional inorganic particles with different properties can be decorated on boron nitride to achieve surface engineering. The binding force between inorganic particles and coated boron nitride is not as strong as covalent bonds, it is more similar to non-covalent modification.[137] In order to benefit the application of BNNP for improving the thermal conductivity of polymer composites, many inorganic particles have been studied for BNNP modification, such as Ag,[138] SiO_2[139] and Fe_3O_4.[140]

Wang *et al.* first performed the ultrasonic-assisted liquid-phase exfoliation of commercial h-BN powder with *N,N*-dimethylformamide (DMF) as solvent, then mixed and stirred the obtained BNNP solution with silver nitrate solution, during which DMF also acted as a reducing agent, and finally achieved silver nanoparticle-deposited BNNP (Figure 2.9).[138] Due to the 'bridging' function of silver nanoparticles (AgNPs), the thermal resistance between AgNP deposited BNNPs was remarkably decreased. The BNNP/AgNPs/epoxy composite has a maximum thermal conductivity of 3.06 W m^{-1} K^{-1}, which is approximately twice that of the BNNP/epoxy composite at the same load rate.

Ahn *et al.* modified the boron nitride surface with silica (SiO_2) by a sol–gel method using tetraethyl orthosilicate (TEOS) as a precursor and hydrochloric acid (HCl) as a catalyst.[139] The h-BN was first mixed with the TEOS solution to form a gel structure. After the heat treatment in two steps, the h-BN/TEOS gel was sintered to eliminate by-products and consolidate the particle/coating adhesion. Subsequent experiments and characterisation results prove that the surface free energy of the surface-modified BN (SiO_2–BN) particles increases and has better interface compatibility with the epoxy-terminated dimethylsiloxane (ETDS) matrix, which is more favourable for improving the thermal conductivity and mechanical properties of the composites.

Kim *et al.* synthesised novel magnetically responsive boron nitride *via* a two-step procedure by decorating with Fe_3O_4.[140] After boron nitride was treated with sodium hydroxide (NaOH), Fe-BN was obtained by coating boron nitride with Fe_3O_4 through a hydrolysis reaction. Then, on the basis of Fe-BN, a layer of poly(tetramethylene glycol) was further introduced to attain P-BN. The orientation of the modified h-BN in the epoxy resin was

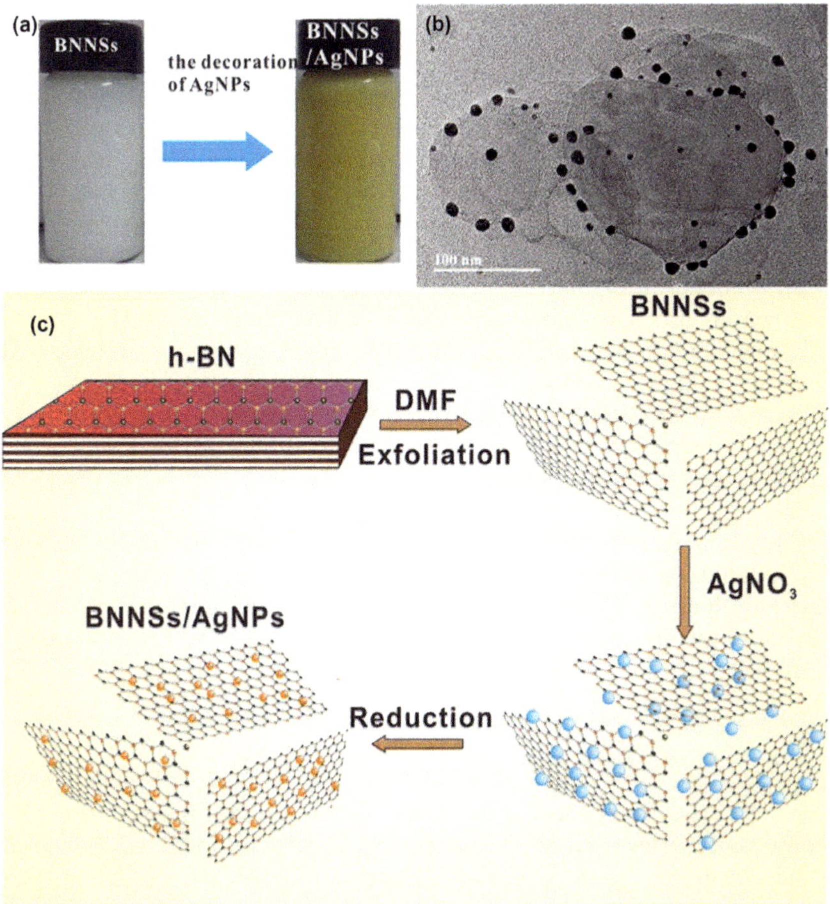

Figure 2.9 (a) Optical image of BNNSs and BNNSs/AgNPs dispersed in DMF. (b) Transmission electron microscopy (TEM) image of BNNSs/AgNPs. (c) Illustration of the preparation process of BNNS/AgNPs nanohybrids. Reproduced from ref. 138 with permission from Springer Nature, Copyright 2016.

controlled by applying a magnetic field during the manufacture of the polymer composite. After orientation of the filler particles, the thermal conductivity increases from 0.832 to 1.448 $W m^{-1} K^{-1}$ (Fe-BN) and 1.577 $W m^{-1} K^{-1}$ (P-BN) at a load of 20 vol%.

Today, the decoration of boron nitride with inorganic particles shows great potential in improving the original properties of h-BN nanomaterials and imparting new properties to h-BN nanomaterials. With further development of functional inorganic particles and more in-depth studies of interaction mechanisms and experimental methods in the future, this method will be applied in more fields.

2.3.2 Covalent Bond Functionalisation

Covalent bond functionalisation refers to the interaction between functional groups and B or N atoms in BNNP being *via* a covalent bond. Similarly to graphene, reactions of BNNP on their surface always involve opening of the polarised bonds or destruction of conjugated structures. Thus, the presence of new bonds should always be paired. When a functional group forms a new bond with B or N atoms, the compensatory group (whether or not it is the same as the functional group) is also attached to the unpaired N or B atoms. Experimental and theoretical studies on covalent bond functionalisation of conjugated carbon systems have shown that bonded functional groups not only are in adjacent positions, but also can be spatially separated, even forming bridging bonds, which can also be applicable to BNNP systems.[141–144] In BNNP systems, B atoms are electron-deficient centres that are reactive with nucleophilic groups, while N atoms are electron-rich centres that can be attacked by electrophilic ones.

Although BNNP and graphene have similar structures, the high chemical stability imparted by the partial ionicity of B–N bonds makes the chemical modification of BNNP much more difficult than graphene. Even so, researchers have succeeded in introducing numerous functional groups into the boron nitride structure, completing the covalent bond functionalisation of BNNP.

2.3.2.1 *Hydroxyl Group (–OH)*

The hydroxyl group (–OH) is the most common and important functional group for covalent functionalisation of BNNP. In most cases, –OH is covalently grafted to the electron-deficient B sites on the surface of BNNP. Hydroxylated boron nitride (OH-BNNP) is widely used not only for direct applications in multiple fields, including thermal conductivity enhancement of polymers, but also as a frequently used precursor for further reaction and functionalisation.[145,146] So far, researchers have reported a number of different ways to form hydroxyl-BNNP, including ultrasonic assisted,[50] hydrothermal reaction,[147] alkali-assisted ball milling,[148] plasma treatment,[149] reserved reaction,[150] oxygen radical reaction,[20,151] steam treatment at high temperature,[21] *etc.*

Coleman's group obtained OH-BNNP by a two-step chemical reaction at high temperature and pressure.[20] First, alkoxy groups were introduced onto the surface of BNNP using an organic peroxide reagent (di-*tert*-butyl peroxide), followed by treatment with a piranha solution (H_2SO_4/H_2O_2). The bonded *tert*-butoxy groups hydrolysed into 'H' groups, and successfully grafted –OH onto the surface and edge of BNNP. Subsequently, Pakdel *et al.* reported a simpler one-step plasma-assisted process to hydroxylate BNNP. Under air plasma conditions, O_2 and H_2O in the air could generate OH radicals to react with BNNS. Due to the high reactivity of the OH and OR groups, these reactants reacted randomly with the B/N sites on the BN surface, with both in-plane and edge positions possibly functionalised.[149]

Unlike the above-described method of directly functionalising BNNP, in many cases, the exfoliation and functionalisation of BNNP are performed simultaneously. Lin *et al.* found that with water as the solvent, the hydroxylated monolayer and a few-layer boron nitride nanoplatelet were obtained by the direct ultrasonic treatment of h-BN powder.[50] The hydrolysis effect attracted the exfoliation, cutting and hydroxylation of BN layers. In addition to sonication-assisted exfoliation, ball milling technology is fashionable in the functionalisation of BNNP. Lee *et al.* proposed a hydroxide-assisted ball milling approach where NaOH aqueous solution was added during the ball milling process.[101] High-speed ball milling provides the shear force required for exfoliation, while NaOH utilises the reaction between –OH and h-BN to cut and hydroxylate the boron nitride. The OH-BNNP products obtained by the synergistic action were dispersed into the styrene butadiene rubber (SBR) matrix. OH-BNNP demonstrated a better effect in improving thermal conductivity of SBR nanocomposites than carbon black and graphite at low loading levels (Figure 2.10).

Xiao *et al.* reported that OH-BNNP can be obtained under high-temperature steam treatment at 850 °C without other impurities.[21] The authors used electron energy loss spectroscopy (EELS) plots to provide direct evidence of BNNP edge hydroxylation (Figure 2.11(a, b)). As an effective filler for enhancing the thermal conductivity and thermal response, the cross-linked poly(*N*-isopropylacrylamide) (PNIPAM) and OH-BNNP were integrated through hydrogen interaction, forming the PNIPAM/OH-BNNS hydrogel (Figure 2.11(c, d)).

Apart from the methods based on h-BN or BNNP as precursors, Weng *et al.* developed an oxygen radical reaction to produce BNNP without BN and its derivatives as origin materials.[151] The *g*-C_3N_4 was the N source, while H_3BO_3 was the B source. The hydroxyl-group-containing B precursor from H_3BO_3 replaced the carbon component in the *g*-C_3N_4 solid at moderate temperatures. Thus, a portion of the –OH groups in the B precursor were successfully retained in the obtained product. Therefore, highly hydroxylated OH–BN derivatives were produced directly without any subsequent treatment.

2.3.2.2 Amino (–NH_2) and Amine (–NHR) Groups

Like hydroxylated boron nitride (OH-BNNP), aminated boron nitride (NH_2-BNNP) is also an important intermediate for many complex functionalisation reactions, such as (NHR-BNNP), as well as being applied directly. Amino (–NH_2) and amine (–NHR) groups belong to nucleophilic groups, by which B sites on the surface of BNNP can be modified. Research on the amino (–NH_2) and amine (–NHR) functionalisation of boron nitride nanomaterials dates back to 2007, and the report from Zettl's team that the –NH_2 group can be covalently attached to the BN nanotube surface.[152]

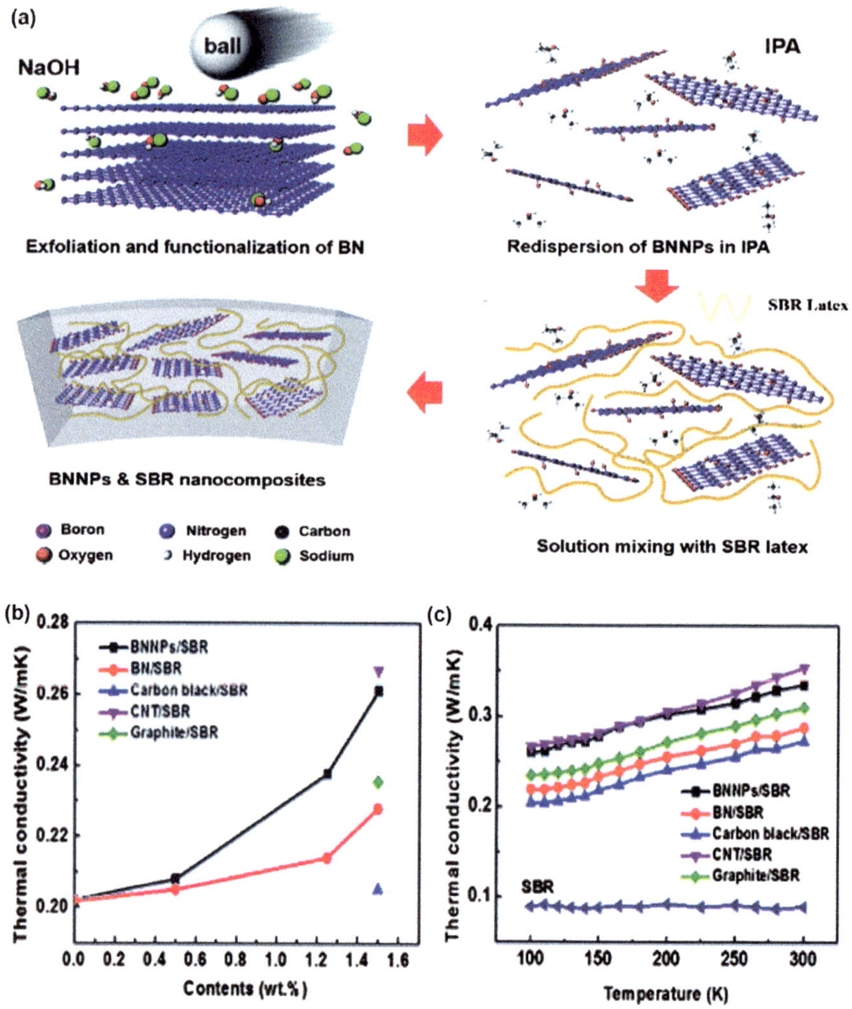

Figure 2.10 (a) Schematic showing the overall processing for preparing BNNP and BNNP/SBR nanocomposites. (b) Thermal conductivities of the elastomer nanocomposites with diverse fillers. (c) Temperature dependence of the thermal conductivity for different elastomer nanocomposites. Reproduced from ref. 101 with permission from the Royal Society of Chemistry.

As a typical example for direct application of NH_2-BNNP, Zeng *et al.* added a kind of surfactant, sodium cholate, during low energy ball milling and subsequent ultrasonic treatment and centrifugation to obtain NF-BNNP.[153] Density-controlled BNNP aerogels with excellent superplastic property were generated after mixing NF-BNNP and 1,4-butanediol diglycidyl ether (BDGE) as the cross-linking agent, and freeze drying. In addition, using ball milling technology, Lei and Chen replaced sodium cholate with urea to complete the exfoliation and amination of boron nitride in one step.[154] The obtained

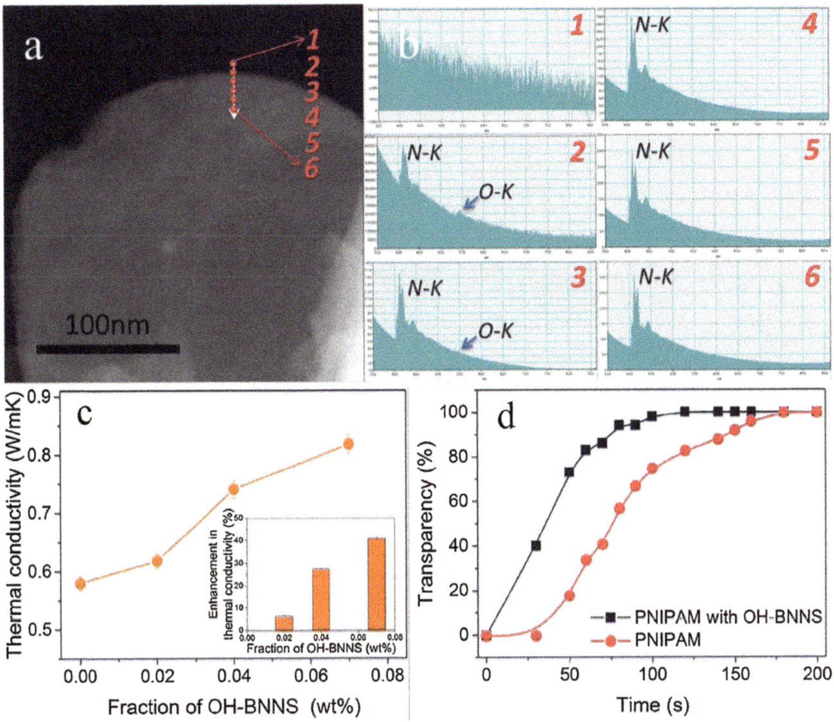

Figure 2.11 (a, b) Electron energy loss spectroscopy (EELS) mapping shows that the hydroxyl groups are predominately located along the edges of the nanoplatelets. (c) Thermal conductivity improvement with increasing ratio of OH-BNNP. (d) Change in transparency after being taken out of 50 °C warm water.
Reproduced from ref. 21 with permission from John Wiley & Sons, Copyright 2015 WILEY-VCH Verlag GmbH & Co. KGaA, Weinheim.

NH_2-BNNS products dispersed evenly and well in water, forming a colloid solution that can be stable for several months, with a concentration up to 30 mg ml^{-1}. By drying naturally in the air, an almost transparent boron nitride membrane is produced, while after complete dehydration by freeze-drying technology, an ultra-light (density of 1.4 mg cm^{-3}) NH_2-BNNP aerogel is produced.

The functionalisation of amine groups (–NHR) is more complicated. Yu *et al.* first introduced BNNP into the amino group by silane coupling agent treatment, and then grafted the hyperbranched polyacrylamide (HBP) on the modified BNNP (Figure 2.12(a, b, c)). The large number of amino groups carried on the surface of HBP-BNNP and the epoxy groups in the epoxy resin formed covalent bonds during the curing process, further forming a strong interfacial interaction. Under the premise of the same processing technology and the same content, the stronger the interface interaction, the better the effect on enhancing thermal conductivity (Figure 2.12(d, e)).[113]

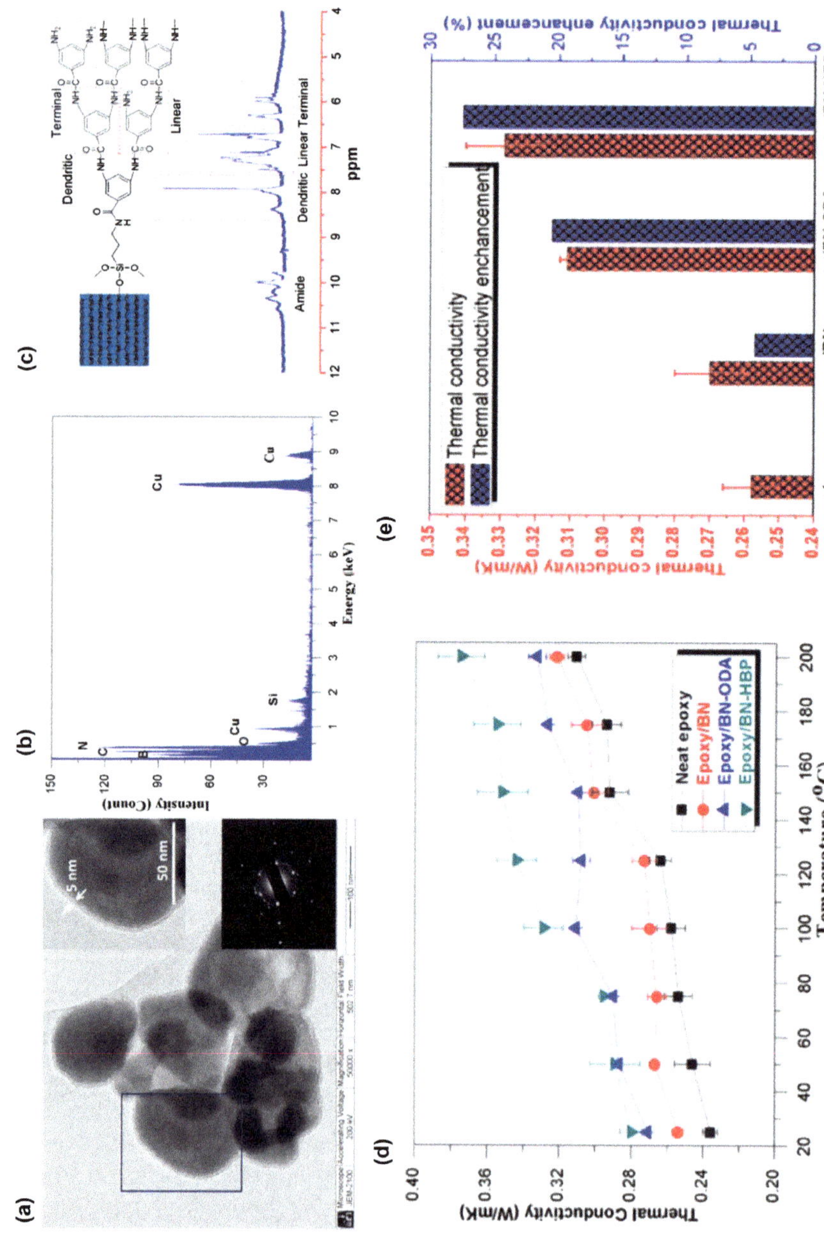
(a)
5 nm
50 nm
(b)
Intensity (Count)
Energy (keV)
N
C
B
O
Cu
Si
Cu
Cu
(c)
Dendritic
Terminal
Linear
Amide
Dendritic Linear Terminal
ppm
(d)
Thermal Conductivity (W/mK)
Temperature (°C)
Neat epoxy
Epoxy/BN
Epoxy/BN-ODA
Epoxy/BN-HBP
(e)
Thermal conductivity (W/mK)
Thermal conductivity enhancement (%)
Thermal conductivity
Thermal conductivity enchancement
neat epoxy
epoxy/BN
epoxy/BN-ODA
epoxy/BN-HBP

Another recent example is that Li *et al.* grafted polyethyleneimine (PEI) onto hydroxylated boron nitride treated with sulfuric acid.[155] On the one hand, PEI on the surface of BN can help to limit the aggregation of filler in the poly(vinyl alcohol) (PVA) polymer matrix, so as to achieve more uniform distribution and build more effective thermal conduction channels. On the other hand, this linear polyelectrolyte and PVA form a bridging structure that can be a phonon transfer pathway, significantly reducing the interfacial thermal resistance. When the surface-functionalised filler particle loading is increased to 50 vol%, the thermal conductivity of the PVA composite strengthens to 14.22 $W m^{-1} K^{-1}$.

2.3.2.3 Alkyl (–R) Groups

For alkyl groups, there are two ways to link with boron nitride particles: directly on B or N atoms, and by forming B–C–N bridged bonds. Sainsbury *et al.* used dibromocarbene (DBC) to exfoliate and functionalise boron nitride nanoplatelets at the same time.[156] In this chemical process, the DBC-BNNP were obtained through dibromo-bridged bicyclo BCN systems between BNNP and dibromocarbene, with BNNP as the reaction substrate, which formed B–CBr_2 ylides to stabilise carbenes. Subsequently, the DBC-BNNP were treated with a highly reactive organolithium reagent, *n*-butyl lithium (*n*-BuLi), with the bromine atoms replaced by butyl groups and finally obtaining the required alkylated BNNP, *i.e.* $(n\text{-}C_4H_9)_2C$-BNNP. The detailed characterisation of intermediates and products in this report confirmed the feasibility of alkylation of BNNP by carbene reaction and alkyl/halogen exchange.

2.3.2.4 Silanisation

A silane-coupling agent refers to the organic silicon monomer with more than two different reaction groups in the molecule. Its chemical molecular formula can be written as follows: $Y(CH_2)_nSiX_3$, where $n = 1$–3, X is a hydrolysable group and Y is an organic functional group. A silane-coupling agent can bond chemically with organic and inorganic substances, building a 'molecular bridge' on the interfaces of inorganic and organic substances and coupling two kinds of materials with different properties together. Therefore, it is often used as a highly effective surfactant and is also widely used in the surface treatment of composite materials to improve their

Figure 2.12 (a) Representative TEM image of HBP-BNNP. Inset: magnified image of the boxed region and corresponding selected area electron diffraction (SAED). (b) Corresponding energy-dispersive X-ray spectroscopy (EDS) spectrum. (c) ^1H NMR spectra of HBP-BNNP in DMSO-d_6. (d) Thermal conductivity of the neat epoxy and epoxy composites as a function of test temperature. (e) Thermal conductivity and thermal conductivity enhancement of the neat epoxy and its composites at 100 °C. Reproduced from ref. 113 with permission from Elsevier, Copyright 2011.

performance, including thermal conductivity.[157] For the application of a silane-coupling agent in BNNP surface engineering, previous in-depth research has been conducted and achieved considerable results.

As mentioned above, the silanised BNNP can be utilised as an intermediate for the subsequent functionalisation reaction, with the characteristics of the silane-coupling agent having multiple active groups.[113] Moreover, the use of silanised BNNP directly in the preparation of composite materials has a remarkable effect in enhancing the thermal conductivity of polymers.[158] The interface binding force between raw BNNP and polymer matrix is generally weak, which leads to the accumulation of filler particles in the processing procedure and poor dispersion levels. At the same time, due to the gap because of the filler agglomeration, and the interface where the compatibility is not high enough, the phonons that transfer heat flux will be strongly scattered, resulting in high thermal resistance, and thus greatly affecting the thermal conductivity of the material. Under these conditions, the silane molecule modified on the surface of the BNNP can function as a bridge, with boron nitride particles attached to one end and polymer matrix surrounding the other end, so that the degree of interface affinity is increased, and the bonding force is enhanced. The nanocomposite produced has better uniformity, as well as phonon scattering at the interface being inhibited and thermal resistance reduced.

Seyhan *et al.* reported a method for silanising BNNP by microfluidisation techniques and vinyl-trimethoxy silane (VTS) reagents.[159] After chemical exfoliation of BNNP through microfluidisation and acid oxidation, hydroxyl groups were introduced onto the surface of BNNP. VTS was hydrolysed to silanol in the BNNP solution, then bonded to the hydroxyl groups on the surface of the BNNS to form hydrogen bonds. After drying, a condensation reaction occurred between adjacent silanol molecules to form a so-called polysiloxane layer bonded to the constituent surface. Silanised BNNP shows good water solubility. At different BNNP contents (1–5 wt%), PP nanocomposites containing silanised BNNP show higher thermal conductivity than those containing silanisation-free BNNP.

Lee *et al.* oxidised boron nitride using sulfuric acid in the presence of a sodium nitrate catalyst, followed by surface silanisation in an aqueous ethanol solution.[160] Compared with raw boron nitride (control group), the silanised boron nitride (experimental group) showed both enhanced thermal conductivity and mechanical properties. The thermal conductivity of epoxy composites in the experimental group was 0.357 $W\,m^{-1}\,K^{-1}$, with a bending strength of 63 MPa and fracture toughness of 1.23 MPa at 10 wt% of filler loading, which was 57%, 57.5% and 10% higher than that of the control group, respectively.

2.3.2.5 *Natural Macromolecules*

Natural macromolecules have also been applied in the functionalisation of boron nitride nanomaterials, which have unique advantages, such as low toxicity, environmental friendliness, low price and convenient access,[161] compared with synthetic chemical reagents. Different studies have been reported on efficient functionalisation of nanofillers using protein[162,163] and polysaccharide.[164]

The abundance of amphiphilic functional groups on protein molecules gives them great potential for decorating filler particles to improve compatibility. Wang *et al.* explored a one-step method to obtain functional BNNP from h-BN powder by using soybean protein isolate (SPI).[165] In long-term aqueous sonication, SPI acted not only as a surfactant in the exfoliation process, but also as an active agent in the modification process. The resulting SPI-BNNP and nanofibres were combined to form high thermal conductivity biobased composite films, whose in-plane thermal conductivity reaches 6.88 $W\,m^{-1}\,K^{-1}$ at low loading (1.0 wt%).

Like proteins, the reactive groups on polysaccharide macromolecules can be chemically bonded to boron nitride to complete the surface engineering of boron nitride. Chen *et al.* reported a method named sugar (sucrose)-assisted mechanochemical exfoliation (SAMCE) (Figure 2.13(a)).[96] Sucrose was added during the ball milling process. Under the shear and impact forces generated by the ball milling, the boron nitride sheets were fractured and

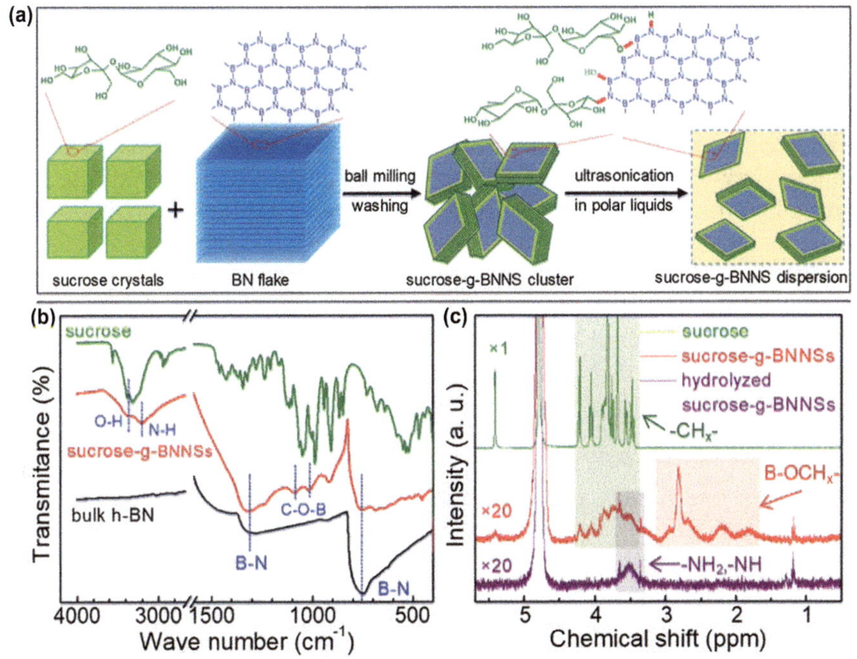

Figure 2.13 (a) Illustration of the SAMCE process. (b) Fourier transform infrared (FTIR) spectra of sucrose-*g*-BNNSs. (c) ^{1}H NMR spectra of sucrose-*g*-BNNSs. Reproduced from ref. 96 with permission from John Wiley & Sons, Copyright 2019 WILEY-VCH Verlag GmbH & Co. KGaA, Weinheim.

produced active B and N sites on the surface. The active N site tends to capture the –H and –R moieties from the sucrose molecule, while the active B sites are more favourable for binding to the –OR and –OH moieties, resulting in the stable sucrose-*g*-BNNP. Subsequent characterisation revealed that the sucrose molecule was indeed grafted onto the BNNP (Figure 2.13(b, c)). Also taking advantage of ball milling techniques, Wang *et al.* used glucose to functionalise boron nitride.[164] Through freeze-drying and thermal cross-linking, functional boron nitride formed a lightweight and thermoconductive aerogel with hydrophobicity, high temperature stability and high thermal conductivity of 9.8 W m^{-1} K^{-1}.

2.3.2.6 Other Groups

In addition to the functional groups already discussed here, there are many other reports on the surface engineering of boron nitride. Organic functional groups include urea, esters (–OCOR), amides (–NHCOR), carbonyl derivatives –COX (X = H, CH_3, OCH_3, OH, and NH_2), *etc.*, and inorganic functional groups, such as hydrogen (–H), fluorine (–F), *etc.*[137,141]

Complex organic functional groups can be directly modified on boron nitride nanoplatelets, but the more common way to achieve this goal is through derivative reactions with hydroxylated (OH-BNNP) and aminated (NH_2-BNNP) intermediates. For example, the –OH group in OH-BNNP can be esterified with perfluorobutyric acid ($CF_3CF_2CF_2COOH$) or thioglycolic acid ($HSCH_2COOH$), or functionalised with isocyanate, to graft new groups onto the surface of boron nitride.[20,50,147] The successful application of similar derivative reactions with OH-BNNP and NH_2-BNNP as intermediates, not only greatly enriches the chemical reaction pathway of complex functional groups grafted on BNNP, but also provides the possibility of BN nanofillers with customised properties for composite materials.

Hydrogen is the simplest structural group used to modify BNNP. Functionalisation of BNNP will change its band gap. The fully hydrogenated BNNP band gap changed from the original 5.0–6.0 eV to 3.0–3.1 eV, and the half-hydrogenated BNNP shows a narrower band gap of 2.24 eV.[166] The fluorination of BNNP can be achieved by hydrothermal reaction.[56] Under high temperature and pressure, F atoms in ammonium fluoride (NH_4F) covalently bonded with B atoms in boron nitride, making the B atoms sp^3-hybridised and finally completed the stripping and fluoridation of BNNP.

Besides exploring experimental methods, researchers have also made many attempts in the functionalisation of BNNP using theoretical calculations. From first principles calculations, Bhattacharya *et al.* theoretically discussed the effect of many other groups (such as –CH_3, –CHO, –CN) on the band gap of BN monolayers, which is between 0.3 and 3.2 eV.[166] At the same time, based on phonon frequency calculations, they predict some chemically functionalised structures can be considered as having intrinsic stability. Tian *et al.* performed density functional theory (DFT) calculations to study the effect of CHO radicals on the structure and properties of boron nitride sheets with different coverages.[167] CHO radical functionalisation will narrow the band gap of BNNP, especially when the number of CHO radicals modified on BNNP is odd.

2.3.2.7 *Heteroatom Doping*

Heteroatom doping is a kind of special functionalisation, referring to B or N atoms in the in-plane layer structure of boron nitride being substituted by heteroatoms, without forming large domains. The doping of special heteroatoms can significantly change the properties of BN nanomaterials and endow them with new properties.[168] Carbon (C) and oxygen (O) atoms are the most common doped heteroatoms.

In the case of carbon-atom doping, Liu *et al.* partially etched boron nitride films prepared by CVD by means of lasers and then a small amount of graphene was grown in the etched portion, with CH_4 as the carbon source at 1000 °C.[169] Oxygen-doped BNNP was reported as being obtained *via* a sol–gel method.[170] Lie *et al.* mixed B_2O_3 and urea in methanol solution to form a complex between the boron and the urea precursors, which converted into few-layer oxygen-doped BNNP after high-temperature calcining.

2.4 High Thermal Conductivity BNNP Polymer Composites

With the rapid development of microelectronics technology and the electrical industry, the size of electronic components and logic circuits is getting smaller and smaller, while the operating frequency is increasing sharply, and the ambient temperature is changing toward high temperature. In order to ensure reliable operation of electronic components for a long time, the requirements for thermal conductivity and heat dissipation of polymer insulation materials are also ever-changing. The traditional method of preparing a thermally conductive polymer composite is to directly incorporate the thermally conductive filler into a polymer matrix. With the content of boron nitride nanosheets increasing, the composite thermal conductivity increases, but the extent of the increase is modest. Until the content is up to a certain level, the adjacent boron nitride nanoplatelets are in contact with each other forming a large number of mutually interconnected heat conduction networks. The thermal conductivity of the composite material can be significantly increased. However, under the high addition of BNNP, the processing performance of the composite material will be noticeably reduced, with serious loss of mechanical properties, and the raw material cost will rise markedly, which severely limits the large-area application of such composite materials.[171–173]

Since the beginning of the 21st century, many methods have been adopted to try to improve the efficiency of enhancing the thermal conductivity, achieving high thermal conductivity composites at low loading of boron nitride filler. In the past decade, scientists have found that enhancing the thermal conductivity efficiency of BNNP can be improved effectively by properly controlling the internal microstructure of the composite materials.[174,175]

2.4.1 The Orientation of BNNP

Compared with random scattered BNNP, oriented BNNP has a higher efficiency for enhancing thermal conductivity. On the one hand, due to the directionality of heat transfer, when BNNP is arranged along a specific orientation in the matrix, it can significantly improve the thermal conductivity of the composite materials in that direction. On the other hand, the oriented BNNP is more likely to achieve a high concentration of the filler locally and overlap between adjacent fillers, forming a heat conduction network at a low level of BNNP addition. The orientation of BNNP in the polymer can be caused by mechanical actions, such as moulding, casting and multistage stretching, and by external fields, such as magnetic and electric fields.

Taking orientation by mechanical action as an example, Zhang *et al.* provided a continuous and powerful shear flow field through multistage stretch extrusion, introducing boron nitride into the high viscosity high-density polyethylene (HDPE) matrix. With boron nitride almost fully orienting along the extrusion plane, the thermal conductivity of the composite reached 3.57 $W m^{-1} K^{-1}$ at a filling amount of 40 wt%.[176] Wang *et al.* self-assembled BNNP and polystyrene (PS) microspheres to form hybrid microspheres through strong electrostatic interaction. The prepared composite microspheres were hot-pressed near the glass transition temperature to selectively distribute BNNP at the deformed PS microsphere interface. The preparation method is shown in Figure 2.14. Thanks to the higher degree of planar orientation of BNNP in the PS matrix, the thermal conductivity of the composite can be as high as 8.0 $W m^{-1} K^{-1}$ when the BNNP load is 13.4 vol%.[177]

Magnetic and electric fields are the most common external field forces applied to BNNP orientation. Du *et al.* used boron nitride co-functionalised by iron oxide and dopamine as a filler to prepare the anisotropic polyvinylidene fluoride (PVDF)/BN composite by orienting BN under a magnetic field.[178] The results showed that the thermal conductivity of the composite increased from 0.35 $W m^{-1} K^{-1}$ to 0.82 $W m^{-1} K^{-1}$ after the modification with dopamine, and further increased to 1.43 $W m^{-1} K^{-1}$ after orientation alignment. At the same time, the material exhibits good tensile strength and impact resistance.

In addition to the magnetic field, the electric field can also significantly promote the orientation of BNNP. It has been reported that under high direct-current voltage, linear assembly of BN nanosheets (LABN) is perpendicular to the membrane plane in the polysiloxane matrix, with high anisotropy.[179] A more complex and sophisticated application of electric fields is electrospinning technology. Chen *et al.* first prepared PVA/BNNS composite fibres with orientation arrangement and internal contact of BNNS through electrostatic spinning, and then rolled up the fibre to form a porous and vertically in-plane-oriented PVA/BNNS fibre array structure. Based on this structure, polydimethylsiloxane (PDMS) was poured by vacuum

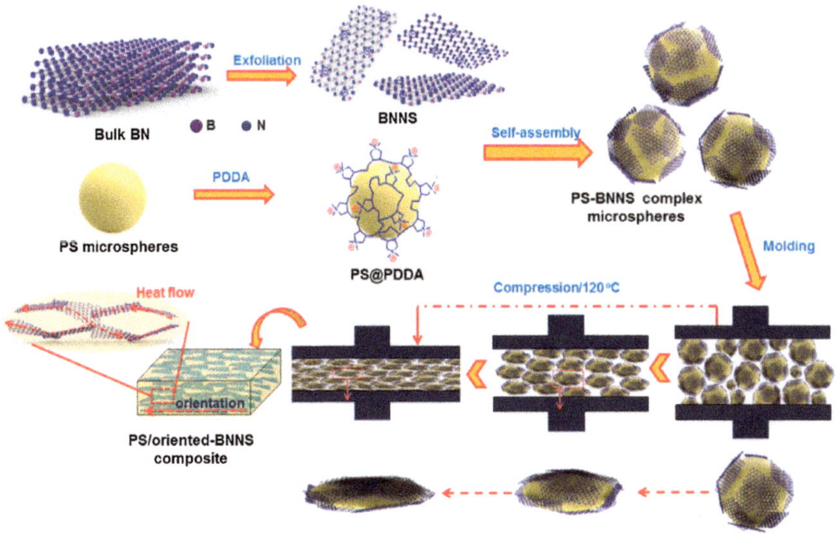

Figure 2.14 Schematic of the formation mechanism of PS-BNNS complex microspheres and PS/oriented BNNS composites.
Reproduced from ref. 177 with permission from American Chemical Society, Copyright 2017.

assistance to make PDMS/PVA/BNNS nanocomposites. At a loading of 15.6 vol%, the nanocomposite shows a super-high in-plane thermal conductivity of 1.94 $W\,m^{-1}\,K^{-1}$ at room temperature, which is nearly nine times higher than pure PDMS, and is a potential thermal interface material.[10]

Besides the examples discussed above, Table 2.2 lists some high thermal conductivity polymer composites with regard to the orientation of boron nitride.

2.4.2 Synergy Effects

The introduction of a thermally conductive filler other than boron nitride into the polymer matrix is a double-edged sword. On the one hand, the addition of new fillers will inevitably bring new interfaces, aggravating the interface thermal resistance problem; on the other hand, the synergy between different thermal conductive fillers can effectively improve the efficiency of enhancing the thermal conductivity. Developing methods to restrain the former effects and promote the latter are key for researchers to address this challenge. There are two main synergistic effects: one is represented by graphene oxide (GO), which can effectively reduce the interfacial thermal resistance due to the matching phonon spectrum and strong interfacial interaction between the fillers and BNNP;[180] the other is represented by metal particles and carbon tubes, which can construct more thermal paths between BNNP through the 'bridge' and 'weld' effects of fillers.[102]

Huang *et al.* prepared graphene oxide-encapsulated boron nitride hybrids (h-BN@GO) *via* electrostatic self-assembly, which were subsequently added

Table 2.2 High thermal conductivity polymer composites and the orientation of boron nitride.

Filler	Matrix	Composite processing procedure	Thermal conductivity ($W m^{-1} K^{-1}$)	References
BN sheets	Polyethylene (PE)	Multistage stretching extrusion	1.21 (30 wt%)	201
BN sheets	Polycarbonate (PC)	Hot-pressing	3.09 (18.5 vol%)	202
BNNP	PVA	Tape-casting on glass panel	4.5 (30 wt%)	203
BNNP	Poly(dimethylsiloxane)	Electric field (50 Hz)	1.6 (15 vol%)	204
BN platelets (functionalised with Fe_3O_4)	Epoxy resin	Magnetic field	5.5 (40 vol%)	205
BN platelets (functionalised with iron oxide nanoparticles)	Epoxy resin	Magnetic field	1.4 (20 vol%)	206
BNNP	PVA	Electrostatic spraying and electrospinning	21.4 (22.2 vol%)	207

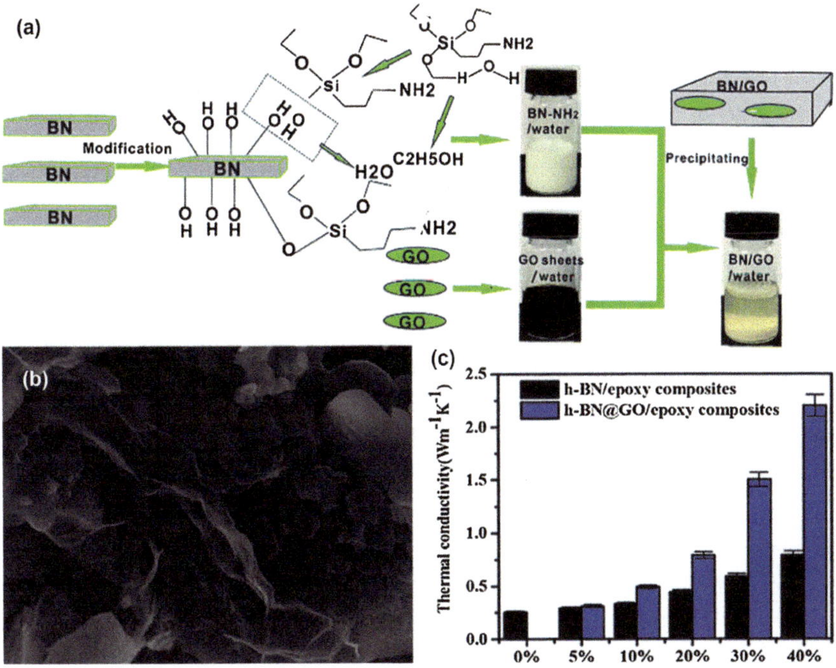

Figure 2.15 (a) Schematic of the preparation of h-BN@GO hybrids. (b) SEM image of h-BN@GO hybrids. (c) Thermal conductivity of h-BN@GO/epoxy and h-BN/epoxy composites with different mass fractions.
Reproduced from ref. 181 with permission from the Royal Society of Chemistry.

into epoxy to fabricate h-BN@GO/epoxy composites (Figure 2.15(a)).[181] When the loading capacity of the filler is 40 wt%, the thermal conductivity of the composite can reach 2.23 W m^{-1} K^{-1}, which is nearly three and ten times the thermal conductivity of the pure h-BN filler composite under the same loading and the pure epoxy resin, respectively (Figure 2.15(b)). Li *et al.* reported a GO/BNNP composite film with an in-plane thermal conductivity up to 11.9 W m^{-1} K^{-1} at a BNNP content of 80 wt%.[182] After exfoliation and surface modification of BNNP by tannic acid, the modified BNNP and GO were directly filtered into a film. In addition, if 1.6 wt% PVA was added to fill the voids in the film, the thermal conductivity of the film could be further improved.

Pak *et al.* studied the hybrid system of multiwalled carbon nanotubes (MWCNT) and h-BN, observing the synergistic effect of the two fillers on enhancing the thermal conductivity of the polymer. The maximum thermal conductivity of the pelletisable, injection mouldable composite material that is produced is 1.74 W m^{-1} K^{-1}.[183] Apart from existing between single nanomaterials and boron nitride, synergy effects can also be found between three or more nanomaterials. Yao *et al.* utilised the synergistic effect of BNNP silicon nanowires (SiCNW) and silver particles (Ag) to prepare a novel, highly thermally conductive composite paper.[184] By reducing the silver ions attached to the surface of BNNP and SiCNW, hybrids of BNNP-Ag and SiCNW-Ag were obtained, and they were collectively mixed with PVA as filler to filter into a film (BNNP-Ag/SiCNW-Ag/PVA). When the weight ratio of BNNP-Ag and SiCNW-Ag is 85 : 10, the in-plane thermal conductivity of the resulting composite paper can be as high as 21.7 $W m^{-1} K^{-1}$.

2.4.3 Three-dimensional Frame Structure

An ordered three-dimensional (3D) structural framework can not only effectively decrease the number of interfaces between filler and polymer matrix, thus reducing the interface thermal resistance, but also serve as a thermal conduction path, letting heat be transferred along the thermal conductivity framework, so as to obtain thermal conductivity composites with high thermal conductivity at relatively low BNNP content. There are two common methods of 3D structure skeleton construction: the template method and the self-assembly method.

2.4.3.1 *Template Method*

The template method is to prepare a three-dimensional structural composite material by taking the porous material with ordered structure as the skeleton, and then fill with the polymer. For example, Wang *et al.* used melamine foam as a skeleton, which allowed BNNP to adhere layer-by-layer through electrostatic interaction, and formed a thermally conductive composite *via* vacuum-assisted injection of epoxy resin.[185] Its thermal conductivity is up to 0.6 $W m^{-1} K^{-1}$ at a BNNP content of 1.1 vol%.

A boron nitride frame can also be synthesised directly. Manuela Loeblein *et al.* used Ni foam as a catalytic substrate, with 3D-BN growing in a split tube furnace from ammonia borane (NH_3–BH_3). After Ni was completely etched away, a high-density ultra-high thermal conductivity 3D-BN foam was acquired.[186] Xue *et al.* took the interconnection network of carbon-based nanorods as a template and successfully prepared two kinds of three-dimensional h-BN (h-BN nanorods and nanoplatelets) co-assembled porous skeletons by an *in situ* carbothermal reduction, CVD substitution reaction.[187] Based on this skeleton structure, various light-weight 3D-BN/polymer porous composite materials can be reliably produced, which have excellent mechanical strength, great dielectric properties, highly isotropic thermal conductivities (~26–51 times that of pristine polymers) and super-low coefficient of thermal expansion, and even recyclability and reusability.

Some special physical phenomena are also utilised to construct 3D skeletons. Researchers have developed a new method for preparing 3D structures, the ice-templating method, based on the discovery that ice crystals have specific orientations as water freezes directionally. Zeng *et al.* prepared a 3D-BNNP/epoxy thermally conductive composite *via* this method.[119] The specific process is as follows: the uniformly mixed BNNP and PVA aqueous solutions were directionally frozen, making BNNP oriented along the ordered ice crystal structure, and a 3D-BNNP aerogel was attained after vacuum freeze drying and annealing treatment, and, finally, epoxy resin was injected. When the BNNS content is 9.29 vol%, the thermal conductivity along the orientation can reach 2.85 W m^{-1} K^{-1}.

In order to realise a better orientation structure of BNNP, Han *et al.* used a unique bidirectional freezing technique to attain a nacre-mimetic 3D-BNNP anisotropic network skeleton, and then poured epoxy resin to prepare nanocomposites.[120] This composite exhibits excellent thermal stability and superior thermal conductivity, with a thermal conductivity of 6.07 W m^{-1} K^{-1} at a load of 15 vol% BNNP. Figure 2.16 shows the preparation process, morphology of the 3D-BNNS network skeleton and thermal conductivity of the composite.

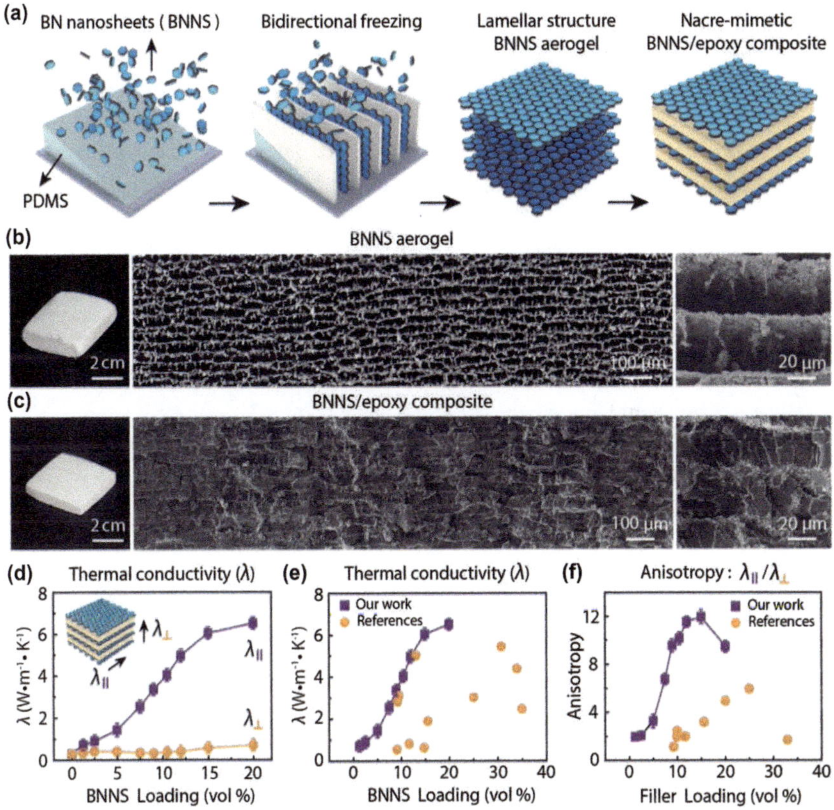

Figure 2.16 (a) Schematic of the fabrication route using a bidirectional freezing technique together with epoxy resin infiltration. (b) Optical and SEM images of the 3D-BNNP aerogel. (c) Optical and SEM images of the 3D-BNNS/epoxy composite. (d) Thermal conductivity of 3D-BNNS/epoxy composite in the parallel (λ∥) and perpendicular (λ⊥) directions. (e, f) Comparison of the 3D-BNNS/epoxy composites with some reported composites for thermal conductivity and anisotropy (λ∥/λ⊥). Reproduced from ref. 120 with permission from John Wiley & Sons, Copyright 2019 WILEY-VCH Verlag GmbH & Co. KGaA, Weinheim.

2.4.3.2 Self-assembly Method

Self-assembly refers to a process in which the basic structural units interact with each other (such as attraction and repulsion, or spontaneous formation of chemical bonds, *etc.*) to form an orderly organisational structure from a disordered system, without the intervention of external conditions. The self-assembly method has become an effective 'bottom-up' approach for building 3D structures.

In 2016, Shao *et al.* prepared a porous aerogel formed by self-assembly of BNNP and GO through hydrothermal reduction, and then introduced polyamine-6 into the aerogel *via in situ* polymerisation to gain thermal conductive composite materials with the highest thermal conductivity of 0.89 W m^{-1} K^{-1}.[188] Chen *et al.* first dissolved cellulose in sodium hydroxide/urea solution and added BNNS to form a cellulose/BNNS sol. A hydrogel with a 3D structure was obtained with epichlorohydrin (ECH) as the cross-linking agent. In this process, BNNP self-assembled onto the surface of cellulose nanofibres and lined up one by one along the cellulose skeleton. Finally, epoxy resin was injected into the freeze-dried aerogel to produce the heat-conducting composite. The entire process is as shown in Figure 2.17.

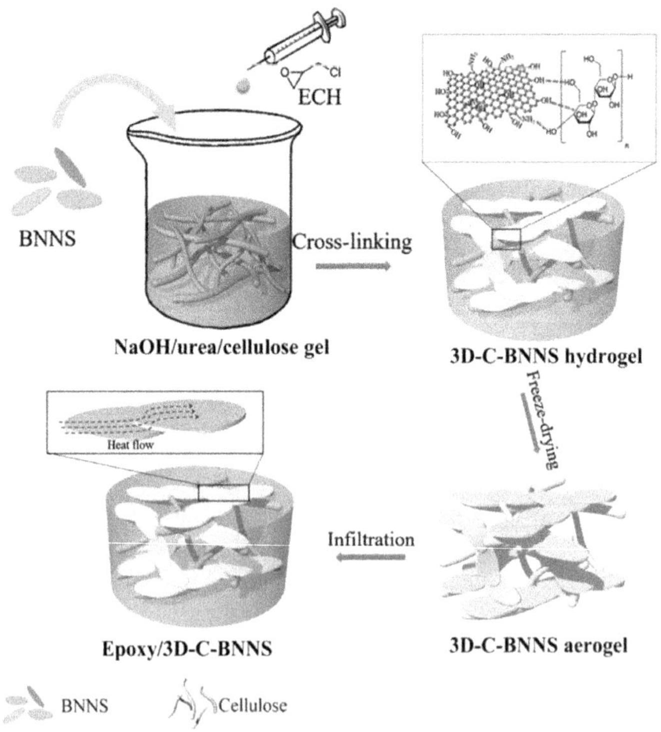

Figure 2.17 Scheme illustrating the preparation process of epoxy/3D-C-BNNS nanocomposites.
Reproduced from ref. 9 with permission from John Wiley & Sons, Copyright 2016 WILEY-VCH Verlag GmbH & Co. KGaA, Weinheim.

At a BNNS content of only 9.6 vol%, the thermal conductivity of the 3D-C-BNNS/epoxy composite can reach 3.13 $W m^{-1} K^{-1}$, which is 1400% higher than that of pure epoxy resin.[9]

2.4.4 Nanocomposite Paper/Film

The demand for thermal conductivity materials in emerging industries, electronics, aerospace and other fields is not limited to ordinary block polymer materials, but also includes ultra-high thermally conductive nanocomposite paper or nanocomposite film.[8,189] In response to this demand, scientists have undertaken a lot of fruitful research in this field in recent years.

In 2014, Zhu *et al.* reported an insulating nanocomposite paper with an in-plane thermal conductivity of up to 145.7 $W m^{-1} K^{-1}$ at 50 wt% BNNP loading, which was made by filtration of a suspension containing two-dimensional BNNP and one-dimensional cellulose nanofibres.[190] In this composite paper, the BNNP was oriented, forming a thermally conductive network, while the one-dimensional cellulose nanofibres acted as stabilisers for the dispersion of BNNP on one hand, and provided mechanical strength to the system on the other. Subsequently, Wu *et al.* modified BNNP by ball milling before mixing it with cellulose nanofibres (CNF), and then fabricated a CNF/BNNP nanocomposite film *via* filtration.[174] This nanocomposite film shows thickness dependence, and its in-plane thermal conductivity reaches 30.25 $W m^{-1} K^{-1}$ at a film thickness of 30 μm and with the BNNP addition of 70 wt%.

Besides cellulose nanofibres, BNNP can also form compounds with polymers to prepare nanocomposite films. Wu *et al.* combined aminated BNNP with poly(diallyldimethylammonium chloride) (PDDA) and prepared a highly anisotropic thermally conductive and high-strength BNNP/PDDA composite film after vacuum-assisted filtration with an in-plane thermal conductivity of 212.8 $W m^{-1} K^{-1}$ and through-plane thermal conductivity of 1.0 $W m^{-1} K^{-1}$.[173] In addition, using aminated BNNP and vacuum-assisted filtration technology, Wang *et al.* took PVA as the matrix to fabricate a layer-by-layer laminate nanostructure composite film (FBN/PVA film).[191] Figure 2.18 shows the photograph and schematic illustration of the FBN/PVA film. The FBN/PVA film exhibits superior in-plane thermal conductivity of 120 $W m^{-1} K^{-1}$ for 90 wt% BNNP loading, which is nearly 100 times higher than that of pure PVA film.

Instead of filtration, Chen *et al.* adopted mould pressing technology to obtain a highly thermally conductive, yet electrically insulating, PVDF and boron nitride nanoplatelets composite film.[23] First, the oriented PVDF/BNNP nanofibres were prepared *via* electrospinning. Second, the cut PVDF/BNNP nanofibres were folded vertically into a criss-cross grid array. Finally, PVDF/BNNP nanocomposite film was produced after meld pressing. When the content of BNNP is 33 wt%, the in-plane thermal conductivity of the resulting film reaches 10.4 $W m^{-1} K^{-1}$ and the thermal resistance between BNNP is about 1.26×10^{-6} $m^2 K W^{-1}$, which is an order of magnitude smaller than that of randomly dispersed BNNP (1.81×10^{-5} $m^2 K W^{-1}$).

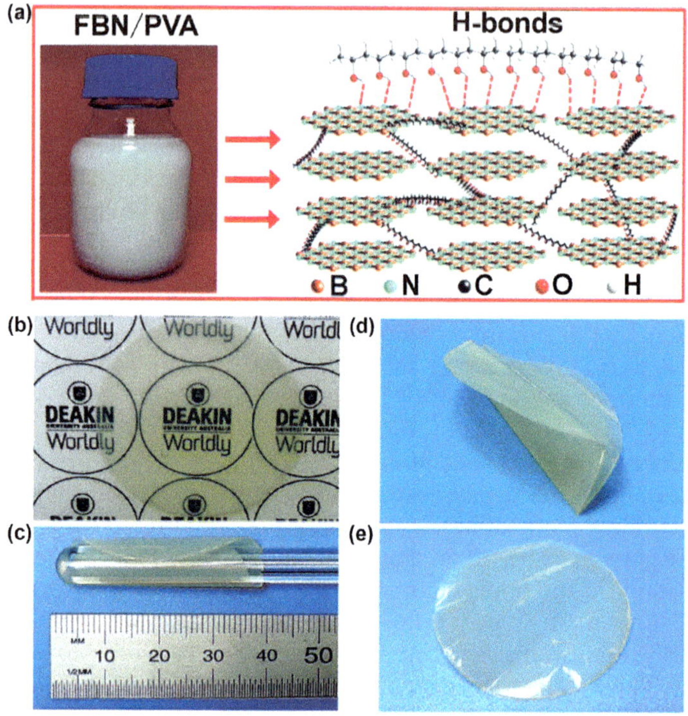

Figure 2.18 (a) Photograph and schematic of the homogeneous mixing of FBN nanosheets and PVA solutions. (b) Photograph of a free-standing FBN/PVA film. (c) Photograph of a piece of an FBN/PVA film that was rolled around a glass rod, and (d) folded into a paper airplane shape. (e) The folded film recovered to its original shape without observable damage. Reproduced from ref. 191 with permission from the Royal Society of Chemistry.

2.4.5 Other Materials

In addition to the methods mentioned in the previous sections, some novel material forms and methods have emerged in high-thermal conductivity polymer-based nanocomposites in recent years.

As an example of these novel material forms, Wang *et al.* developed a high-temperature-stable and highly thermally conductive textile *via* green non-toxic electrospinning (Figure 2.19).[192] Both the polyimide precursor and BNNP were chemically modified to be water soluble. The entire mixing process was carried out in aqueous solution without any toxic organic solvents. The polyimide/amino-functionalised boron nitride (PI/FBN) composite textile has light weight and flexibility, as well as a high thermal conductivity of 13.1 $W\,m^{-1}\,K^{-1}$ at 300 °C, which means that it has broad prospects for application in the fields of space suits and electronic packaging.

As an example of the novel methods, Gao *et al.* reported a fast and scalable 3D printing method to fabricate thermally conductive composite fibres of PVA

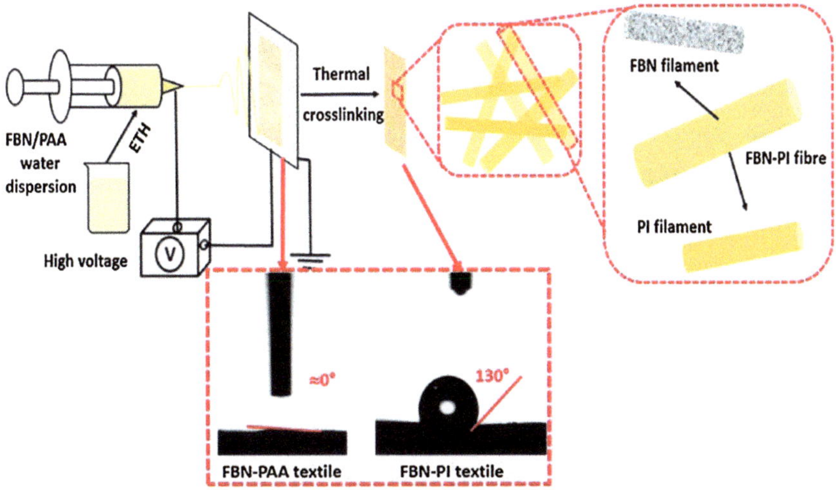

Figure 2.19 Schematic of the 'green' electrospinning process for producing the FBN-PI textile. Reproduced from ref. 192 with permission from the Royal Society of Chemistry.

and BNNP.[193] In the production process, 3D printing realised the uniform dispersion of BNNP, while hot drawing improved the orientation alignment of BNNP. The composite fibres show high mechanical strength and high thermal conductivity, which can be used in human-wearable textiles to improve the heat transmission performance of the textiles. Through experimental verification, the BNNS/PVA composite fibres achieve a better cooling effect, which is increased by 55% compared to commercial cotton fibres.

2.5 Summary and Outlook

The ongoing changes and rapid development in fields such as electronic components, power systems, communication equipment and others place a demand for higher requirements on the thermal conductivity of polymer insulating materials. Due to its extraordinary properties, including mechanical properties and chemical stability, and especially its ultra-high thermal conductivity (experimental values of 300–360 $W m^{-1} K^{-1}$) and excellent insulation properties (10^{17} Ω cm at room temperature), the application of two-dimensional BNNP in polymer composites has aroused intense interest in both academia and industry.

Challenges lie in finding effective methods to produce high quality BNNP and strategies to promote the dispersion and interfacial interactions with the polymers. BNNP has been produced *via* two methods: 'top-down' and 'bottom-up', *i.e.* exfoliation from h-BN and synthesis from B and N precursors, respectively. The top-down methods include the scotch tape method, ball milling and other mechanical exfoliation, chemical exfoliation and sonication-assisted liquid-phase exfoliation in various solvents. CVD is the most representative and commonly used bottom-up method. In addition to CVD, BNNP can also be synthesised directly by chemical reaction.

Thermal conduction is a complex process, involving interface thermal resistance, phonon scattering and the formation of heat conduction pathways. The efficiency and effectiveness of BNNP in enhancing polymer thermal conductivity are affected by complex factors, such as morphology, crystal structure integrity, filler/matrix interfacial compatibility and the dispersion and orientation of BNNP.

Surface engineering of BNNP is an effective way to improve the compatibility between BNNP and the polymer matrix. A variety of functional groups can be introduced to BNNP surfaces through different interaction forces, including covalent bonding and non-covalent interactions, such as Lewis acid–base interactions, π–π stacking interactions and hydrogen bonding. Different chemical groups and substances have been used to modify BNNP, including hydroxyl (–OH), amino (–NH_2), amine (–NHR), alkyl (–R), esterification (–OCOR), amidation (–NHCOR), carbonyl derivatives (–COX), halogen (–X) and inorganic particles, silane coupling agent, natural macromolecules and heteroatom dopants.

In order to attain polymer composites with higher thermal conductivity at lower BNNP loadings, scientists have developed a number of ways to control the internal microstructure of the materials. Oriented BNNP is more effective in enhancing thermal conductivity than randomly scattered BNNP, which can be achieved by processing methods or by applying an external field; by constructing a three-dimensional network structure to build more efficient heat pathways; and by utilising synergistic effects between different thermally conductive fillers to reduce thermal resistance. In addition to conventional bulk polymer composites, other types of highly thermally conductive nanocomposites, such as polymer papers/films and fibres/textiles are also reported.

Moving forward there are six critical issues that need to be addressed for BNNP/polymer composites.

1. BNNP structure and properties. Improving the theoretical study and comprehensive characterisation of BNNP, including, but not limited to, morphology, structure, physical properties and chemical reactivity.
2. BNNP production. Developing facile, efficient, large-scale methods of producing high-quality BNNP, solving the shortage of raw materials for BNNP-based polymer composites.
3. Thermal conduction theory. The phonon theory fails to explain many experimental phenomena. Improving the understanding of the theoretical system of thermal conduction; thermal resistance of materials, contact thermal resistance and interface thermal resistance; establishing a computational model with practical guidance for experiments.
4. BNNP surface engineering. Realise the property-customised design of functionalised BNNP.
5. BNNP distribution. An even distribution of BNNP filler in polymers cannot effectively form heat conduction paths at low content, while

excessive content may deteriorate the mechanical and processing properties of the material. Hence, the mutual contact, orientation or ordered heat conduction network of BNNP needs to be considered for enhancing the thermal conductivity of polymer composites.

6. Polymer composite material production. The current methods for preparing high thermal conductivity polymer composites are still limited to the laboratory scale. It is of great scientific and practical significance to develop new methods that are simple, feasible and can be industrialised.

Acknowledgements

The authors gratefully acknowledge the National Natural Science Foundation of China (No. 51877132) for support.

References

1. A. L. Moore and L. Shi, *Mater. Today*, 2014, **17**, 163.
2. H. Chen, V. V. Ginzburg, J. Yang, Y. Yang, W. Liu, Y. Huang, L. Du and B. Chen, *Prog. Polym. Sci.*, 2016, **59**, 41.
3. N. Burger, A. Laachachi, M. Ferriol, M. Lutz, V. Toniazzo and D. Ruch, *Prog. Polym. Sci.*, 2016, **61**, 1.
4. W. Li, Y. Liu and G. Wu, *Carbon*, 2015, **95**, 545.
5. C. Xue, H. Bai, P. F. Tao, J. W. Wang, N. Jiang and S. L. Wang, *Mater. Des.*, 2016, **108**, 250.
6. J. Yang, J. Sprengard, L. Ju, A. Hao, M. Saei, R. Liang, G. J. Cheng and C. Xu, *Carbon*, 2016, **108**, 38.
7. X. Li, L. Shao, N. Song, L. Shi and P. Ding, *Composites, Part A*, 2016, **88**, 305.
8. M. Loeblein, A. Bolker, S. H. Tsang, N. Atar, C. Uzan-Saguy, R. Verker, I. Gouzman, E. Grossman and E. H. Teo, *Small*, 2015, **11**, 6425.
9. J. Chen, X. Huang, Y. Zhu and P. Jiang, *Adv. Funct. Mater.*, 2017, **27**, 1604754.
10. J. Chen, X. Huang, B. Sun, Y. Wang, Y. Zhu and P. Jiang, *ACS Appl. Mater. Interfaces*, 2017, **9**, 30909.
11. Y. Yao, X. Zhu, X. Zeng, R. Sun, J. B. Xu and C. P. Wong, *ACS Appl. Mater. Interfaces*, 2018, **10**, 9669.
12. Y. Chen, J. Zou, S. J. Campbell and G. Le Caer, *Appl. Phys. Lett.*, 2004, **84**, 2430.
13. C. Zhi, Y. Bando, C. Tang, H. Kuwahara and D. Golberg, *Adv. Mater.*, 2009, **21**, 2889.
14. M. S. R. N. Kiran, K. Raidongia, U. Ramamurty and C. N. R. Rao, *Scr. Mater.*, 2011, **64**, 592.
15. S. Kemaloglu, G. Ozkoc and A. Aytac, *Thermochim. Acta*, 2010, **499**, 40.
16. D. Pacilé, J. C. Meyer, Ç. Ö. Girit and A. Zettl, *Appl. Phys. Lett.*, 2008, **92**, 133107.

17. Y. Lin and J. W. Connell, *Nanoscale*, 2012, **4**, 6908.
18. J. N. Coleman, M. Lotya, A. O'Neill, S. D. Bergin, P. J. King, U. Khan, K. Young, A. Gaucher, S. De, R. J. Smith, I. V. Shvets, S. K. Arora, G. Stanton, H. Y. Kim, K. Lee, G. T. Kim, G. S. Duesberg, T. Hallam, J. J. Boland, J. J. Wang, J. F. Donegan, J. C. Grunlan, G. Moriarty, A. Shmeliov, R. J. Nicholls, J. M. Perkins, E. M. Grieveson, K. Theuwissen, D. W. McComb, P. D. Nellist and V. Nicolosi, *Science*, 2011, **331**, 568.
19. E. Ćavar, R. Westerström, A. Mikkelsen, E. Lundgren, A. S. Vinogradov, M. L. Ng, A. B. Preobrajenski, A. A. Zakharov and N. Mårtensson, *Surf. Sci.*, 2008, **602**, 1722.
20. T. Sainsbury, A. Satti, P. May, Z. Wang, I. McGovern, Y. K. Gun'ko and J. Coleman, *J. Am. Chem. Soc.*, 2012, **134**, 18758.
21. F. Xiao, S. Naficy, G. Casillas, M. H. Khan, T. Katkus, L. Jiang, H. Liu, H. Li and Z. Huang, *Adv. Mater.*, 2015, **27**, 7196.
22. M. Tanimoto, T. Yamagata, K. Miyata and S. Ando, *ACS Appl. Mater. Interfaces*, 2013, **5**, 4374.
23. J. Chen, X. Huang, B. Sun and P. Jiang, *ACS Nano*, 2019, **13**, 337.
24. H. Shen, J. Guo, H. Wang, N. Zhao and J. Xu, *ACS Appl. Mater. Interfaces*, 2015, **7**, 5701.
25. L. F. Dobrzhinetskaya, R. Wirth, J. Yang, H. W. Green, I. D. Hutcheon, P. K. Weber and E. S. Grew, *Am. Mineral.*, 2014, **99**, 764.
26. A. Pakdel, Y. Bando and D. Golberg, *Chem. Soc. Rev.*, 2014, **43**, 934.
27. K. Kim and J. Kim, *Ceram. Int.*, 2014, **40**, 5181.
28. K. Watanabe, T. Taniguchi, T. Niiyama, K. Miya and M. Taniguchi, *Nat. Photonics*, 2009, **3**, 591.
29. E. K. Sichel, R. E. Miller, M. S. Abrahams and C. J. Buiocchi, *Phys. Rev. B*, 1976, **13**, 4607.
30. Y. Pan, K. Huo, Y. Hu, J. Fu, Y. Lu, Z. Dai, Z. Hu and Y. Chen, *Small*, 2005, **1**, 1199.
31. A. Gupta, T. Sakthivel and S. Seal, *Prog. Mater. Sci.*, 2015, **73**, 44.
32. C. R. Dean, A. F. Young, I. Meric, C. Lee, L. Wang, S. Sorgenfrei, K. Watanabe, T. Taniguchi, P. Kim, K. L. Shepard and J. Hone, *Nat. Nanotechnol.*, 2010, **5**, 722.
33. Z. Cui, Z. Cao, R. Ma, A. V. Dobrynin and D. H. Adamson, *ACS Appl. Mater. Interfaces*, 2015, **7**, 16913.
34. C. Lee, Q. Li, W. Kalb, X. Z. Liu, H. Berger, R. W. Carpick and J. Hone, *Science*, 2010, **328**, 76.
35. K. S. Novoselov, A. K. Geim, S. V. Morozov, D. Jiang, Y. Zhang, S. V. Dubonos, I. V. Grigorieva and A. A. Firsov, *Science*, 2004, **306**, 666.
36. K. Watanabe, T. Taniguchi and H. Kanda, *Nat. Mater.*, 2004, **3**, 404.
37. J. Li, G. Gui and J. Zhong, *J. Appl. Phys.*, 2008, **104**, 094311.
38. K. Yoichi, *Science*, 2007, **5840**, 932.
39. L. Britnell, R. V. Gorbachev, R. Jalil, B. D. Belle, F. Schedin, M. I. Katsnelson, L. Eaves, S. V. Morozov, A. S. Mayorov, N. M. Peres, A. H. Neto, J. Leist, A. K. Geim, L. A. Ponomarenko and K. S. Novoselov, *Nano Lett.*, 2012, **12**, 1707.

40. L. H. Li, Y. Chen, G. Behan, H. Zhang, M. Petravic and A. M. Glushenkov, *J. Mater. Chem.*, 2011, **21**, 11862.
41. T. A. Chen, C. P. Chuu, C. C. Tseng, C. K. Wen, H. P. Wong, S. Pan, R. Li, T. A. Chao, W. C. Chueh, Y. Zhang, Q. Fu, B. I. Yakobson, W. H. Chang and L. J. Li, *Nature*, 2020, **579**, 219.
42. N. Alem, R. Erni, C. Kisielowski, M. D. Rossell, W. Gannett and A. Zettl, *Phys. Rev. B*, 2009, **80**, 155425.
43. A. N. Streletskii, D. G. Permenov, B. B. Bokhonov, I. V. Kolbanev, A. V. Leonov, I. V. Berestetskaya and K. A. Streletzky, *J. Alloys Compd.*, 2009, **483**, 313.
44. Z. P. Xia and Z. Q. Li, *J. Alloys Compd.*, 2007, **436**, 170.
45. H. Yurdakul, Y. Göncü, O. Durukan, A. Akay, A. T. Seyhan, N. Ay and S. Turan, *Ceram. Int.*, 2012, **38**, 2187.
46. X. Chen, J. F. Dobson and C. L. Raston, *Chem. Commun.*, 2012, **48**, 3703.
47. Y. Hernandez, V. Nicolosi, M. Lotya, F. M. Blighe, Z. Sun, S. De, I. T. McGovern, B. Holland, M. Byrne, Y. K. Gun'Ko, J. J. Boland, P. Niraj, G. Duesberg, S. Krishnamurthy, R. Goodhue, J. Hutchison, V. Scardaci, A. C. Ferrari and J. N. Coleman, *Nat. Nanotechnol.*, 2008, **3**, 563.
48. S. Zhao, S. Xie, Z. Zhao, J. Zhang, L. Li and Z. Xin, *ACS Sustainable Chem. Eng.*, 2018, **6**, 7652.
49. W. Liu, P. Li, W. Wang, D. Zhu, Y. Chen, S. Pen, E. Paek and D. Mitlin, *ACS Nano*, 2018, **12**, 12255.
50. Y. Lin, T. V. Williams, T.-B. Xu, W. Cao, H. E. Elsayed-Ali and J. W. Connell, *J. Phys. Chem. C*, 2011, **115**, 2679.
51. K. Varoon, X. Zhang, B. Elyassi, D. D. Brewer, M. Gettel, S. Kumar, J. A. Lee, S. Maheshwari, A. Mittal, C. Y. Sung, M. Cococcioni, L. F. Francis, A. V. McCormick, K. A. Mkhoyan and M. Tsapatsis, *Science*, 2011, **334**, 72.
52. S. D. Bergin, Z. Sun, D. Rickard, P. V. Streich, J. P. Hamilton and J. N. Coleman, *ACS Nano*, 2009, **3**, 2340.
53. K. G. Zhou, N. N. Mao, H. X. Wang, Y. Peng and H. L. Zhang, *Angew. Chem. Int. Ed.*, 2011, **50**, 10839.
54. Y. Wang, Z. Shi and J. Yin, *J. Mater. Chem.*, 2011, **21**, 11371.
55. X. Li, X. Hao, M. Zhao, Y. Wu, J. Yang, Y. Tian and G. Qian, *Adv. Mater.*, 2013, **25**, 2200.
56. M. Du, X. Li, A. Wang, Y. Wu, X. Hao and M. Zhao, *Angew. Chem. Int. Ed.*, 2014, **53**, 3645.
57. N. Wang, G. Yang, H. Wang, C. Yan, R. Sun and C.-P. Wong, *Mater. Today*, 2018, **27**, 33.
58. M. J. Rand and J. F. Roberts, *J. Electrochem. Soc.*, 1968, **115**, 423.
59. W. Auwaerter, H. U. Suter, H. Sachdev and T. Greber, *ChemInform*, 2004, **35**, 343.
60. A. S. Rozenberg, Y. A. Sinenko and N. V. Chukanov, *J. Mater. Sci.*, 1993, **28**, 5528.
61. S. Middleman, *Mater. Sci. Eng., A*, 1993, **163**, 135.

62. G. Kim, A. R. Jang, H. Y. Jeong, Z. Lee, D. J. Kang and H. S. Shin, *Nano Lett.*, 2013, **13**, 1834.
63. L. Song, L. Ci, H. Lu, P. B. Sorokin, C. Jin, J. Ni, A. G. Kvashnin, D. G. Kvashnin, J. Lou, B. I. Yakobson and P. M. Ajayan, *Nano Lett.*, 2010, **10**, 3209.
64. Y. Shi, C. Hamsen, X. Jia, K. K. Kim, A. Reina, M. Hofmann, A. L. Hsu, K. Zhang, H. Li, Z. Y. Juang, M. S. Dresselhaus, L. J. Li and J. Kong, *Nano Lett.*, 2010, **10**, 4134.
65. K. H. Lee, H. J. Shin, J. Lee, I. Y. Lee, G. H. Kim, J. Y. Choi and S. W. Kim, *Nano Lett.*, 2012, **12**, 714.
66. A. Nag, K. Raidongia, K. P. S. S. Hembram, R. Datta, U. V. Waghmare and C. N. R. Rao, *ACS Nano*, 2010, **4**, 1539.
67. Z. Zhao, Z. Yang, Y. Wen, Y. Wang and E. Dickey, *J. Am. Ceram. Soc.*, 2011, **94**, 4496.
68. W.-Q. Han, H.-G. Yu and Z. Liu, *Appl. Phys. Lett.*, 2011, **98**, 203112.
69. X. Wang, C. Zhi, L. Li, H. Zeng, C. Li, M. Mitome, D. Golberg and Y. Bando, *Adv. Mater.*, 2011, **23**, 4072.
70. X. Wang, A. Pakdel, C. Zhi, K. Watanabe, T. Sekiguchi, D. Golberg and Y. Bando, *J. Phys.: Condens. Matter*, 2012, **24**, 314205.
71. C. Ageorges, L. Ye, Y.-W. Mai and M. Hou, *Composites, Part A*, 1998, **29**, 921.
72. S. Jia, Z. Wang, N. Ding, Y. L. Elaine Wong, X. Chen, G. Qiu and T. W. Dominic Chan, *Anal. Chim. Acta*, 2016, **936**, 123.
73. I. Jo, M. T. Pettes, J. Kim, K. Watanabe, T. Taniguchi, Z. Yao and L. Shi, *Nano Lett.*, 2013, **13**, 550.
74. W. Meng, Y. Huang, Y. Fu, Z. Wang and C. Zhi, *J. Mater. Chem. C*, 2014, **2**, 10049.
75. Y. Yao, X. Zeng, G. Pan, J. Sun, J. Hu, Y. Huang, R. Sun, J. B. Xu and C. P. Wong, *ACS Appl. Mater. Interfaces*, 2016, **8**, 31248.
76. J. R. Choi, S. Yu, H. Jung, S. K. Hwang, R. H. Kim, G. Song, S. H. Cho, I. Bae, S. M. Hong, C. M. Koo and C. Park, *Nanoscale*, 2015, **7**, 1888.
77. M. Chau, B. A. F. Kopera, V. R. Machado, S. M. Tehrani, M. A. Winnik, E. Kumacheva and M. Retsch, *Mater. Horiz.*, 2017, **4**, 236.
78. C. Zeng, S. Lu, L. Song, X. Xiao, J. Gao, L. Pan, Z. He and J. Yu, *RSC Adv.*, 2015, **5**, 35773.
79. K. J. Tielrooij, N. C. H. Hesp, A. Principi, M. B. Lundeberg, E. A. A. Pogna, L. Banszerus, Z. Mics, M. Massicotte, P. Schmidt, D. Davydovskaya, D. G. Purdie, I. Goykhman, G. Soavi, A. Lombardo, K. Watanabe, T. Taniguchi, M. Bonn, D. Turchinovich, C. Stampfer, A. C. Ferrari, G. Cerullo, M. Polini and F. H. L. Koppens, *Nat. Nanotechnol.*, 2018, **13**, 41.
80. Y. Zhou, X. Zhang and M. Hu, *Nano Lett.*, 2017, **17**, 1269.
81. C. Sevik, H. Sevincli, G. Cuniberti and T. Cagin, *Nano Lett.*, 2011, **11**, 4971.
82. Z. Yu, Z. Y. Ong, Y. Pan, Y. Cui, R. Xin, Y. Shi, B. Wang, Y. Wu, T. Chen, Y. W. Zhang, G. Zhang and X. Wang, *Adv. Mater.*, 2016, **28**, 547.

83. J. H. Seol, I. Jo, A. L. Moore, L. Lindsay, Z. H. Aitken, M. T. Pettes, X. Li, Z. Yao, R. Huang, D. Broido, N. Mingo, R. S. Ruoff and L. Shi, *Science*, 2010, **328**, 213.
84. N. Bonini, J. Garg and N. Marzari, *Nano Lett.*, 2012, **12**, 2673.
85. S. Lee, D. Broido, K. Esfarjani and G. Chen, *Nat. Commun.*, 2015, **6**, 6290.
86. H. W. Kim, W. Ko, J. Ku, I. Jeon, D. Kim, H. Kwon, Y. Oh, S. Ryu, Y. Kuk, S. W. Hwang and H. Suh, *Nat. Commun.*, 2015, **6**, 7528.
87. H. Chakraborty, S. Mogurampelly, V. K. Yadav, U. V. Waghmare and M. L. Klein, *Nanoscale*, 2018, **10**, 22148.
88. W. Yu, Y. Qi, Y. Zhou, L. Chen, H. Du and H. Xie, *J. Appl. Polym. Sci.*, 2016, **133**, 43242.
89. L. Yang, Y. Yang, Q. Zhang, Y. Zhang, Y. Jiang, Z. Guan, M. Gerboth, J. Yang, Y. Chen, D. Greg Walker, T. T. Xu and D. Li, *Nanoscale*, 2016, **8**, 17895.
90. S. Xu, H. Liu, Q. Li, Q. Mu and H. Wen, *J. Mater. Chem. C*, 2016, **4**, 872.
91. C. Xia, A. C. Garcia, S. Q. Shi, Y. Qiu, N. Warner, Y. Wu, L. Cai, H. R. Rizvi, N. A. D'Souza and X. Nie, *Sci. Rep.*, 2016, **6**, 34726.
92. S. Wu, R. B. Ladani, J. Zhang, E. Bafekrpour, K. Ghorbani, A. P. Mouritz, A. J. Kinloch and C. H. Wang, *Carbon*, 2015, **94**, 607.
93. Y. Wang, C. Yang, Y.-W. Mai and Y. Zhang, *Carbon*, 2016, **102**, 311.
94. T. Wang, H. Sun, J. Long, Y.-Z. Wang and D. Schiraldi, *ACS Sustainable Chem. Eng.*, 2016, **4**, 2601.
95. J. A. Spechler, T.-W. Koh, J. T. Herb, B. P. Rand and C. B. Arnold, *Adv. Funct. Mater.*, 2015, **25**, 7428.
96. S. Chen, R. Xu, J. Liu, X. Zou, L. Qiu, F. Kang, B. Liu and H. M. Cheng, *Adv. Mater.*, 2019, **31**, 1804810.
97. O. Cometto, B. Sun, S. H. Tsang, X. Huang, Y. K. Koh and E. H. Teo, *Nanoscale*, 2015, **7**, 18984.
98. H. Fang, S.-L. Bai and C. P. Wong, *Composites, Part A*, 2017, **100**, 71.
99. X. Huang, S. Wang, M. Zhu, K. Yang, P. Jiang, Y. Bando, D. Golberg and C. Zhi, *Nanotechnology*, 2015, **26**, 015705.
100. H. Jiang, Z. Wang, H. Geng, X. Song, H. Zeng and C. Zhi, *ACS Appl. Mater. Interfaces*, 2017, **9**, 10078.
101. O. S. Kwon, D. Lee, S. P. Lee, Y. G. Kang, N. C. Kim and S. H. Song, *RSC Adv.*, 2016, **6**, 59970.
102. F. Wang, Y. Yao, X. Zeng, T. Huang, R. Sun, J. Xu and C.-P. Wong, *RSC Adv.*, 2016, **6**, 41630.
103. W. Dai, J. Yu, Z. Liu, Y. Wang, Y. Song, J. Lyu, H. Bai, K. Nishimura and N. Jiang, *Composites, Part A*, 2015, **76**, 73.
104. C. Zhang, R. Huang, Y. Wang, Z. Wu, S. Guo, H. Zhang, J. Li, C. Huang, W. Wang and L. Li, *J. Mater. Chem. A*, 2018, **6**, 20663.
105. J. Yu, X. Huang, L. Wang, P. Peng, C. Wu, X. Wu and P. Jiang, *Polym. Chem.*, 2011, **2**, 1380.
106. H. Wu, W. Zhao, H. Hu and G. Chen, *J. Mater. Chem.*, 2011, **21**, 8626.

107. C. Yu, Q. Zhang, J. Zhang, R. Geng, W. Tian, X. Fan and Y. Yao, *ACS Appl. Nano Mater.*, 2018, **1**, 4875.
108. N. Yang, C. Xu, J. Hou, Y. Yao, Q. Zhang, M. E. Grami, L. He, N. Wang and X. Qu, *RSC Adv.*, 2016, **6**, 18279.
109. X. Wu, H. Liu, Z. Tang and B. Guo, *Compos. Sci. Technol.*, 2016, **123**, 179.
110. Z. Su, H. Wang, X. Ye, K. Tian, W. Huang, J. He, Y. Guo and X. Tian, *Composites, Part A*, 2018, **111**, 12.
111. X. Huang, C. Zhi, P. Jiang, D. Golberg, Y. Bando and T. Tanaka, *Adv. Funct. Mater.*, 2013, **23**, 1824.
112. H. Wu and M. R. Kessler, *ACS Appl. Mater. Interfaces*, 2015, **7**, 5915.
113. J. Yu, X. Huang, C. Wu, X. Wu, G. Wang and P. Jiang, *Polymer*, 2012, **53**, 471.
114. X.-B. Wang, Q. Weng, X. Wang, X. Li, J. Zhang, F. Liu, X.-F. Jiang, H. Guo, N. Xu, D. Golberg and Y. Bando, *ACS Nano*, 2014, **8**, 9081.
115. Y.-F. Zhang, D. Han, Y.-H. Zhao and S.-L. Bai, *Carbon*, 2016, **109**, 552.
116. N. Song, X. Hou, L. Chen, S. Cui, L. Shi and P. Ding, *ACS Appl. Mater. Interfaces*, 2017, **9**, 17914.
117. H. S. Lim, J. W. Oh, S. Y. Kim, M.-J. Yoo, S.-D. Park and W. S. Lee, *Chem. Mater.*, 2013, **25**, 3315.
118. Z. Lin, Y. Liu, S. Raghavan, K. S. Moon, S. K. Sitaraman and C. P. Wong, *ACS Appl. Mater. Interfaces*, 2013, **5**, 7633.
119. X. Zeng, Y. Yao, Z. Gong, F. Wang, R. Sun, J. Xu and C. P. Wong, *Small*, 2015, **11**, 6205.
120. J. Han, G. Du, W. Gao and H. Bai, *Adv. Funct. Mater.*, 2019, 1900412.
121. L. K. Randeniya, H. Shi, A. S. Barnard, J. Fang, P. J. Martin and K. K. Ostrikov, *Small*, 2013, **9**, 3993.
122. K. Wu, P. Liao, R. Du, Q. Zhang, F. Chen and Q. Fu, *J. Mater. Chem. A*, 2018, **6**, 11863.
123. R. Rizvi, E. P. Nguyen, M. D. Kowal, W. H. Mak, S. Rasel, M. A. Islam, A. Abdelaal, A. S. Joshi, S. Zekriardehani, M. R. Coleman and R. B. Kaner, *Adv. Mater.*, 2018, **30**, e1800200.
124. A. A. Balandin, *Nat. Mater.*, 2011, **10**, 569.
125. B. Mortazavi, M. Potschke and G. Cuniberti, *Nanoscale*, 2014, **6**, 3344.
126. D. Kim, Y. Hwangbo, L. Zhu, A. E. Mag-Isa, K. S. Kim and J. H. Kim, *Nanoscale*, 2013, **5**, 11870.
127. W. Zhou, Q. Chen, X. Sui, L. Dong and Z. Wang, *Composites, Part A*, 2015, **71**, 184.
128. N. Song, D. Jiao, S. Cui, X. Hou, P. Ding and L. Shi, *ACS Appl. Mater. Interfaces*, 2017, **9**, 2924.
129. Z. Liu, Q. Liu, Y. Huang, Y. Ma, S. Yin, X. Zhang, W. Sun and Y. Chen, *Adv. Mater.*, 2008, **20**, 3924.
130. Y. Lin, T. V. Williams and J. W. Connell, *J. Phys. Chem. Lett.*, 2010, **1**, 277.
131. W.-Q. Han, L. Wu, Y. Zhu, K. Watanabe and T. Taniguchi, *Appl. Phys. Lett.*, 2008, **93**, 223103.
132. Y. T. Liu, X. M. Xie and X. Y. Ye, *Chem. Commun.*, 2013, **49**, 388.

133. P. Ma and J. T. Spencer, *J. Mater. Sci.*, 2014, **50**, 313.
134. D. Yang, Y. Ni, X. Kong, D. Gao, Y. Wang, T. Hu and L. Zhang, *Compos. Sci. Technol.*, 2019, **177**, 18.
135. J. Wang, D. Zhao, X. Zou, L. Mao and L. Shi, *J. Mater. Sci.: Mater. Electron.*, 2017, **28**, 12984.
136. X. Zeng, L. Ye, S. Yu, H. Li, R. Sun, J. Xu and C. P. Wong, *Nanoscale*, 2015, **7**, 6774.
137. Z. Zheng, M. Cox and B. Li, *J. Mater. Sci.*, 2017, **53**, 66.
138. F. Wang, X. Zeng, Y. Yao, R. Sun, J. Xu and C. P. Wong, *Sci. Rep.*, 2016, **6**, 19394.
139. K. Ahn, K. Kim and J. Kim, *Ceram. Int.*, 2015, **41**, 9488.
140. K. Kim, H. Ju and J. Kim, *Polymer*, 2016, **91**, 74.
141. Q. Weng, X. Wang, X. Wang, Y. Bando and D. Golberg, *Chem. Soc. Rev.*, 2016, **45**, 3989.
142. Y.-Z. Tan, S.-Y. Xie, R.-B. Huang and L.-S. Zheng, *Nat. Chem.*, 2009, **1**, 450.
143. Q.-H. Weng, Q. He, T. Liu, H.-Y. Huang, J.-H. Chen, Z.-Y. Gao, S.-Y. Xie, X. Lu, R.-B. Huang and L.-S. Zheng, *J. Am. Chem. Soc.*, 2010, **132**, 15093.
144. I. V. Kuvychko, A. V. Streletskii, N. B. Shustova, K. Seppelt, T. Drewello, A. A. Popov, S. H. Strauss and O. V. Boltalina, *J. Am. Chem. Soc.*, 2010, **132**, 6443.
145. Z. Wang, Q. Li, Z. Chen, H. Li, S. Zheng and X. Tang, *Polym. Compos.*, 2018, **40**, 2346.
146. L. Wu, K. Wu, D. Liu, R. Huang, J. Huo, F. Chen and Q. Fu, *J. Mater. Chem. A*, 2018, **6**, 7573.
147. C. Y. Zhi, Y. Bando, T. Terao, C. C. Tang, H. Kuwahara and D. Golberg, *Chem. - Asian J.*, 2009, **4**, 1536.
148. D. Lee, B. Lee, K. H. Park, H. J. Ryu, S. Jeon and S. H. Hong, *Nano Lett.*, 2015, **15**, 1238.
149. A. Pakdel, Y. Bando and D. Golberg, *ACS Nano*, 2014, **8**, 10631.
150. Q. Weng, B. Wang, X. Wang, N. Hanagata, X. Li, D. Liu, X. Wang, X. Jiang, Y. Bando and D. Golberg, *ACS Nano*, 2014, **8**, 6123.
151. Q. Weng, B. Wang, X. Wang, N. Hanagata, X. Li, D. Liu, X. Wang, X. Jiang, Y. Bando and D. Golberg, *ACS Nano*, 2014, **8**, 6123.
152. T. Ikuno, T. Sainsbury, D. Okawa, J. M. J. Fréchet and A. Zettl, *Solid State Commun.*, 2007, **142**, 643.
153. X. Zeng, L. Ye, S. Yu, R. Sun, J. Xu and C.-P. Wong, *Chem. Mater.*, 2015, **27**, 5849.
154. W. Lei, V. N. Mochalin, D. Liu, S. Qin, Y. Gogotsi and Y. Chen, *Nat. Commun.*, 2015, **6**, 8849.
155. H. Li, Y. Gao, P. Zhu, X. Du, X. Yu, L. Ma, G. Li, R. Sun and C. P. Wong, *Adv. Mater. Interfaces*, 2019, 1900787.
156. T. Sainsbury, A. O'Neill, M. K. Passarelli, M. Seraffon, D. Gohil, S. Gnaniah, S. J. Spencer, A. Rae and J. N. Coleman, *Chem. Mater.*, 2014, **26**, 7039.

157. Y. Xie, C. A. S. Hill, Z. Xiao, H. Militz and C. Mai, *Composites, Part A*, 2010, **41**, 806.
158. D. S. Muratov, D. V. Kuznetsov, I. A. Il'inykh, I. N. Burmistrov and I. N. Mazov, *Compos. Sci. Technol.*, 2015, **111**, 40.
159. A. T. Seyhan, Y. Göncü, O. Durukan, A. Akay and N. Ay, *J. Solid State Chem.*, 2017, **249**, 98.
160. J.-H. Lee, H. Shin and K. Y. Rhee, *Composites, Part B*, 2019, **157**, 276.
161. A. Salama, N. Shukry, A. El-Gendy and M. El-Sakhawy, *Ind. Crops Prod.*, 2017, **95**, 170.
162. Y. Li, J. Ji, Y. Wang, R. Li and W.-H. Zhong, *J. Mater. Sci.*, 2018, **53**, 8653.
163. A. Khakalo, I. Filpponen and O. J. Rojas, *Compos. Sci. Technol.*, 2018, **160**, 145.
164. J. Wang, D. Liu, Q. Li, C. Chen, Z. Chen, P. Song, J. Hao, Y. Li, S. Fakhrhoseini, M. Naebe, X. Wang and W. Lei, *ACS Nano*, 2019, **13**, 7860.
165. Z. Wang, Y. Wen, S. Zhao, W. Zhang, Y. Ji, S. Zhang and J. Li, *Ind. Crops Prod.*, 2019, **137**, 239.
166. A. Bhattacharya, S. Bhattacharya and G. P. Das, *Phys. Rev. B*, 2012, **85**, 035415.
167. Y. Tian, X.-F. Pan, Y.-J. Liu and J.-X. Zhao, *Appl. Surf. Sci.*, 2014, **295**, 137.
168. R. S. Singh, *AIP Adv.*, 2015, **5**, 117150.
169. Z. Liu, L. Ma, G. Shi, W. Zhou, Y. Gong, S. Lei, X. Yang, J. Zhang, J. Yu, K. P. Hackenberg, A. Babakhani, J. C. Idrobo, R. Vajtai, J. Lou and P. M. Ajayan, *Nat. Nanotechnol.*, 2013, **8**, 119.
170. W. Lei, H. Zhang, Y. Wu, B. Zhang, D. Liu, S. Qin, Z. Liu, L. Liu, Y. Ma and Y. Chen, *Nano Energy*, 2014, **6**, 219.
171. Z. Zhu, C. Li, E. Songfeng, L. Xie, R. Geng, C.-T. Lin, L. Li and Y. Yao, *Compos. Sci. Technol.*, 2019, **170**, 93.
172. C. Yu, W. Gong, W. Tian, Q. Zhang, Y. Xu, Z. Lin, M. Hu, X. Fan and Y. Yao, *Compos. Sci. Technol.*, 2018, **160**, 199.
173. Y. Wu, Y. Xue, S. Qin, D. Liu, X. Wang, X. Hu, J. Li, X. Wang, Y. Bando, D. Golberg, Y. Chen, Y. Gogotsi and W. Lei, *ACS Appl. Mater. Interfaces*, 2017, **9**, 43163.
174. K. Wu, J. Fang, J. Ma, R. Huang, S. Chai, F. Chen and Q. Fu, *ACS Appl. Mater. Interfaces*, 2017, **9**, 30035.
175. T. Morishita and H. Okamoto, *ACS Appl. Mater. Interfaces*, 2016, **8**, 27064.
176. X. Zhang, J. Zhang, L. Xia, C. Li, J. Wang, F. Xu, X. Zhang, H. Wu and S. Guo, *ACS Appl. Mater. Interfaces*, 2017, **9**, 22977.
177. X. Wang and P. Wu, *ACS Appl. Mater. Interfaces*, 2017, **9**, 19934.
178. C. Du, M. Li, M. Cao, S. Song, S. Feng, X. Li, H. Guo and B. Li, *ACS Appl. Mater. Interfaces*, 2018, **10**, 34674.
179. H.-B. Cho, T. Nakayama, T. Suzuki, S. Tanaka, W. Jiang, H. Suematsu and K. Niihara, *J. Nanomater.*, 2011, **2011**, 1.
180. Y. Yao, X. Zeng, F. Wang, R. Sun, J.-B. Xu and C.-P. Wong, *Chem. Mater.*, 2016, **28**, 1049.

181. T. Huang, X. Zeng, Y. Yao, R. Sun, F. Meng, J. Xu and C. Wong, *RSC Adv.*, 2016, **6**, 35847.
182. P. Li, H. Shen, Z. Qian, X. Yang, N. Zhao, C. Zhu and J. Xu, *J. Mater. Sci.*, 2017, **53**, 4189.
183. S. Y. Pak, H. M. Kim, S. Y. Kim and J. R. Youn, *Carbon*, 2012, **50**, 4830.
184. Y. Yao, X. Zeng, R. Sun, J. B. Xu and C. P. Wong, *ACS Appl. Mater. Interfaces*, 2016, **8**, 15645.
185. X. Wang and P. Wu, *Chem. Eng. J.*, 2018, **348**, 723.
186. M. Loeblein, S. H. Tsang, M. Pawlik, E. J. Phua, H. Yong, X. W. Zhang, C. L. Gan and E. H. Teo, *ACS Nano*, 2017, **11**, 2033.
187. Y. Xue, X. Zhou, T. Zhan, B. Jiang, Q. Guo, X. Fu, K. Shimamura, Y. Xu, T. Mori, P. Dai, Y. Bando, C. Tang and D. Golberg, *Adv. Funct. Mater.*, 2018, **28**, 1801205.
188. L. Shao, L. Shi, X. Li, N. Song and P. Ding, *Compos. Sci. Technol.*, 2016, **135**, 83.
189. X. Zeng, J. Sun, Y. Yao, R. Sun, J. B. Xu and C. P. Wong, *ACS Nano*, 2017, **11**, 5167.
190. H. Zhu, Y. Li, Z. Fang, J. Xu, F. Cao, J. Wan, C. Preston, B. Yang and L. Hu, *ACS Nano*, 2014, **8**, 3606.
191. J. Wang, Y. Wu, Y. Xue, D. Liu, X. Wang, X. Hu, Y. Bando and W. Lei, *J. Mater. Chem. C*, 2018, **6**, 1363.
192. J. Wang, Q. Li, D. Liu, C. Chen, Z. Chen, J. Hao, Y. Li, J. Zhang, M. Naebe and W. Lei, *Nanoscale*, 2018, **10**, 16868.
193. T. Gao, Z. Yang, C. Chen, Y. Li, K. Fu, J. Dai, E. M. Hitz, H. Xie, B. Liu, J. Song, B. Yang and L. Hu, *ACS Nano*, 2017, **11**, 11513.
194. M. Sajjad, G. Morell and P. Feng, *ACS Appl. Mater. Interfaces*, 2013, **5**, 5051.
195. M. Chubarov, H. Pedersen, H. Högberg, S. Filippov, J. A. A. Engelbrecht, J. O'Connel and A. Henry, *Physica B*, 2014, **439**, 29.
196. A. S. Nazarov, V. N. Demin, E. D. Grayfer, A. I. Bulavchenko, A. T. Arymbaeva, H.-J. Shin, J.-Y. Choi and V. E. Fedorov, *Chem. - Asian J.*, 2012, **7**, 554.
197. N. Kostoglou, K. Polychronopoulou and C. Rebholz, *Vacuum*, 2015, **112**, 42.
198. N. Kostoglou, J. Lukovic, B. Babic, B. Matovic, D. Photiou, G. Constantinides, K. Polychronopoulou, V. Ryzhkov, B. Grossmann, C. Mitterer and C. Rebholz, *Mater. Des.*, 2016, **110**, 540.
199. K. N. Kudin, G. E. Scuseria and B. I. Yakobson, *Phys. Rev. B*, 2001, **64**, 235406.
200. R. Kumar, G. Rajasekaran and A. Parashar, *Nanotechnology*, 2016, **27**, 085706.
201. X. Zhang, L. Shen, H. Wu and S. Guo, *Compos. Sci. Technol.*, 2013, **89**, 24.
202. N. Sun, J. Sun, X. Zeng, P. Chen, J. Qian, R. Xia and R. Sun, *Composites, Part A*, 2018, **110**, 45.

203. B.-H. Xie, X. Huang and G.-J. Zhang, *Compos. Sci. Technol.*, 2013, **85**, 98.
204. H.-B. Cho, T. Nakayama, H. Suematsu, T. Suzuki, W. Jiang, K. Niihara, E. Song, N. S. A. Eom, S. Kim and Y.-H. Choa, *Compos. Sci. Technol.*, 2016, **129**, 205.
205. K. Kim and J. Kim, *Int. J. Therm. Sci.*, 2016, **100**, 29.
206. C. Yuan, B. Xie, M. Huang, R. Wu and X. Luo, *Int. J. Heat Mass Transfer*, 2016, **94**, 20.
207. J. Chen, H. Wei, H. Bao, P. Jiang and X. Huang, *ACS Appl. Mater. Interfaces*, 2019, **11**, 31402.

CHAPTER 3

Transition Metal Carbide (MXene)–Polymer Nanocomposites

JIE KONG,*[a] YAN SONG,[a] LEI WANG[a] AND BEN BIN XU[b]

[a] Shaanxi Key Laboratory of Macromolecular Science and Technology, School of Chemistry and Chemical Engineering, Northwestern Polytechnical University, Xi'an 710072, China; [b] Department of Mechanical and Construction Engineering, Faculty of Engineering and Environment, Northumbria University, Newcastle upon Tyne NE1 8ST, UK
*Email: kongjie@nwpu.edu.cn

3.1 Introduction

Two-dimensional (2D) materials have received increasing attention because of their unique structures and related electrical, optical, mechanical, and thermal properties.[1] Since Novoselov and Geim *et al.* used a mechanical stripping method in 2004 to successfully strip graphene from graphite, people have had great interest in various two-dimensional nanomaterials. Graphene has a single atomic layer and exhibits excellent properties, such as ultra-high carrier mobility (~10 000 $cm^2 V^{-1} s^{-1}$),[2] large theoretical specific surface area (2630 $m^2 g^{-1}$),[3] excellent transmittance (~97.7%),[4] high Young's modulus (~1 TPa),[5] and excellent thermal conductivity (3000–5000 $W m^{-1} K^{-1}$) at room temperature.[6] These unexpected physical and photoelectric properties of graphene have aroused interest in exploring new two-dimensional nanomaterials. MXene is another kind of new two-dimensional nanomaterial that has attracted much attention. In addition to possessing the conductivity of

Inorganic Materials Series No. 6
Two-dimensional Inorganic Nanomaterials for Conductive Polymer Nanocomposites
Edited by Chaoying Wan, Xingyi Huang and Chris Bowen

Published by the Royal Society of Chemistry, www.rsc.org

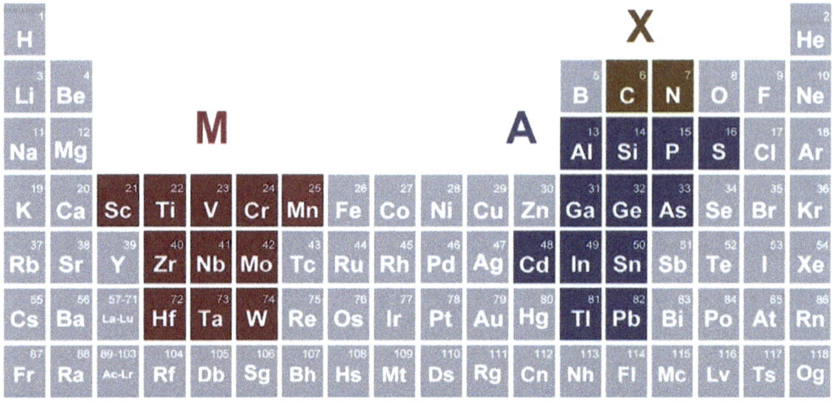

Figure 3.1 Periodic table illustrating elements forming MAX phase materials.

transition metals, MXene also has a hydrophilic surface modified by functional groups; this combination makes MXene more suitable than graphite for use as a conductive substrate and conductive filler.[7] In 2011, MXene was first prepared as $Ti_3C_2T_x$ *via* selective chemical etching of aluminium from a MAX phase ceramic, that is, Ti_3AlC_2.[8] Today, MXenes refer to transition metal carbides, nitrides, or carbonitrides with the formula $M_{n+1}X_nT_x$ ($n = 1 \sim 3$), where M represents transition metal elements, X is carbon and/or nitrogen, and T_x represents surface functional groups (such as –OH, –F, or =O).[9] The constituent elements of MAX phase materials are highlighted in the periodic table as shown in Figure 3.1.

There are many elements available, which makes it possible to synthesise two-dimensional materials with rich composition and structure. Also, a series of solid solutions with a MAX phase can also be synthesised *via* heat treatment of the M, A, or X positions, and it is expected that this will further expand their chemical composition.[10,11] Because the interlayer force of MAX phase materials is much higher than the van der Waals forces between graphite and transition metal sulfides, the two-dimensional structure is not easily obtained *via* simple mechanical stripping. However, because of the high reactivity of A atoms and the significant difference in chemical bonding between M–X and M–A in crystal structures, it is possible to selectively etch substrates. For example, with selective etching of the A-position (such as aluminium), the corresponding $M_{n+1}X_n$ lamellae can be obtained, as presented in Figure 3.2.[7] Subsequently, these materials are saturated with =O, –F, –OH, and other functional groups in fluorine-containing compounds, to form the final $M_{n+1}X_nT_x$ compound. At present, nearly 20 MXene structures (such as $Ti_3C_2T_x$, Ti_2CT_x, $Ta_4C_3T_x$, $(V_{0.5}Cr_{0.5})_3C_2T_x$, $TiNbCT_x$, Ti_3CNT_x, Nb_2CT_x, and V_2CT_x) have been reported as a result of etching MAX phases in HF, NH_4HF_2, LiF, and HCl mixed solutions.[9,12–14]

Similar to graphene, MXene has a high specific surface area and high electrical conductivity, in addition to having a flexible and adjustable composition and a controllable nanoscale layer thickness. MXene has shown

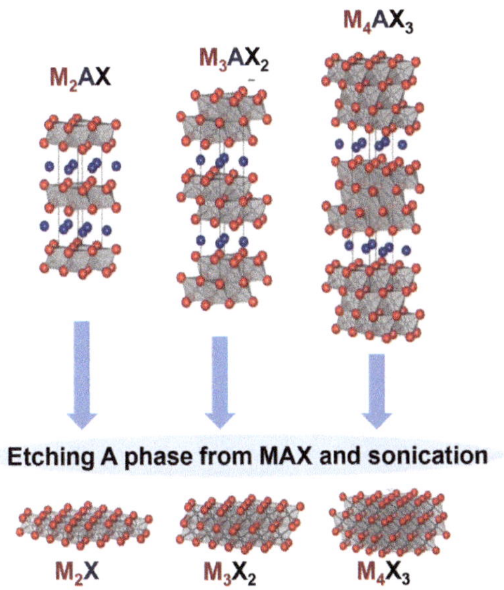

Figure 3.2 Structure of MAX phases and the corresponding MXenes.[7] Reproduced from ref. 7 with permission from John Wiley & Sons, Copyright 2013 WILEY-VCH Verlag GmbH & Co. KGaA, Weinheim.

great potential in applications, such as energy storage, gas adsorption and separation, sensors, conductive fillers, and polymer matrix nanocomposites.

3.2 Preparation Methods of MXenes

Similar to the lamellar structure of graphite, the structure of a MAX phase can also be regarded as the bonding of a transition metal (M), C/N compound layer, and element A layer. There are covalent bonds, metal bonds, and ionic bonds between M and X, whereas there are only metal bonds between M and A. Because of the presence of these strong chemical bonds, the stripping process of MAX is particularly difficult compared to graphite. A lot of research has been devoted to developing synthesis methods for new MXenes. The so-called 'top-down' and 'bottom-up' methods have been used to produce new MXenes. For the 'top-down' approach, a layered structure of the original precursor is necessary, whereas the 'bottom-up' strategy mainly relies on a chemical reaction of the precursor under a specific condition. That is, MXene is prepared *via* the selective etching of the 'A' layer by controlling the reaction conditions, based on the weak bond energy between M–A and with the goal of retaining the M–X layer to the greatest extent. Then, the 'A' layer is completely stripped *via* ultrasound. In the 'bottom-up' method, etching and spilling in solution or eutectic salt generates multilayered MXenes. Normally, an aqueous HF solution, NH_4HF_2 solution, NH_4F, LiF–HCl mixed solution, and a eutectic salt base of fluoride compounds have been employed. The composition of etching agents strongly influences the microstructure and properties of the

MXenes. $Ti_3C_2T_x$ MXene can be obtained *via* the selective etching of a layer of aluminium atoms of the Ti_3AlC_2 precursor in HF solution.

3.2.1 HF Acid Etching

As mentioned above, in the ternary layers of the MAX phase, the M–A bond is a metal bond. The binding energy of the above chemical bonds is much higher than the van der Waals forces. Interestingly, the layer of aluminium atoms can be selectively corroded and stripped *via* HF acid etching. In this process, the etching temperature and time should be strictly controlled to avoid damage to the MXene matrix. Moreover, hydrogen generation is an exothermal process, and so ventilation and experimental safety should be considered in detail.

Naguib *et al.*[12] used 50% HF solution to etch Ti_3AlC_2 (MAX) powders under continuous magnetic stirring. When the pH of the supernatant fluid was between 4 and 6 as a result of washing and centrifugal treatment, accordion-like MXene was obtained, as shown in Figure 3.3(a).[15] In the first step, the Al atoms were peeled off to form two-dimensional Ti_3C_2. Then, two or three reaction steps occurred simultaneously to form MXene containing –OH and –F groups. To shorten the etching time, the powder size can be decreased or the reaction temperature can be increased. The chemical equations of the whole process are described in Scheme 3.1. Hydrogen and side reactions are generated in the corrosion process, leading to impure MXene. Thus, the selective corrosion flow chart of Ti_3AlC_2 is shown in Figure 3.3(b).[15]

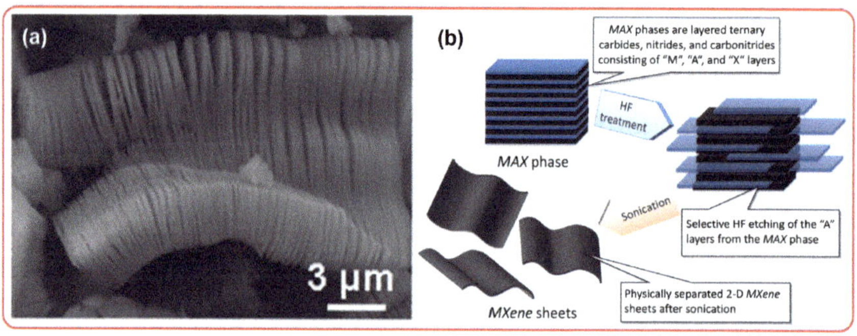

Figure 3.3 (a) SEM image of accordion-like Ti_3AlC_2 after HF acid etching. (b) Schematic describing the synthesis process of MXenes from MAX phases.[15] Reproduced from ref. 15 with permission from American Chemical Society, Copyright 2012.

$$Ti_3AlC_2 + 3HF \longrightarrow AlF_3 + 3/2H_2 + Ti_3C_2$$

$$Ti_3C_2 + 2H_2O \longrightarrow Ti_3C_2(OH)_2 + H_2$$

$$Ti_3C_2 + 2HF \longrightarrow Ti_3C_2F_2 + H_2$$

Scheme 3.1 Reaction mechanism of MXene synthesised by HF acid etching.

3.2.2 Mixed Solution Etching

Because of the strong corrosivity of HF, improper operation may cause serious harm to the human body. Therefore, it is essential to develop a mild etchant and preparation method. The LiF–HCl mixed solution is another *in situ* etching agent that is widely used in the synthesis of high-quality MXenes. It is environmentally friendly and conducive to the formation of pre-inserted layers during Li^+ etching; this increases the spacing between MXene materials and weakens the interaction between layers. This makes it easy to use ultrasonic vibration delamination to obtain MXene with high electronic conductivity. Ghidiu *et al.*[16] were the first to obtain $Ti_3C_2T_x$ *via* etching of Ti_3AlC_2 with LiF–HCl solution, where T represents –OH, –F, or –O– groups. This MXene has excellent hydrophilic and film-forming characteristics because of its special terminal groups, and it has high conductivity (an electronic conductivity of 1500 $S\,cm^{-1}$). Therefore, it has very quickly attracted the attention of researchers. In addition, a high concentration (>6 M) of HCl and a high molar ratio of LiF : Ti_3AlC_2 (>5 : 1) also make it easier to achieve the delamination of $Ti_3C_2T_x$.

As a result of the high bonding energy of $Ti_{n+1}AlN_n$, it is difficult to etch layers of atomic aluminium from nitrogen-containing MAX materials, even in HF acid solution. However, Ti_2N-layered MXene can be successfully obtained in a KF and HCl mixed solution. Soundiraraju *et al.*[17] soaked Ti_2AlN powders in the mixed solution of KF and HCl at room temperature, followed by washing with dimethyl sulfoxide (DMSO) and ultrasonic treatment to obtain Ti_2NT_x. Two-dimensional Ti_2NT_x material with a few layers was obtained. The reaction route is described in Scheme 3.2.

3.2.3 Molten Salt Etching

A molten salt etching approach can be used to avoid the splitting of MXene in a strong acid environment. Urbankowski *et al.*[18] mixed Ti_4AlN_3 powders with KF, NaF, and LiF. At 550 °C, the nitride Ti_4N_3 formed, and it could be further stratified by using tetrabutylammonium hydroxide (TBAOH) under ultrasonication. Hamil *et al.*[19] used NH_4HF_2 to etch Ti_3AlC_2 at room temperature to obtain Ti_3C_2. NH_4HF_2 is milder and less dangerous than HF. As described in Scheme 3.3, Ti_3C_2 can be generated from Ti_3AlC_2 and NH_4HF_2, where NH_3 and

$$2Ti_2AlN + 6KF + 6HCl \longrightarrow 2Ti_2NT_x + K_3AlF_6 + AlCl_3 + 3KCl + 3H_2$$

$$2Ti_2AlN + 6KF + 6HCl \longrightarrow 2Ti_2NT_x + 2AlF_3 + 6KCl + 3H_2$$

Scheme 3.2 Reaction mechanism of MXene synthesised by mixed solution etching.

$$Ti_3AlC_2 + 3NH_4HF_2 \longrightarrow (NH_4)_3AlF_6 + Ti_3C_2 + 3/2H_2$$

$$Ti_3C_2 + aNH_4HF_2 + bH_2O \longrightarrow (NH_3)_c(NH_4)_dTi_3C_2(OH)_xF_y$$

Scheme 3.3 Reaction mechanism of MXene synthesised by molten salt etching.

NH_4^+ are present between layers of $Ti_3C_2T_x$. Thus, the interlayer gap of this type of MXene can be large.

As mentioned above, each etching method has specific applicable conditions. HF etching is the fastest and most common etching method for MXene. It can be used to etch Ti_3AlC_2, Ti_3SiC_2, and other popular MAX materials, but it is not suitable for etching nitride MAX materials because of its poor stability. Moreover, HF acid is highly corrosive and is associated with safety issues. In a mixed etching solution, HF acid is generated *via in situ* reaction on the surface of MAX materials, and the lower corrosive agent can be used to etch nitride-containing MAX materials. Currently, molten salt etching is the safest etching method that can be used to etch nitride MAX materials, although it requires a higher etching temperature and has a low etching yield.[20]

3.2.4 Exfoliation of MXene Nanosheets

Several methods for etching the 'A' phase in MAX have been mentioned above, but lamellar nanosheets of MXene have more potential applications. MXene nanosheets are usually obtained by peeling off layered MXenes. According to first-principles calculations, groups (such as =O, –OH, or –F) on the surface of layered MXene can significantly weaken the interlayer interaction force, but it is still 2–6 times larger than that of layered graphite and MoS_2 with van der Waals forces as the main force.[21] Therefore, it is difficult to obtain single-layer MXene nanosheets *via* simple mechanical stripping. At present, the intercalation stripping method is the most commonly used to prepare MXene nanosheets. The basic principle is that the organic molecular intercalation agent is inserted into layered MXenes to increase the interlayer spacing and so reduce the interlayer force. Next, they were layered *via* ultrasonic treatment and oscillating forces to obtain a single layer and few layers of a MXene nanosheet colloidal solution.

An appropriate intercalation agent is the key factor for preparing high-quality MXene nanosheets. There are two kinds of commonly used intercalators: polar organic molecules (such as DMSO,[22,23] hydrazine,[16] urea,[24] and isopropylamine[25]) and organic base molecules (such as tetramethylammonium hydroxide (TMAOH),[26] tetrabutylammonium hydroxide (TBAOH), choline hydroxide, and *n*-butylamine[27]). Different intercalation agents are suitable for specific MXene phases; for example, DMSO, urea, and hydrazine can be used for the intercalation reaction of Ti_3C_2; DMSO can also be inserted into $Mo_2TiC_2T_x$. Hydrazine can also be intercalated into Ti_3CN and $(Ti, Nb)_2C$ interlayers. TBAOH can be used for the large-scale delamination of MXenes that contain Nb, V, Ta, and Mo; however, it cannot make $Ti_3C_2T_x$ layered, and this is probably because of its larger molecular size.[28] The smaller TMAOH molecules can be intercalated into the $Ti_3C_2T_x$ layers and make them laminate. DMSO is the first intercalation agent for the preparation of MXene nanosheets. However, its boiling point is high, and it can easily remain in MXene nanodispersion solution after intercalation. The key point is that it is difficult to remove DMSO, and it easily causes delaminated MXene to adhere.

In addition to the important role of an intercalation agent, the lamellar formation of MXenes is also a result of the presence of hydrophilic functional groups on the surface of MXenes. These groups cause the layered structure to form a stable colloidal solution without a surfactant, in water.[29]

In addition, some metal cations (such as Li^+, Na^+, K^+, Mg^{2+}, and Al^{3+}) can also be inserted into the interlayer of MXenes to increase the interlayer distance of the MXenes.[30] Wang *et al.*[31] inserted Na^+ into the interlayer of layered MXene *via* etching with an aqueous HF solution and grafted $-SO_3H$ groups on the surface to achieve uniform dispersion of Ti_3C_2 nanosheets. Therefore, when etching the MAX precursor, etching–intercalation can be completed at the same time when a mixture of fluoride molten salt and strong acid was used as the etchant. Then, high-quality MXene nanosheets can be obtained *via* stripping. On the basis of the amphoteric property of Al in the Ti_3AlC_2 phase, Xuan *et al.*[32] used the organic base TMAOH as the etchant and intercalation agent to react with Al and to cause it to be gradually de-embedded. At the same time, TMA^+ was embedded in the interlayer of Ti_3AlC_2 to increase the interlayer distance. After that, Ti_3C_2 nanosheets were peeled off *via* oscillation. This study provides a new way to simplify the preparation of MXene nanosheets.

3.3 Structure and Properties of MXenes

3.3.1 Structure

Similar to the MAX phase ($M_{n+1}AX_n$), MXene has a hexagonal crystal structure, in which M atoms are in a hexagonal close packing arrangement, with X atoms filling the spaces in between. The general formula ($M_{n+1}X_n$) is consistent with that of MAX. Without surface groups, MXene has very high chemical activity, and so it is easy to react with solution to form MXene with specific groups ($M_{n+1}X_nT_x$, where T_x mainly represents –F, –OH, and =O groups) in the solution etching process.[33] Tang *et al.*[34] constructed a microstructure model of Ti_3C_2 and $Ti_3C_2X_2$ (X = F and OH). For Ti_3C_2 with no functional groups, the atomic structure is mainly in the form of Ti(1)–C–Ti(2)–C–Ti(1), whereas $Ti_3C_2X_2$ (X = F, OH) with functional groups can be divided into three types, depending on the distribution of the functional group X. These three types are as follows. (1) In the first type, the X functional groups are distributed on both sides of Ti_3C_2 lamellae with the X group directly above the Ti(2) atom and adjacent to three C atoms at the same time. (2) In the second type, the X functional groups are on both sides of Ti_3C_2 with the X group directly above the C atom. (3) The third type can be regarded as a combination of the first two types, with the X group on one side facing the Ti(2) atom and on the other side facing the C atom. In addition, X groups (including O, *etc.*) are randomly distributed on both sides of Ti_3C_2; there is no specific sequence, and there is no correlation between groups.[35] Hope *et al.*[36] found that there can be a single surface group between Ti_3C_2 layers, or that –OH and –F can be present at the same time. They also believe that the relationship between –F and –OH proves the existence of interlayer structural water. Because of this, it is difficult to move this interlayer. Also, the 2D ^{1}H NMR spectrum shows that there are no two adjacent –OH between Ti_3C_2 lamellae, and the amount of –OH is much less than that of –F and –O. Wang *et al.*[35] used high-angle annular dark-field imaging and directly observed the Ti atom arrangement of Ti_3C_2X. The interlayer spacing of Ti_3C_2X was measured to be about 0.98 nm, and the results are consistent with X-ray diffraction (XRD) results. However, C, F, and O atoms cannot be observed in the dark-field image, so they used an annular bright field image to successfully discover the presence of C, F, and O and the relative position with the Ti atom, which is consistent with previous research.[37]

Because the surface of the prepared MXene contains a large number of functional groups, the instability of these functional groups and the interaction with MXene lead to large differences in the stability of MXene when it is prepared under different conditions. Therefore, it is very important to study the stability of MXene, for its use in different fields. Zhou *et al.*[24] pyrolysed the prepared Ti_3C_2 at different temperatures (500 °C, 900 °C, and 1200 °C in Ar). XRD and thermogravimetric–differential thermal analysis (TG–DTA) showed that –OH and –F were gradually desorbed from the surface of Ti_3C_2, and this was accompanied by the formation of by-products (AlF_3 and TiO_2 anatase). Meanwhile, the composition and structure of Ti_3C_2 changed with an increase in temperature. The results obtained by Li *et al.*,[38] who used heat treatment on Ti_3C_2 under Ar protection, were similar to those mentioned above. In addition, for comparison, they also conducted an experiment in an O_2 atmosphere. The study found that at a higher temperature (200 °C), Ti_3C_2 reacts with O_2 to form a small amount of cubic TiO_2 anatase, and when the temperature was increased to 1000 °C, almost all of the Ti_3C_2 transformed into anatase and connected into a network structure. Because of the diversity of the functional groups, MXene is endowed with outstanding optical, electrical, and magnetic properties.

3.3.2 Electron Migration Performance

Studies of the electronic, dielectric, magnetic, thermoelectric, optical, and mechanical properties of MXenes are mainly at the stage of theoretical prediction and experimental verification. Among them, the electronic properties have been studied the most. Kurtoglu *et al.*[39] used first-principles calculations to prove that pure MXene $Ti_{n+1}X_n$ possessed high metallic electrical conductivity. The MAX phase is staggered layers of an MX atomic layer and an 'A' atomic layer; and the MX_6 octahedron is separated by an 'A' layer. Therefore, when the 'A' atomic layer is corroded, changes in the electron arrangement of the MX layer are inevitable, resulting in changes in the electronic properties. Through first-principles calculations, Shein *et al.*[40] found that Ti_1 changed from a 3D state to a Ti_1–Ti_1 metallic bonded state after the Ti–Al bond was destroyed. Therefore, the near-Fermi energy state density of $Ti_{n+1}N_n/Ti_{n+1}C_n$, which has metallic characteristics, was approximately 2.5–4.5 times higher than that of its MAX phase.

The M-A metal bonds were broken as a result of the stripping of the atomic layer 'A', where the atoms of covalent bonds were firmly confined. The layers were relatively independent with only weak van der Waals interactions, and so the atoms can only move within the layer rather than out of the layer. However, Michael *et al.*[13] showed that the electronic conductivity decreases with an increase in n, which is the number of atomic layers of M_2X, M_3X_2, and M_4X_3 (3, 5, and 7, respectively). Moreover, the electronic conductivity of MXene where X = N is better than that of X = C. Meanwhile, Berdiyorov[41] also used first-principles calculations to show that when the surface functional groups of Ti_3C_2 were different, the electronic conductivity changed. From high to low, the electronic conductivity follows the order of $Ti_3C_2F_2 > Ti_3C_2(OH)_2 > Ti_3C_2 > Ti_3C_2O_2$. However, Tang *et al.*[42] obtained different results using density function theory (DFT). Ti_3C_2 showed the best electronic conductivity, whereas $Ti_3C_2F_2$ and $Ti_3C_2(OH)_2$ were semiconductors, each with a narrow band gap.

Similarly, the elements and surface groups of MXenes have a great influence on their performance. Theoretically, MXenes that do not have functional groups on the surface show metallic properties, and the electron density is close to the Fermi level.[40,42] For $Ti_{n+1}C_n$ and $Ti_{n+1}N_n$, the near-Fermi energy state density was approximately 2.5–4.5 times higher than that of the parent phases, $Ti_{n+1}AlC_n$ and $Ti_{n+1}AlN_n$.[40,43] Naguib *et al.*[12] found that the width of the gap for both $Ti_3C_2(OH)_2$ and $Ti_3C_2F_2$ was very large; with values of 0.05 eV and 0.1 eV, respectively. Khazaei *et al.*[44] showed that the Seebeck coefficient of both Ti_2CO_2 and $Sc_2C(OH)$ was about 1000 V K^{-1} (100 K), and the Seebeck coefficient of $SrTiO_3$ was about 850 V K^{-1} (90 K). $Ti_3C_2T_x$ transformed into anatase TiO_2 and C (with TiO_2 embedded in amorphous C in the form of nanocrystalline material) when $Ti_3C_2T_x$ was oxidised in air, CO_2, and water.[45] Wang *et al.*[46] calcined $Ti_3C_2T_x$ at 600 °C for 1 h and found that the surface functional groups of $Ti_3C_2T_x$ were greatly decreased and that the electrical conductivity increased from 850 S cm^{-1} to 2410 S cm^{-1}.

In the MAX phase, the element A is connected to M and X *via* metallic bonds. The breaking of chemical bonds leads to rearrangement of electrons. When the A phase is selectively etched, it greatly improves the mobility of the electrons and affects the electronic properties of MXene. Zha *et al.*[47] used first-principles calculations to compute the electron properties of Sc_2CT_2 (T = F and OH). Electron migration was strongly anisotropic. Shein *et al.*[40] showed that MXene exhibited metal-like properties. According to DFT, the pure MXene phase shows metal-like conductivity, whereas MXene with functional groups on the surface shows semiconductor properties. The directional arrangement of the functional groups on the two-dimensional plane and the different types of functional groups change the electronic structure of MXene.[48] Zha *et al.*[49] calculated the electrical properties of oxy-functionalised M_2CO_2 (M = Ti, Zr, and Hf) MXenes. Hf_2CO_2 was a semiconductor with a band gap of 1.657 eV and high anisotropic carrier mobility. At room temperature, the hole mobility of Hf_2CO_2 was in a sawtooth direction with values in the range of 13.5×10^3–17.6×10^3 cm^2 V^{-1} s^{-1}.

3.3.3 Thermal Stability

The thermal stability of materials is of great importance for many applications. Lattice energy is one of the key factors that affect crystal stability. HF acid can be used to etch the MAX phase, usually resulting in the functionalisation of the surface of the synthesised MXene with oxygen, fluorine, or hydroxide. The lattice energy of MXene is negative, according to the first-principles analysis.[40] This indicates that MXene can be stable under normal temperature and pressure. The results of Barsoum's work support this conclusion.[13] From comparing the stability of $Ti_{n+1}C_n$ and $Ti_{n+1}N_n$, it is found that the stability increases with an increase in the value of n, and the formation energy of MXene decreases with an increase of n. The formation energy of $Ti_{n+1}C_n$ is lower than that of $Ti_{n+1}N_n$. The stability of carbon-containing MXene is better than that of nitrogen-containing MXene.[13] Thermogravimetric analysis showed that $Ti_3C_2T_x$ remains stable at 800 °C under the protection of argon. However, it is completely oxidised at 1000 °C. Barsoum *et al.*[12] studied the stability of $Ti_3C_2T_x$ under the 200 kV irradiation of an electron beam and found that a nanosheet layer of $Ti_3C_2T_x$ was more stable than graphene. That is, it has higher resistance to electron irradiation and resistance to photophores. Theoretically, it is predicted that MXene possesses moderate oxidation resistance.[50] Gogotsi's group discussed the oxidation behaviour of $Ti_3C_2T_x$ MXenes under different conditions.[45] The results showed that at a high temperature of 1150 °C in an atmosphere of air, $Ti_3C_2T_x$ was oxidised in just 5 seconds to form a mixture of TiO_2/C. In a hydrothermal environment, a CO_2 atmospheric environment (50–500 °C) can also oxidise $Ti_3C_2T_x$ MXene. Table 3.1 shows elastic constants, lattice constants, and electron state density data of MXenes near the Fermi surface.

3.3.4 Thermal Conductivity

Similar to the MAX phase, MXenes possess high thermal conductivity. The thermal conductivity of the MAX phase is even higher than that of most semiconductors, low-dimensional materials, and metals. At room temperature,

Table 3.1 Structure characteristics and properties of different MXenes.

Layer	a (Å)	$a_{MX}-a_{MAX}$ (%)	r (Å)	$r_{MX}-r_{MAX}$ (%)	C_{11} (GPa)	DOS at E_f (eV^{-1} $atom^{-1}$)
Ti_2C	3.007	−1.8	2.291	0.9	636 (312)	2.63
V_2C	2.869	−0.6	2.165	−2.8	718 (338)	1.45
Cr_2C	2.787	−1.1	2.080	0.0	690 (340)	2.55
Zr_2C	3.238	−1.8	2.522	0.6	594 (261)	1.77
Hf_2C	3.239	−1.6	2.592	2.0	658 (291)	2.25
Ta_2C	3.138	−0.5	2.491	−1.9	788 (334)	0.89
Ti_3C_2	3.071	0.4	4.605	−1.9	523 (368)	2.19
Ta_3C_3	3.196	1.2	5.025	−4.1	575 (417)	1.44
Ti_4C_3	3.066	0.1	7.142	−0.2	512 (403)	1.83
Ta_4C_3	3.172	−0.6	7.804	0.3	633 (437)	0.56

the thermal conductivity of 5 μm armchair Sc_2CF_2 materials reaches 472 W mK^{-1}, which is much higher than that of the traditional zig-zag Sc_2CF_2, which has a thermal conductivity of 178 W mK^{-1}. For $Sc_2C(OH)_2$, which contains hydroxyl groups, the thermal conductivity is much lower than Sc_2CF_2. However, Sc_2CT_2 (T = F, OH) shows excellent thermal conductivity and electron mobility. In general, it is expected that MXenes can be applied in highly thermal conductive electronic devices in the future.[47]

3.3.5 Magnetic Properties

A theoretical study found that MXenes that have an exposed surface have good magnetism, whereas MXenes that have functional groups on their surface (such as $TiCO_2$, $TiCF_2$, $TiCH_2$, and $TiC(OH)_2$) generally show no magnetism.[43] Khazaei *et al.* used theoretical calculations and found that both Cr_2CT_x and Cr_2NT_x are magnetic at room temperature, they studied the formation and electronic properties of MXene functionalised with F, OH, V, Cr, Zr, Nb, Ta, and M_2N (M = Ti, Cr, and Zr).[44] It was theoretically deduced that functionalised Cr_2C and Cr_2N MXenes have magnetic properties. $Ti_{n+1}C_n$ and $Ti_{n+1}N_n$ are also magnetic, but most MXenes are not magnetic because they have a large number of functional groups. Currently, only Cr_2C and Cr_2N-based MXenes show good ferromagnetism when they possess functional groups.[43]

MXenes can also show anti-ferromagnetism and ferromagnetism.[40] Also, non-magnetic MXene can become magnetic under stress. Therefore, the magnetic properties of MXene materials can be regulated by controlling the type and number of functional groups in MXene. Xie *et al.* found that both $Ti_{n+1}C_n$ and $Ti_{n+1}N_n$ ($n = 1 \sim 9$) have magnetic properties. However, the magnetism disappeared for functionalised MXene. For Cr_2C and Cr_2N MXene, the presence or absence of surface groups does not affect their magnetism.[51] In addition, the magnetic properties of MXenes that are obtained *via* different calculation methods have also been verified. The magnetic properties of MXenes can be calculated using LDA theory and the PBEsol method. LDA theory showed that Ta_2C MXene was not magnetic and that Ta_4C_3 MXene was anti-ferromagnetic. Meanwhile, the results calculated using the PBEsol method were the opposite; specifically, the results showed that Ta_2C MXene was ferromagnetic and that Ta_4C_3 MXene was not magnetic.[52] In addition, the presence of stress has a great influence on the magnetic properties of MXenes, and this may lead to magnetic transformation.

Some MXenes have also shown good magnetism. Mashtalir *et al.*[53] used DFT to study the dehydrogenation/hydrogenation mechanism of $Sc_2C(OH)_2$/Sc_2CO_2 super-crystalline cells. Three different super-crystalline cells (2×2×1, 3×3×1, and 4×4×1) were used to obtain intermediate $Sc_2C(OH)_xO_{2-x}$, which changes the hydrogen content ($0.0625 \leq x \leq 1.94$). If $Sc_2C(OH)_xO_{2-x}$ was used as a nanosheet, the intrinsic spin polarisation semiconductor characteristics could be observed. The Ti_3C_2 monolayer could be used as a magnetic metal, and doping with nitrogen atoms significantly reduced the spin magnetic moment. On the other hand, when the two surfaces of the Ti_3C_2 monolayer are wrapped by external groups, the magnetism will automatically annihilate. More importantly, both doping and surface modification can reduce the Curie temperature of the Ti_3C_2 monolayer. This provides a new way to study the Ti_3C_2 monolayer for high temperature magnetism applications.

3.3.6 Mechanical Properties

$Ti_3C_2(OH)_2$ has been compared to typical transition metal carbides. Specifically, theoretical calculations have been used to estimate the elastic modulus of monolayer $Ti_3C_2(OH)_2$ to be ~300 GPa, which is close to the value of graphene, but is lower than those of most oxides and layered clays. From Kurtoglu's first-principles calculations,[39] MXene has been found to have a high elastic modulus in the direction of the base plane, and the value is higher than that of most oxides and layered clays. The bending strength is higher than that of graphite at the same thickness. MXene films are flexible, strong, and foldable after repeated treatment.[54] Polyvinyl alcohol can bind tightly to Ti_3C_2 to form a staggered layered material that has four times higher mechanical properties than monolayer MXene films.[55] Kurtoglu *et al.*[56] calculated the mechanical properties of various MXenes, such as T_2C, Ti_3C_2, Ti_4C_3, V_2C, Cr_2C, Zr_2C, Hf_2C, and Ta_2C. Similar to MAX phase ceramics, MXenes have a high elastic modulus when they are extended on the base plane. The values of the bending strength of Ti_4C_3 and Ta_4C_3 are about ten times greater than those of graphene because stripping a bit of the atomic layer enhances the binding energy of M–X bonds.

3.3.7 Adsorption Performance

Because of the two-dimensional structure of MXenes, the light transmission and flexibility of MXenes are much better than those of MAX. The lamellae of MXenes are only a few atomic layers thick and have a large specific surface area. The typical lamellar structure endows MXenes with potentially good adsorption performance. Alkali MXene ($Ti_3C_2(OH/ONa)_xF_{2-x}$) prepared *via* selective corrosion stripping and alkalisation embedding show excellent adsorption performance for divalent lead ions compared to a high concentration of cationic divalent calcium and magnesium. A kinetics experiment demonstrated that the adsorption process reached a balance within 120 s. Both experiment and theoretical calculations proved that the adsorption characteristics of MXenes are significantly related to the hydroxyl-activated T site.[57] Such excellent adsorption performance of MXenes enables MXenes to be widely used in the field of environmental protection.

3.3.8 Hydrogen Storage Performance

As a clean energy, hydrogen has many advantages, such as a high calorific value, small volume, and no pollution. However, hydrogen has storage and transportation problems. In theory, Sc_2C as an MXene phase is expected to have a high hydrogen storage capacity per unit weight (~9.0 wt%).

Normally, 5.5 wt% is an important storage capacity index, as set by the US Department of Energy.[58] On the basis of hydrogen adsorption sites, hydrogen storage of Sc_2C can be divided into three modes: chemical adsorption, physical adsorption, and Kubas adsorption. The values of the corresponding binding energy are 4.703 eV, 0.087 eV, and 0.164 eV, respectively. The weight storage capacity of Kubas-adsorbed and interacting hydrogen is 3.6 wt%.

3.4 Applications of MXenes

3.4.1 Supercapacitors

Because of the unique two-dimensional structure and surface chemical composition, MXene materials exhibit good metallic conductivity, hydrophilicity, and excellent mechanical flexibility. MXene has a transition metal oxide active surface, and this enables rapid reversible oxidation–reduction reactions to occur on the surface when MXene is used as an electrode material for supercapacitors. Two-dimensional nanochannels provide fast ion migration paths, and these give MXene excellent specific electric capacity and rate performance.[35–37] Among the discovered MXenes, $Ti_3C_2T_x$ and Ti_2CT_x are relatively easy to prepare and possess good electrical conductivity. Therefore, $Ti_3C_2T_x$ and Ti_2CT_x are suitable electrode materials for supercapacitors, lithium-ion batteries, and sodium-ion batteries.

As an electrode material in supercapacitors, electrolyte ions can quickly enter into MXene nanosheets through nanochannels. Highly reversible redox reactions occur on the active surface of a transition metal atom layer, thus providing embedded pseudocapacitance. At the same time, the carbon atom layer or nitrogen atom layer in MXene can rapidly transfer electrons, providing good conductivity for electrodes. Therefore, MXene shows a good rate performance, even at a higher charge–discharge current density. The volume specific capacitance of an independent $Ti_3C_2T_x$ electrode in neutral and alkaline electrolyte is 300–400 F cm^{-3} (see Figure 3.4).[30] This value exceeds the current best carbon-based double-layer capacitors, which use graphene-based electrodes (350 F cm^{-3}).[59] Cyclic voltammetry for $Ti_3C_2T_x$ in different electrolyte solutions was compared, and the results show that although the shape of the cyclic voltammetry curves varies slightly with the cations, there are no obvious redox peaks, and the cyclic voltammetry curve is nearly rectangular, indicating good capacitance.[16,22]

In addition, MXenes can be used in planar supercapacitor devices. As shown in Figure 3.5(a), a flexible planar microcapacitor can be made by scraping $Ti_3C_2T_x$ MXenes on A4 paper, and such a microcapacitor can be used in wearable intelligent electronic devices.[60] An MXene suspension can be used as conductive and electrochemical inks to prepare conductive electrodes on A4 paper or polyethylene terephthalate (PET) substrates using impression printing,[61] printing,[62] and writing.[63] The micro-supercapacitor that is fabricated *via* this method shows good flexibility and high capacity under different bending

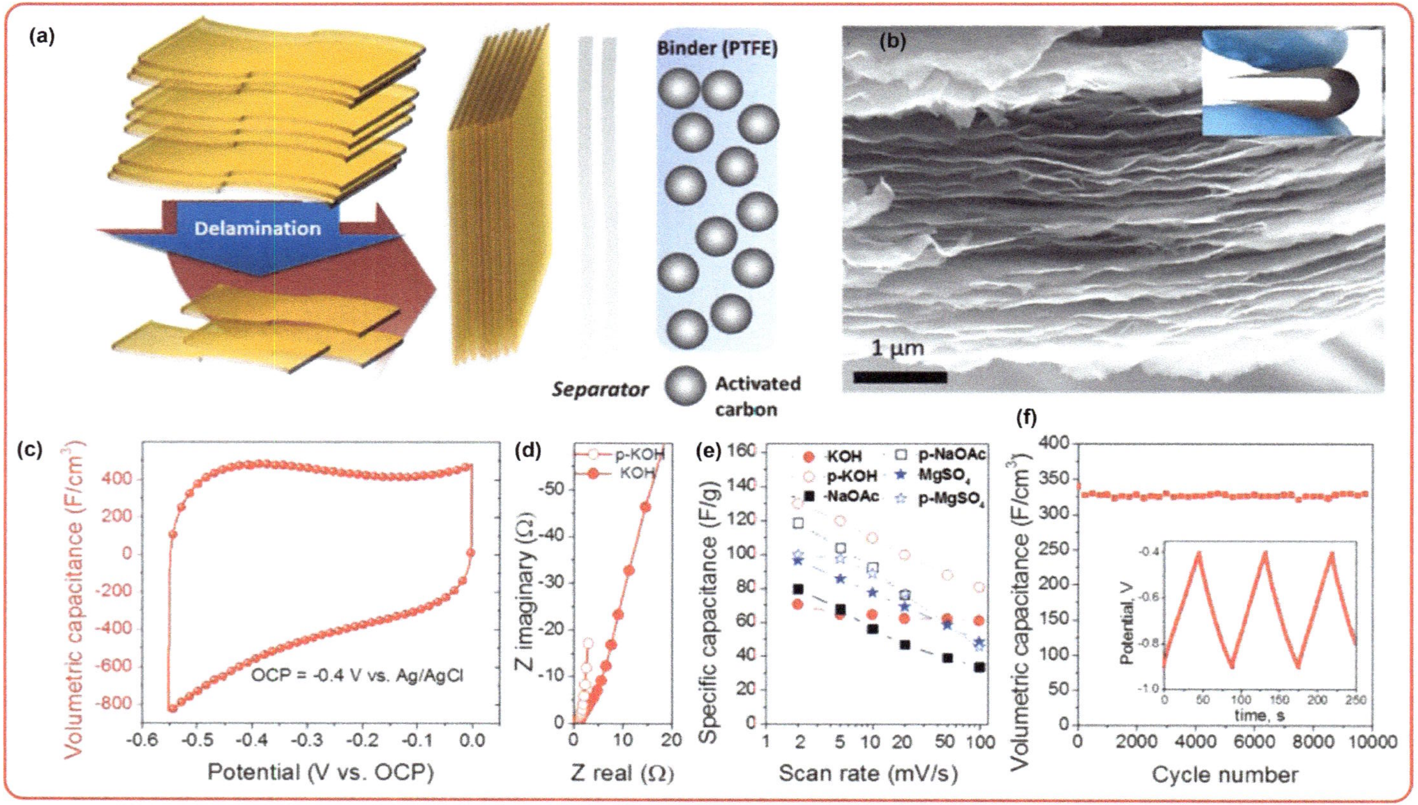

Figure 3.4 Electrochemical performance of $Ti_3C_2T_x$ electrodes in alkaline electrolytes.[30] Reproduced from ref. 30 with permission from AAAS, Copyright 2013.

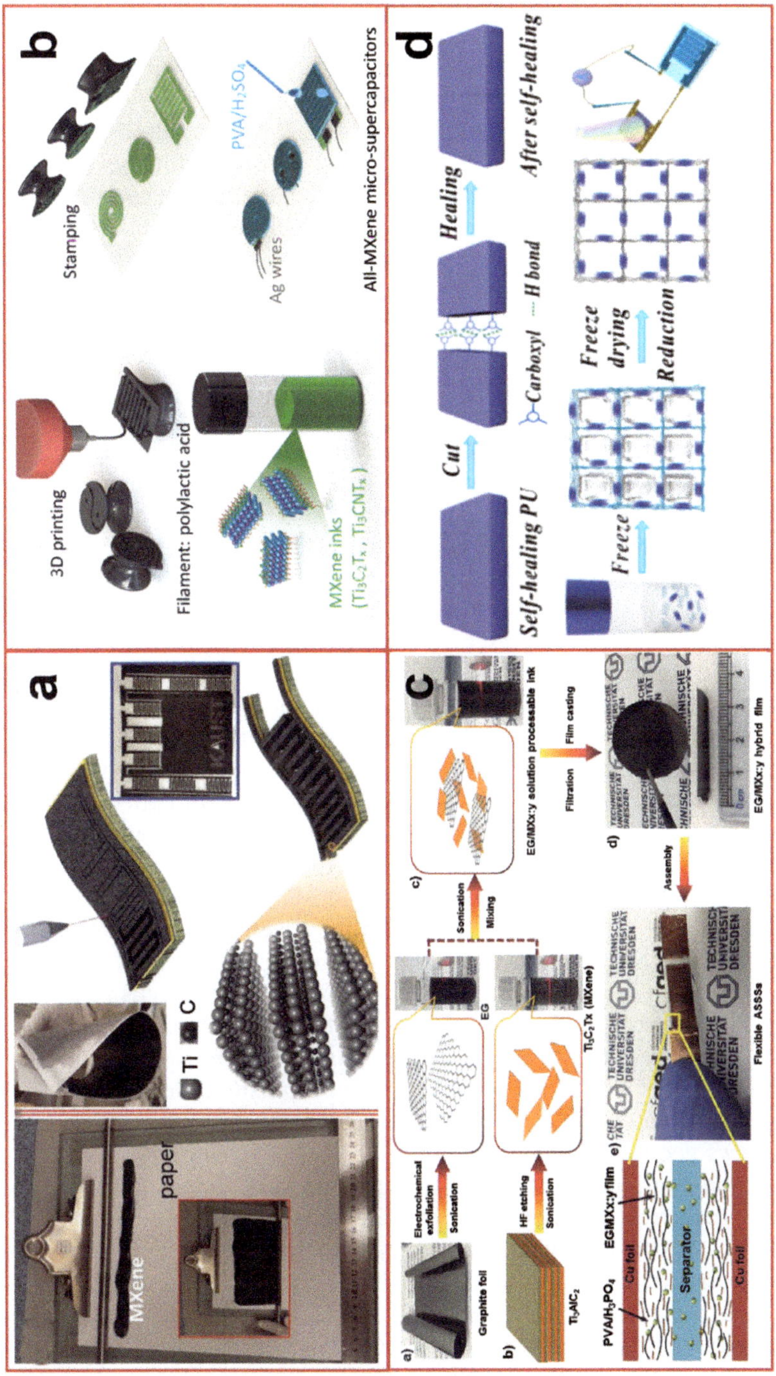

Figure 3.5 (a) Fabrication of MXene-on paper coplanar micro-supercapacitor. Reproduced from ref. 60 with permission from John Wiley & Sons, Copyright 2016 WILEY-VCH Verlag GmbH & Co. KGaA, Weinheim. (b) Stamping of flexible, coplanar micro-supercapacitors using MXene inks. Reproduced from ref. 61 with permission from John Wiley & Sons, Copyright 2018 WILEY-VCH Verlag GmbH & Co. KGaA, Weinheim. (c) Flexible all-solid-state supercapacitors with high volumetric capacitances boosted by solution-processable MXene and electrochemically exfoliated graphene. Reproduced from ref. 67 with permission from John Wiley & Sons, Copyright 2016 WILEY-VCH Verlag GmbH & Co. KGaA, Weinheim. (d) Schematic of fabrication of highly self-healable 3D micro-supercapacitor with MXene–graphene composite aerogel.[47,69] Reproduced from ref. 69 with permission from American Chemical Society, Copyright 2018.

conditions (Figure 3.5(b)). A flexible bendable capacitor device can also be made by bonding MXene to a carbon fibre substrate using PEDOT-PSS as an adhesive.[64] All-MXene and substrate-free micro-supercapacitors can also be prepared.[65] MXene nanochips can also be pumped into self-supporting membranes to directly form flexible micro-supercapacitors.[66]

In addition to the active materials mentioned above, MXenes can also be combined with other electrochemical materials to further optimise the performance of the capacitor. As shown in Figure 3.5(c), MXene with an ultra-high bulk capacity can be mixed with graphene *via* an ultra-flexible electrochemical method to self-assemble into flexible electrodes. A volume capacity of 216 F cm^{-3} can be obtained at a current density of 0.1 A cm^{-3}. The capacity can be maintained above 82% after 2500 cycles even under bending conditions.[67] In addition, MXenes that have negative charge and modified redox graphene with positive charge can assemble into flexible membrane electrodes *via* electrostatic adsorption.[68] A three-dimensional network or foams enable rapid diffusion of electrolyte to the electrode, and this is helpful for enhancing the rate performance. Yue *et al.* used lyophilisation to prepare composite aerogels made from MXene and graphene.[69] Because of the large specific surface area and good electrical conductivity, the composite aerogels showed excellent electrochemical performance. The micro-supercapacitor can be prepared *via* laser etching. It is interesting that a self-healing effect of the capacitor can be achieved, as presented in Figure 3.5(d).

Metal oxides with high electrical capacity are also helpful for increasing the energy density of MXene-based capacitor devices. Liu *et al.* mixed MnO_2 nanosheets with a suspension of $Ti_3C_2T_x$ nanosheets to form flexible composite membranes *via* vacuum filtration.[70] For the assembled symmetrical capacitor devices, the capacitance was 390 F g^{-1}, and the electrochemical stability was retained well after the device was bent 180°; this is several times higher than that of MnO_2 or $Ti_3C_2T_x$. In addition to metal oxides, layered double hydroxide (LDH) with high capacitance and MXene have also been used to fabricate flexible devices. Zhao *et al.* used negatively charged MXene and positively charged LDH to assemble a heterostructure film *via* electrostatic interactions. The flexible capacitor device has an energy density of 45.8 Wh kg^{-1}.[71] In addition, a composite with electrochemical activity and conductive polymers can improve the electrochemical performance and mechanical properties of electrodes. For example, a composite film electrode formed *via in situ* polymerisation of polypyrrole (PPy) on the surface of an MXene film shows good electrochemical and mechanical stability.[53,72]

MXene has good conductivity and hydrophilicity and also has adjustable optoelectronic properties depending on the thickness of the film. Thus, MXene can be used as a transparent supercapacitor electrode.[73,74] Zhang *et al.* used suspension coating and vacuum heat treatment to prepare a transparent supercapacitor of 2D titanium carbide.[73] The capacitor has a higher volume capacity when the transmittance of light is 93%, and this is suitable for flexible transparent electronic devices.

3.4.2 Electromagnetic Interference Shielding

High electronic conductivity of materials is necessary to obtain high electromagnetic interference (EMI) shielding efficiency. In 2016, Gogotsi *et al.* pioneered research on EMI shielding performance of MXenes. Some MXenes have a high electronic conductivity of 4600 S m^{-1}, and their two-dimensional lamellar structures induce multiple reflections of electromagnetic waves inside the material. When the thickness is 45 μm, the electromagnetic shielding performance can reach 92 dB. Even with a thickness of 2.5 μm, the electromagnetic shielding performance can also reach above 50 dB.[75] Figure 3.6(a) shows a comparison of the shielding effectiveness of MXenes and composites, and the electromagnetic shielding mechanism model of MXenes is also presented.

The electromagnetic shielding effect of $Ti_3C_2T_x$ MXene originates from both the electronic conductivity and the two-dimensional lamellar structure. As shown in Figure 3.6(b), when an electromagnetic wave is incident on the surface of the $Ti_3C_2T_x$ MXene sheet, some electromagnetic waves are immediately reflected and some interact with the high electron density $Ti_3C_2T_x$ MXene through the lamellar structure because of the abundance of free electrons on the surface, and this results in current and ohmic loss in terms of the attenuation of electromagnetic energy. After passing through the first layer of $Ti_3C_2T_x$ MXene (marked 'I' in Figure 3.6(b)), the remaining electromagnetic wave encounters the next layer (marked 'II'), and this attenuation process repeats back and forth. At the same time, layer II can act as a reflecting surface and cause multiple internal reflections. The electromagnetic waves can be reflected back and forth between layers (I, II, III, *etc.*) until the electromagnetic wave is fully absorbed.[75] Therefore, the multilayer structure of $Ti_3C_2T_x$ MXene induces a high shielding performance. In addition, the Ti in MXene forms a local dipole with the terminal groups (=O, –F, and –OH) in an alternating electric field. F has strong electronegativity and can induce polarisation to further enhance the effectiveness of the electromagnetic shielding.

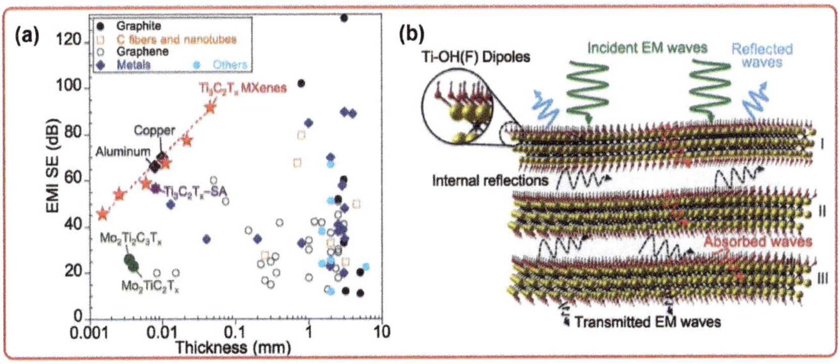

Figure 3.6 (a) EMI SE *versus* thickness of different materials. (b) Proposed EMI shielding mechanism.[75]
Reproduced from ref. 75 with permission from AAAS, Copyright 2016.

Sun *et al.*[76] used an electrostatic self-assembly to prepare nanocomposites composed of $Ti_3C_2T_x$ and polystyrene spheres. An electronic conductivity of 1081 $S\,m^{-1}$ and an electromagnetic shielding effectiveness of 54 dB were achieved over the whole X-band. When the amount of added $Ti_3C_2T_x$ was 1.9 vol%, the maximum value of the electromagnetic shielding effectiveness was 62 dB. The electromagnetic shielding materials are summarised in Table 3.2.

Liu *et al.*[77] used hydrazine hydrate to induce MXene film foaming and obtained light, hydrophobic, and durable MXene foam, as shown in Figure 3.7. The electromagnetic shielding effectiveness can reach 70 dB. For comparison, it was only 53 dB for the $Ti_3C_2T_x$ film without foaming. The presence of a porous structure is conducive to multiple attenuations of electromagnetic waves. One-dimensional nanocellulose and two-dimensional MXene can also be combined to form a composite paper, which is similar to a pearl-layered structure that is flexible and has excellent electromagnetic shielding performance. Because of the 'mortar and brick' structure formed from cellulose nanocellulose and MXene, the toughness of the composite paper is improved. The tensile strength is 135.4 MPa, and the electronic conductivity is 739.4 $S\,m^{-1}$. It shows excellent electromagnetic shielding performance.[78]

Because MXene has a high density, it is difficult for MXene to meet light weight requirements. Therefore, MXene aerogels have been developed. Pradeep *et al.*[79] reported a $Ti_3C_2T_x$/carbon nanotube (CNT) aerogel that has high electrical conductivity and clcctromagnctic shiclding. As reinforcing fillers, CNTs improve the mechanical properties (such as compression modulus) of porous $Ti_3C_2T_x$ aerogels. The aerogel is lightweight and has high conductivity, good mechanical stability, and strong electromagnetic interference resistance. In the X-band, the EMI shielding effectiveness (SE) values of MXene and MXene/CNT with a thickness of 3 mm are 63.56 dB, 78.56 dB, 87.34 dB, and 103.99 dB. The excellent electromagnetic shielding performance of MXene/CNT aerogels is mainly attributed to the three-dimensional porous structure, which has high electronic conductivity and a uniform lamellar structure.

3.4.3 Electromagnetic Wave Absorption

For electromagnetic (EM) wave-absorbing materials, the key requirements are low density, wide absorption ability, and high absorption efficiency. Two-dimensional materials are widely used in designing new highly efficient EM-absorbing materials, and an MXene two-dimensional material has potential in this field. MXene has several inherent advantages when it is used as a microwave-absorbing material. First, the intense HF etching reaction in the MXene synthesis process causes MXene materials to have a large number of defects, and this can help enhance the absorption of electromagnetic waves. Second, the active surface functional groups of MXene materials can improve and tune the microwave absorption performance. Finally, the two-dimensional structure of an MXene that has a large spacing between layers can reflect and scatter electromagnetic waves in multiple layers, and this enhances the absorption of incident electromagnetic waves.

Table 3.2 Comparison of EMI SE of different EMI shielding materials reported in references.

		Content		EMI SE	Conductivity	Thick-ness	Frequency		
Filler	Matrix	vol%	wt%	dB	$S\ m^{-1}$	mm	GHz	Method	Ref.
RGO	PEI	1.38	10	13	2.20×10^{-3}	2.3	8.2–12.4 (X-band)	Water vapour-induced phase separation	96
	Epoxy	8.80	15	21	≈10	—		*In situ* process	97
	PS	5.6	30	29	1.25	2.5		Compression moulding + salt leaching	98
	PU	4.70	10	57.7	0.06	60		Dip-coating	99
TGO	PMMA	1.8	5.0	19	3.11	2.4	8–12	Solution mixing + batch foaming	100
		4.23	11.8	30	20	3.4		Solution blending	101
3D Graphene	PDMS	0.36	0.8	20	200	1	8–12	CVD + RTM	102
Aligned RGO	Epoxy	0.89	2	38	1	>0.1	0.4–4	*In situ* process	103
Multiwalled CNT	PC	3.39	5	25	—	1.85	X-band	Melt blending	104
	PP	7.5	10.8	35	$\approx10^{4}$	1	8–12		105
	WPU	7.2	76.2	50	44.6	4.5	X-band	Solution blending + directional freezing	106
	ABS	10.5	15	50	123	1.1		Solution blending	107
Single-walled CNT	PU	14.3	20	17	2.20×10^{-2}	2			108
	Epoxy	10.5	15	20	20	1.5	0.5–1.5	*In situ* process	109
	PMMA	14.3	20	40	2	4.5	8–12	Solution blending	110
CNT sponge	Epoxy	1.34	2	40	516	2	8–12	CVD + VARTM	111
Carbon fibre	ABS	8.93	15	35	66	1.1	X-band	Solution blending	107
	PP	10	16.6	25	10	3.2		Melt blending	112
Expanded graphite	SEBS	9.96	20	12	24	5	X-band	Melt blending	113
Carbon nanowires/graphene	PDMS	13.0	25.3	36	340	1.6	X-band	CVD + RTM	114
RGO/Fe_3O_4	PVC	3.4	10	13	7.7×10^{-4}	1.8	8.2–12	Melt blending	115
	PVA	8.54	35	15	—	0.3	X-band	Solution blending	116
Carbon fibre/Fe_3O_4	Epoxy	5.55	20	18	<0.2	13	1–18	*In situ* process	117
$RGO/\gamma\text{-}Fe_2O_3$	PVA	10.8	53	20.3	3	0.36	X-band	Solution mixing	118
TGO/Fe_3O_4	PS	4.13	13	30	21	4	8–12	Solution blending	119
Ag nanowires	PS	2.5	21.2	31.85	1.90×10^{3}	0.8	X-band	Solution processing	120
Ag nanowires	PANI	14	43.4	35	5.30×10^{5}	0.04	X-band	*In situ* polymerisation	121
$Ti_3C_2T_x$	PS	1.9	6.2	62	1081	2	X-band	Self-assembly	76
		1.2	4	40	602				
		0.72	2.4	33	50				

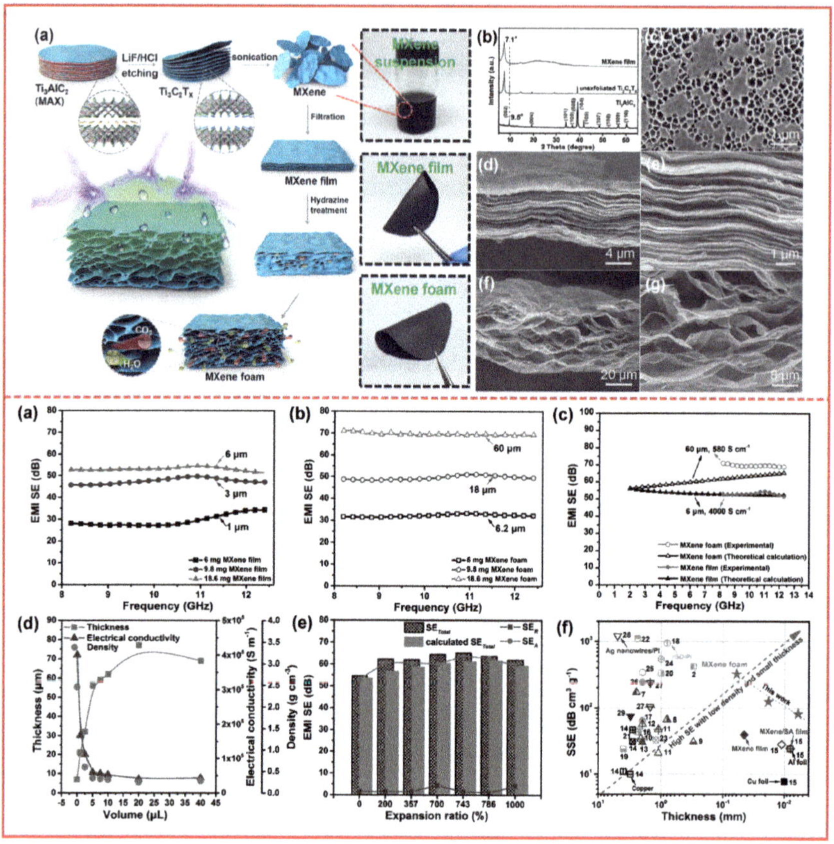

Figure 3.7 Schematic of the fabrication of the hydrophobic and flexible MXene foam and some performance of MXene foam.[77]
Reproduced from ref. 77 with permission from John Wiley & Sons, Copyright 2017 WILEY-VCH Verlag GmbH & Co. KGaA, Weinheim.

The real and imaginary parts of the complex dielectric constant of Ti_3AlC_2 powders in the K_u band are 5 and 7, respectively. The real and imaginary parts of the etched Ti_3C_2 nanolayered MXene material increased to 11 and 14, respectively. The increase in the real part is related to charges and defects on the surface after etching. The increase in the imaginary part is related to polarisation relaxation of Ti_3C_2 after etching. The reflection loss of Ti_3AlC_2 is very poor; however, the minimum reflection coefficient of etched Ti_3C_2 can be decreased to -17 dB in the K_u band.[80] However, the high conductivity leads to a poor impedance match. The EM wave cannot be incident in the bulk material. Therefore, MXene is usually combined with other dielectric or magnetic materials to achieve impedance matching as much as possible.

Both dielectric loss and magnetic loss are necessary to achieve highly efficient EM-absorbing materials. Therefore, magnetic particles such as Fe_3O_4 can be added between multiple layers of MXenes to significantly improve complex permeability and magnetic loss. In this way, MXene EM wave-absorbing materials have the ability to achieve both dielectric loss and magnetic loss under the action of electromagnetic waves. Because of the larger surface area of multilayered MXene material, it is easier to produce interfacial polarisation relaxation. Magnetic particles can be filled into the interlayer gap of MXene *via in situ* reaction or the hydrothermal method. Liu *et al.* use the hydrothermal method to prepare $TiO_2/Ti_3C_2T_x/Fe_3O_4$, which is a two-dimensional EM wave-absorbing material.[81,82] Introducing Fe_3O_4 effectively improved the microwave absorption performance of an MXene complex. Meanwhile, nanosized Fe_3O_4 with multiple interfaces can effectively improve the interfacial polarisation effect. On the other hand, it can also allow the EM wave to be incident on the surface. The layered $Ti_3C_2T_x$ matrix provides a large number of interfaces to promote charge transfer. When the thickness of the sample was 1.9 mm, the minimum reflection coefficient was −57.3 dB at 10.1 GHz, and where the reflection coefficient is less than −10 dB, the effective absorption band is 2 GHz (9.1–11.1 GHz). In addition to the dielectric loss and magnetic loss, the excellent absorbing properties of $TiO_2/Ti_3C_2T_x/Fe_3O_4$ result from the conductive network in MXene as well as from the interfacial polarisation and relaxation at the layered interface.

Han *et al.*[83] studied the influences that heat treatment and surface structure changes of $Ti_3C_2T_x$ have on the dielectric properties and EM wave-absorption properties. The complex permittivity can be significantly tuned by removing the Al layer through HF etching. The multilayer $Ti_3C_2T_x$ that was etched by HF was annealed for 2 h in argon at 800 °C. As a result, the Ti layer on the surface was oxidised to form a local sandwich structure that was composed of anatase, rutile TiO_2 nanoparticles, and amorphous carbon and was supported by a layered $Ti_3C_2T_x$ sheet. The reduction of hydroxyl groups and the formation of amorphous carbon improved the conductivity of MXene. At the same time, the dissipation of EM waves was enhanced by the dielectric dipole interaction and the associated relaxation at multiple interfaces as a result of the formation of the low dielectric constant of anatase and rutile TiO_2. At 11.6 GHz, the minimum reflection coefficient of the annealed $Ti_3C_2T_x$ MXene that was mixed with 50 wt% paraffin was −48.4 dB with a thickness of 1.7 mm. The total shielding effectiveness (SE) of the $Ti_3C_2T_x$/wax composite is mostly attributed to electromagnetic wave absorption rather than electromagnetic wave reflection, even though $Ti_3C_2T_x$ has the free electron density of a metalloid.

3.4.4 Lithium-ion Batteries

Two-dimensional MXenes that have a large specific surface area and ion channels show a wide range of applications in the field of energy storage.[84–86] In the field of portable electronics, lithium-ion batteries (LIBs) are the leading contender for commercial battery packs of electric vehicles. Most LIBs use commercial graphite anodes, which have limited energy density. MXene has excellent conductivity, a low working voltage range (−0.2 to 0.6 V *vs.* Li/Li^+), and a low diffusion energy barrier. In contrast to typical diffusion-limited battery electrode materials, the diffusion of ions in MXene layers occurs at a very high rate without an obviously reduced energy storage capacity. Levi *et al.*[87] showed that water molecules in MXene promoted rapid electrochemical cation intercalation between partially expanded layers. Similar to ion adsorption at the solid–liquid interface, this caused significant expansion and shrinkage. This electrochemical capacitance paradox was caused by the shallow and deep cation adsorption sites, which were full of ions and water molecules. MXene sheets that have more =O groups on the surface showed higher capacity than MXene sheets with other surface groups.[88,89] HCl/LiF etching may be the best way to prepare MXene for LIBs although the yield is relatively low. In this case, controlling the surface chemical property of MXene is very important.

Tang *et al.*[34] used DFT calculation and predicted the theoretical lithium-ion capacity (320 mA h g^{-1}) of an ideal Ti_3C_2 monolayer without any end groups. If the monolayer completely terminated in –OH or –F, the theoretical capacity was 67 and 130 mA h g^{-1}, respectively. Li^+ can be adsorbed and diffused on the surface of a single layer of Ti_3C_2, $Ti_3C_2(OH)_2$, and $Ti_3C_2F_2$. At the same time, because of the high adsorption energy and a large amount of charge transfer, Li^+ adsorbed onto the surface of MXene *via* strong Coulomb interactions with single-layer Ti_3C_2, $Ti_3C_2(OH)_2$, and $Ti_3C_2F_2$. Adsorption of Li^+ can lead to geometric changes of I-$Ti_3C_2F_2$ and I-$Ti_3C_2(OH)_2$. From theoretical calculations, the transfer rate order of Li^+ in the single-layer chip mentioned above was determined to be Ti_3C_2 > I-$Ti_3C_2F_2$ > I-$Ti_3C_2(OH)_2$. This is because the lowest energy barrier and the shortest transport channels are on the surface of Ti_3C_2. Inspired by the first use of etched multilayer Ti_2CT_x as an anode material for LIBs, the reversible capacity was recorded as 70 mA h g^{-1} in the half-cell structure even after 200 cycles at 10 °C. Therefore, more MXenes (such as $Ti_3C_2T_x$, $Mo_2TiC_2T_x$, Nb_2CT_x, V_2CT_x, $Nb_4C_3T_x$, and Mo_2CT_x) can also be considered to be potential anodes for LIBs.[30,90,91]

Mashtalir *et al.*[22] prepared a mixture of single-layer and multilayer $Ti_3C_2T_x$ with a narrow size distribution. DMSO was inserted between the multilayer $Ti_3C_2T_x$ that was etched by HF. Then, an independent $Ti_3C_2T_x$ electrode was assembled without binder *via* membrane filtration. The $Ti_3C_2T_x$ electrode showed an excellent rate performance that was as high as 410 mA h g^{-1} at 1C and 110 mA h g^{-1} at 36C after 700 cycles. The additional lithiated Li layers may form on $Ti_3C_2T_x$, resulting in a capacity that is higher than the theoretical value. Despite the inherent moderate gravimetric capacity (320 mA h g^{-1}), the practical high density of the $Ti_3C_2T_x$ electrode, which has a capacity as high as 5.9 mA h cm^{-2} at 3C, can be prepared *via* HF cold pressing etching of multilayer $Ti_3C_2T_x$ with conductive carbon black. The electrode is superior to the current commercial graphite electrodes.[92]

To address the inherent moderate gravimetric capacity, Luo *et al.*[93] synthesised a PVP–Sn^{4+}@$Ti_3C_2T_x$ nanocomplex with a synergistic 'pillar effect'. The spontaneous expansion of anchored Sn during alloying with Li^+ promoted separation of the $Ti_3C_2T_x$ interlayer and allowed more Li^+ intercalation. When this material was tested in a half-cell, it maintained a reversible capacity of 544 mA h g^{-1} after 200 cycles at 500 mA g^{-1}. It is observed

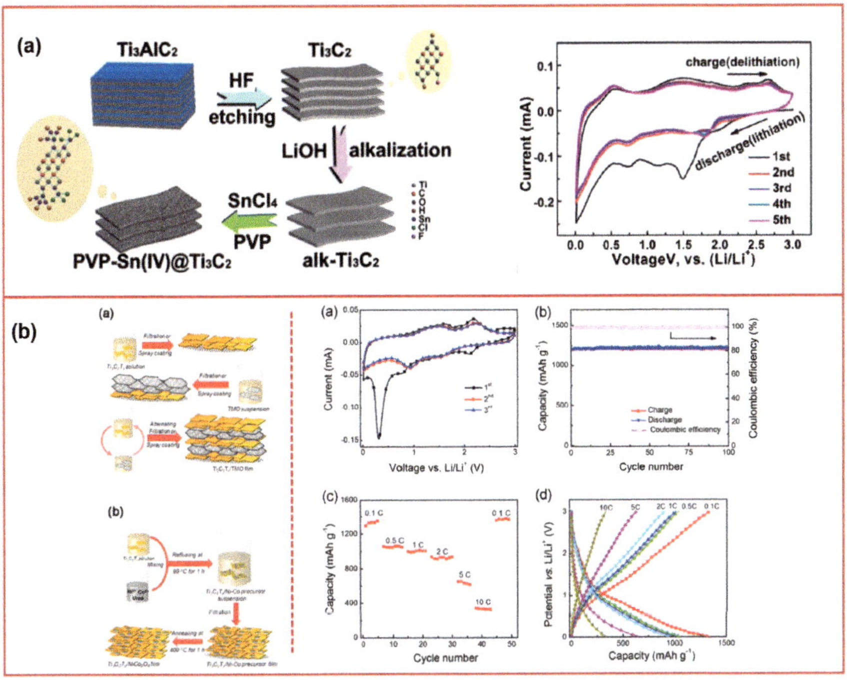

Figure 3.8 (a) Sn^{4+}-ion-decorated highly conductive Ti_3C_2MXene: promising lithium-ion anodes with enhanced volumetric capacity and cyclic performance. Reproduced from ref. 93 with permission from American Chemical Society, Copyright 2016. (b) 2D titanium carbide and transition metal oxides hybrid electrodes for Li-ion storage.[93,94] Reproduced from ref. 94 with permission from Elsevier, Copyright 2016.

that the charge–discharge curve of $Ti_3C_2T_x$ is more like a supercapacitor than LIBs, as seen in Figure 3.8(top). Therefore, $Ti_3C_2T_x$ may be more compatible in lithium-ion capacitors or hybrid devices. Zhao *et al.* alternated filtration, spraying, and *in situ* growth to prepare $Ti_3C_2T_x$/planar $NiCo_2O_4$ hybrid films.[94] The first two methods provide flexibility for the preparation of $Ni_2Co_2O_4$ and $Ti_3C_2T_x$ hybrid films with different mass ratios. With an increase in the $Ni_2Co_2O_4$ content, the mechanical properties changed. During *in situ* growth, Ni^{2+} and Co^{2+} cation catalysts lead to partial oxidation of $Ti_3C_2T_x$ lamellae and reduce the relative conductivity of hybrid films. After 100 cycles of the 50 wt% $Ti_3C_2T_x$/planar $Ni_2Co_2O_4$ hybrid film (3 μm), a reversible capacity of 1010 mA h g^{-1} was maintained at 1C, indicating excellent performance. This is attributed to the higher degree of disorder in $Ti_3C_2T_x$ and to the plane of $Ni_2Co_2O_4$ nanoflakes in the film as a result of spraying and a large number of $Ti_3C_2T_x$ (50 wt%). At the same time, the permeability of the electrolyte improved, and the transmission of electrons and ions was promoted, as shown in Figure 3.8(bottom).

Recently, two kinds of Mo-based MXenes have been used as anode materials for LIBs. Anasori *et al.*[90] reported that the electrochemical behaviour of

$Mo_2TiC_2T_x$ and $Ti_3C_2T_x$ were similar after 160 cycles. The reversible capacity was 145 mA h g^{-1} at 1C. Mo_2CT_x that was mixed with 8 wt% MWCNTs had a reversible capacity of 250 mA h g^{-1} at 5 A g^{-1} and 76 mA h g^{-1} at 10 A g^{-1} after 1000 cycles.[95] In addition, Ti-, Nb-, V-, and Mo-based MXenes were also studied for their future potential use in anodes of LIBs.[25,91] It should be noted that device manufacturing additives (such as current collectors, conductive agents, and polymer adhesives) can inevitably affect the final volume capacity. The weight ratio of the electrode material (normalised to total electrode mass) is considered to be negatively affected by a higher thickness or mass loading. To date, most work has focused on a half-cell structure of LIB electrodes. If full-cells are assembled to determine energy and power density, MXenes will play a more crucial role in practical applications.

3.5 Summary and Outlook

Two-dimensional structured MXene was pioneered by Yury Gogotsi in 2011, as a new type of transition metal carbide and carbonitride. The A phase in MAX is selectively etched by HF acid, and the surface of as-produced exfoliated MXenes was functioned with =O, –F, and –OH groups. Because of the new intercalation structure, along with simultaneously being electronically conducting and hydrophilic, MXene has potential applications in many fields, such as lubricants, environmental pollution control, energy storage, catalysis, sensors, antibacterial substances, and electromagnetic shielding. To date, a number of MXenes, such as, Ti_3C_2, Ti_2C, Nb_2C, V_2C, $(Ti_{0.5}Nb_{0.5})_2C$, $(V_{0.5}Cr_{0.5})_3C_2$, Ti_3C_n, Ta_4C_3, and Nb_4C_3 have been well developed, with high yield and controllable synthesis strategies of the MXenes; however, there are still some challenges to be solved. For example, MXene thin films that are etched with hazardous HF or HF–HCl are multilayered structures and agglomerate easily. In addition, using aqueous HF as an etchant introduces lots of functional groups on the surface of MXene, which may limit its performance. Therefore, the development of MXenes with clean surfaces is a future trend. It is possible to use anhydrous etchants to avoid the formation of –OH, –O, and other functional groups on the surface of MXene; alternatively, MXene can be heat-treated in an inert gas or reducing gas at high temperature. Moreover, most of the etching of phase A is Al except for Mo_2C. The etching of A elements needs to be explored to cover all possible transition metals carbides to form MAX phase. If other A elements can be removed *via* etching, it is possible to synthesise more abundant MXenes. At present, the theoretical research on MXenes has mainly focused on DFT methods and classical and abiotic molecular dynamics. The surface chemical properties need to be well investigated, understood, and ultimately controlled. In addition to single layer and stacked, more MXene-based nanostructures (such as nanotubes and nanorods) need to be further studied in the terms of geometry, electronic structure, and characteristics. In summary, great progress has been achieved recently in the preparation, structure, performance, and application of MXenes, and future applications of functional polymer-based nanocomposites are envisaged.

References

1. H. Zhang, *ACS Nano*, 2015, **9**, 9451–9469.
2. A. K. G. K. S. Novoselov, S. V. Morozov, D. Jiang, Y. Zhang, S. V. Dubonos, I. V. Grigorieva and A. A. Firsov, *Science*, 2004, **306**, 666–669.
3. S. P. Meryl, D. Stoller, Y. Zhu, J. An and R. S. Ruoff, *Nano Lett.*, 2008, **8**, 3498–3502.
4. P. B. R. R. Nair, A. N. Grigorenko, K. S. Novoselov, T. J. Booth, T. Stauber, N. M. R. Peres and A. K. Geim, *Science*, 2008, **320**, 1308.
5. X. W. Changgu Lee, J. W. Kysar and J. Hone, *Science*, 2008, **321**, 385–388.
6. S. G. Alexander, A. Balandin, W. Bao, I. Calizo, D. Teweldebrhan, F. Miao and C. N. Lau, *Nano Lett.*, 2008, **8**, 902.
7. M. Naguib, V. N. Mochalin, M. W. Barsoum and Y. Gogotsi, *Adv. Mater.*, 2014, **26**, 992–1005.
8. X. Zhang, J. Xu, H. Wang, J. Zhang, H. Yan, B. Pan, J. Zhou and Y. Xie, *Angew. Chem., Int. Ed.*, 2013, **52**, 4361–4365.
9. R. Meshkian, L.-Å. Näslund, J. Halim, J. Lu, M. W. Barsoum and J. Rosen, *Scr. Mater.*, 2015, **108**, 147–150.
10. B. Anasori, M. Dahlqvist, J. Lu, L. Hultman, P. Eklund, J. Rosén, J. Halim, E. J. Moon, B. C. Hosler and S. J. May, *J. Appl. Phys.*, 2015, **118**, 094304.
11. J. Yang, M. Naguib, M. Ghidiu, L. M. Pan, J. Gu, J. Nanda, J. Halim, Y. Gogotsi and M. W. Barsoum, *J. Am. Ceram. Soc.*, 2016, **99**, 660–666.
12. M. Naguib, M. Kurtoglu, V. Presser, J. Lu, J. Niu, M. Heon, L. Hultman, Y. Gogotsi and M. W. Barsoum, *Adv. Mater.*, 2011, **23**, 4248–4253.
13. M. Naguib, O. Mashtalir, J. Carle, V. Presser, J. Lu, L. Hultman, Y. Gogotsi and M. W. Barsoum, *ACS Nano*, 2012, **6**, 1322–1331.
14. M. Naguib, J. Halim, J. Lu, K. M. Cook, L. Hultman, Y. Gogotsi and M. W. Barsoum, *J. Am. Chem. Soc.*, 2013, **135**, 15966–15969.
15. M. Naguib, O. Mashtalir, J. Carle, V. Presser, J. Lu, L. Hultman, Y. Gogotsi and M. W. Barsoum, *ACS Nano*, 2012, **6**, 1322–1331.
16. M. Ghidiu, M. R. Lukatskaya, M. Q. Zhao, Y. Gogotsi and M. W. Barsoum, *Nature*, 2014, **516**, 78–81.
17. B. Soundiraraju and B. K. George, *ACS Nano*, 2017, **11**, 8892–8900.
18. P. Urbankowski, B. Anasori, T. Makaryan, D. Er, S. Kota, P. L. Walsh, M. Zhao, V. B. Shenoy, M. W. Barsoum and Y. Gogotsi, *Nanoscale*, 2016, **8**, 11385–11391.
19. J. Halim, M. R. Lukatskaya, K. M. Cook, J. Lu, C. R. Smith, L.-Å. Näslund, S. J. May, L. Hultman, Y. Gogotsi and P. Eklund, *Chem. Mater.*, 2014, **26**, 2374–2381.
20. L. Verger, V. Natu, M. Carey and M. W. Barsoum, *Trends. Chem.*, 2019, **1**, 656–669.
21. T. Hu, M. Hu, Z. Li, H. Zhang, C. Zhang, J. Wang and X. Wang, *Phys. Chem. Chem. Phys.*, 2016, **18**, 20256–20260.
22. O. Mashtalir, M. Naguib, V. N. Mochalin, Y. Dall'Agnese, M. Heon, M. W. Barsoum and Y. Gogotsi, *Nat. Commun.*, 2013, **4**, 1716.

23. K. Maleski, V. N. Mochalin and Y. Gogotsi, *Chem. Mater.*, 2017, **29**, 1632–1640.
24. Z. Li, L. Wang, D. Sun, Y. Zhang, B. Liu, Q. Hu and A. Zhou, *Mater. Sci. Eng., B*, 2015, **191**, 33–40.
25. O. Mashtalir, M. R. Lukatskaya, M. Q. Zhao, M. W. Barsoum and Y. Gogotsi, *Adv. Mater.*, 2015, **27**, 3501–3506.
26. M. Alhabeb, K. Maleski, B. Anasori, P. Lelyukh, L. Clark, S. Sin and Y. Gogotsi, *Chem. Mater.*, 2017, **29**, 7633–7644.
27. M. Naguib, R. R. Unocic, B. L. Armstrong and J. Nanda, *Dalton Trans.*, 2015, **44**, 9353–9358.
28. B. Anasori, M. R. Lukatskaya and Y. Gogotsi, *Nat. Rev. Mater.*, 2017, **2**, 1–17.
29. S. Venkateshalu and A. N. Grace, *Appl. Mater. Today*, 2020, **18**, 100509.
30. M. R. Lukatskaya, O. Mashtalir, C. E. Ren, Y. Dall'Agnese, P. Rozier, P. L. Taberna, M. Naguib, P. Simon, M. W. Barsoum and Y. Gogotsi, *Science*, 2013, **341**, 1502–1505.
31. H. Wang, J. Zhang, Y. Wu, H. Huang, G. Li, X. Zhang and Z. Wang, *Appl. Surf. Sci.*, 2016, **384**, 287–293.
32. J. Xuan, Z. Wang, Y. Chen, D. Liang, L. Cheng, X. Yang, Z. Liu, R. Ma, T. Sasaki and F. Geng, *Angew. Chem., Int. Ed.*, 2016, **55**, 14569–14574.
33. S. Yang, P. Zhang, F. Wang, A. G. Ricciardulli, M. R. Lohe, P. W. M. Blom and X. Feng, *Angew. Chem., Int. Ed.*, 2018, **57**, 15491–15495.
34. Q. Tang, Z. Zhou and P. Shen, *J. Am. Chem. Soc.*, 2012, **134**, 16909–16916.
35. X. Wang, X. Shen, Y. Gao, Z. Wang, R. Yu and L. Chen, *J. Am. Chem. Soc.*, 2015, **137**, 2715–2721.
36. M. A. Hope, A. C. Forse, K. J. Griffith, M. R. Lukatskaya, M. Ghidiu, Y. Gogotsi and C. P. Grey, *Phys. Chem. Chem. Phys.*, 2016, **18**, 5099–5102.
37. I. R. Shein and A. L. Ivanovskii, *Comput. Mater. Sci.*, 2012, **65**, 104–114.
38. Y. D. Jingxiao Li, C. Huo, S. Wang and C. Cui, *Ceram. Int.*, 2015, **41**, 2631–2635.
39. M. Kurtoglu, M. Naguib, Y. Gogotsi and M. W. Barsoum, *MRS Commun.*, 2012, **2**, 133–137.
40. I. Shein and A. Ivanovskii, *Comput. Mater. Sci.*, 2012, **65**, 104–114.
41. G. Berdiyorov, *Europhys. Lett.*, 2015, **111**, 67002.
42. Q. Tang, Z. Zhou and P. Shen, *J. Am. Chem. Soc.*, 2012, **134**, 16909–16916.
43. Y. Xie and P. Kent, *Phys. Rev. B*, 2013, **87**, 235441.
44. M. Khazaei, M. Arai, T. Sasaki, C. Y. Chung, N. S. Venkataramanan, M. Estili, Y. Sakka and Y. Kawazoe, *Adv. Funct. Mater.*, 2013, **23**, 2185–2192.
45. M. Naguib, O. Mashtalir, M. R. Lukatskaya, B. Dyatkin, C. Zhang, V. Presser, Y. Gogotsi and M. W. Barsoum, *Chem. Commun.*, 2014, **50**, 7420–7423.
46. H. Wang, Y. Wu, J. Zhang, G. Li, H. Huang, X. Zhang and Q. Jiang, *Mater. Lett.*, 2015, **160**, 537–540.

47. X.-H. Zha, J. Zhou, Y. Zhou, Q. Huang, J. He, J. S. Francisco, K. Luo and S. Du, *Nanoscale*, 2016, **8**, 6110–6117.
48. M. Khazaei, A. Ranjbar, M. Ghorbani-Asl, M. Arai, T. Sasaki, Y. Liang and S. Yunoki, *Phys. Rev. B*, 2016, **93**, 205125.
49. X.-H. Zha, Q. Huang, J. He, H. He, J. Zhai, J. S. Francisco and S. Du, *Sci. Rep.*, 2016, **6**, 27971.
50. L.-Y. Gan, D. Huang and U. Schwingenschlögl, *J. Mater. Chem. A*, 2013, **1**, 13672–13678.
51. Y. Xie and P. R. C. Kent, *Phys. Rev. B*, 2013, **87**, 235441.
52. N. J. Lane, M. W. Barsoum and J. M. Rondinelli, *Europhys. Lett.*, 2013, **101**, 57004.
53. O. Mashtalir, M. Naguib, V. N. Mochalin, Y. Dall'Agnese, M. Heon, M. W. Barsoum and Y. Gogotsi, *Nat. Commun.*, 2013, **4**, 1716.
54. M. Ghidiu, M. R. Lukatskaya, M.-Q. Zhao, Y. Gogotsi and M. W. Barsoum, *Nature*, 2014, **516**, 78.
55. Z. Ling, C. E. Ren, M.-Q. Zhao, J. Yang, J. M. Giammarco, J. Qiu, M. W. Barsoum and Y. Gogotsi, *Proc. Natl. Acad. Sci.*, 2014, **111**, 16676–16681.
56. I. Shein and A. Ivanovskii, *Superlattices Microstruct.*, 2012, **52**, 147–157.
57. Q. Peng, J. Guo, Q. Zhang, J. Xiang, B. Liu, A. Zhou, R. Liu and Y. Tian, *J. Am. Chem. Soc.*, 2014, **136**, 4113–4116.
58. Q. Hu, H. Wang, Q. Wu, X. Ye, A. Zhou, D. Sun, L. Wang, B. Liu and J. He, *Int. J. Hydrogen Energy*, 2014, **39**, 10606–10612.
59. Y. Tao, X. Xie, W. Lv, D. M. Tang, D. Kong, Z. Huang, H. Nishihara, T. Ishii, B. Li, D. Golberg, F. Kang, T. Kyotani and Q. H. Yang, *Sci. Rep.*, 2013, **3**, 2975.
60. N. Kurra, B. Ahmed, Y. Gogotsi and H. N. Alshareef, *Adv. Energy Mater.*, 2016, **6**, 1601372.
61. C. J. Zhang, M. P. Kremer, A. SeralAscaso, S. Park, N. McEvoy, B. Anasori, Y. Gogotsi and V. Nicolosi, *Adv. Funct. Mater.*, 2018, **28**, 1705506.
62. S. Xu, Y. Dall'Agnese, G. Wei, C. Zhang, Y. Gogotsi and W. Han, *Nano Energy*, 2018, **50**, 479–488.
63. E. Quain, T. S. Mathis, N. Kurra, K. Maleski, K. L. Van Aken, M. Alhabeb, H. N. Alshareef and Y. Gogotsi, *Adv. Mater. Technol.*, 2019, **4**, 1800256.
64. J. Zhang, S. Seyedin, Z. Gu, W. Yang, X. Wang and J. M. Razal, *Nanoscale*, 2017, **9**, 18604–18608.
65. Y. Peng, B. Akuzum, N. Kurra, M. Zhao, M. Alhabeb, B. Anasori, E. C. Kumbur, H. N. Alshareef, M. Ger and Y. Gogotsi, *Energy Environ. Sci.*, 2016, **9**, 2847–2854.
66. H. Huang, H. Su, H. Zhang, L. Xu, X. Chu, C. Hu, H. Liu, N. Chen, F. Liu, W. Deng, B. Gu, H. Zhang and W. Yang, *Adv. Electron. Mater.*, 2018, **4**, 1800179.
67. H. Li, Y. Hou, F. Wang, M. R. Lohe, X. Zhuang, L. Niu and X. Feng, *Adv. Energy Mater.*, 2017, **7**, 1601847.

68. J. Yan, C. E. Ren, K. Maleski, C. B. Hatter, B. Anasori, P. Urbankowski, A. Sarycheva and Y. Gogotsi, *Adv. Funct. Mater.*, 2017, **27**, 1701264.
69. Y. Yue, N. Liu, Y. Ma, S. Wang, W. Liu, C. Luo, H. Zhang, F. Cheng, J. Rao, X. Hu, J. Su and Y. Gao, *ACS Nano*, 2018, **12**, 4224–4232.
70. W. Liu, Z. Wang, Y. Su, Q. Li, Z. Zhao and F. Geng, *Adv. Energy Mater.*, 2017, **7**, 1602834.
71. R. Zhao, M. Wang, D. Zhao, H. Li, C. Wang and L. Yin, *ACS Energy Lett.*, 2017, **3**, 132–140.
72. M. Zhu, Y. Huang, Q. Deng, J. Zhou, Z. Pei, Q. Xue, Y. Huang, Z. Wang, H. Li, Q. Huang and C. Zhi, *Adv. Energy Mater.*, 2016, **6**, 1600969.
73. C. F. Zhang, B. Anasori, A. Seral-Ascaso, S. H. Park, N. McEvoy, A. Shmeliov, G. S. Duesberg, J. N. Coleman, Y. Gogotsi and V. Nicolosi, *Adv. Mater.*, 2017, **29**, 1702678.
74. P. Salles, E. Quain, N. Kurra, A. Sarycheva and Y. Gogotsi, *Small*, 2018, **14**, 1802864.
75. M. A. Faisal Shahzad, C. B. Hatter, B. Anasori, S. M. Hong, C. M. Koo and Y. Gogotsi, *Science*, 2016, **353**, 1137–1140.
76. R. Sun, H. Zhang, J. Liu, X. Xie, R. Yang, Y. Li, S. Hong and Z. Yu, *Adv. Funct. Mater.*, 2017, **27**, 1702807.
77. J. Liu, H. B. Zhang, R. Sun, Y. Liu, Z. Liu, A. Zhou and Z. Z. Yu, *Adv. Mater.*, 2017, **29**, 1702367.
78. W. T. Cao, F. F. Chen, Y. J. Zhu, Y. G. Zhang, Y. Y. Jiang, M. G. Ma and F. Chen, *ACS Nano*, 2018, **12**, 4583–4593.
79. P. Sambyal, A. Iqbal, J. Hong, H. Kim, M. K. Kim, S. M. Hong, M. Han, Y. Gogotsi and C. M. Koo, *ACS Appl. Mater. Interfaces*, 2019, **11**, 38046–38054.
80. Y. Qing, W. Zhou, F. Luo and D. Zhu, *Ceram. Int.*, 2016, **42**, 16412–16416.
81. P. Liu, V. M. H. Ng, Z. Yao, J. Zhou and L. B. Kong, *Mater. Lett.*, 2018, **229**, 286–289.
82. P. Liu, Z. Yao, V. M. H. Ng, J. Zhou, L. B. Kong and K. Yue, *Composites, Part A*, 2018, **115**, 371–382.
83. M. Han, X. Yin, H. Wu, Z. Hou, C. Song, X. Li, L. Zhang and L. Cheng, *ACS Appl. Mater. Interfaces*, 2016, **8**, 21011–21019.
84. M. Naguib, M. Kurtoglu, V. Presser, J. Lu, J. Niu, M. Heon, L. Hultman, Y. Gogotsi and M. W. Barsoum, *Adv. Mater.*, 2011, **23**, 4248–4253.
85. H. Jiang, Z. Wang, Q. Yang, M. Hanif, Z. Wang, L. Dong and M. Dong, *Electrochim. Acta*, 2018, **290**, 695–703.
86. X. Xie, M.-Q. Zhao, B. Anasori, K. Maleski, C. E. Ren, J. Li, B. W. Byles, E. Pomerantseva, G. Wang and Y. Gogotsi, *Nano Energy*, 2016, **26**, 513–523.
87. M. D. Levi, M. R. Lukatskaya, S. Sigalov, M. Beidaghi, N. Shpigel, L. Daikhin, D. Aurbach, M. W. Barsoum and Y. Gogotsi, *Adv. Energy Mater.*, 2015, **5**, 1400815.
88. C. Eames and M. S. Islam, *J. Am. Chem. Soc.*, 2014, **136**, 16270–16276.

89. Y. Xie, M. Naguib, V. N. Mochalin, M. W. Barsoum, Y. Gogotsi, X. Yu, K. W. Nam, X. Q. Yang, A. I. Kolesnikov and P. R. Kent, *J. Am. Chem. Soc.*, 2014, **136**, 6385–6394.
90. B. Anasori, Y. Xie, M. Beidaghi, J. Lu, B. C. Hosler, L. Hultman, P. R. Kent, Y. Gogotsi and M. W. Barsoum, *ACS Nano*, 2015, **9**, 9507–9516.
91. M. Naguib, J. Halim, J. Lu, K. M. Cook, L. Hultman, Y. Gogotsi and M. W. Barsoum, *J. Am. Chem. Soc.*, 2013, **135**, 15966–15969.
92. S. J. Kim, M. Naguib, M. Zhao, C. Zhang, H. Jung, M. W. Barsoum and Y. Gogotsi, *Electrochim. Acta*, 2015, **163**, 246–251.
93. J. Luo, X. Tao, J. Zhang, Y. Xia, H. Huang, L. Zhang, Y. Gan, C. Liang and W. Zhang, *ACS Nano*, 2016, **10**, 2491–2499.
94. M.-Q. Zhao, M. Torelli, C. E. Ren, M. Ghidiu, Z. Ling, B. Anasori, M. W. Barsoum and Y. Gogotsi, *Nano Energy*, 2016, **30**, 603–613.
95. J. Halim, S. Kota, M. R. Lukatskaya, M. Naguib, M.-Q. Zhao, E. J. Moon, J. Pitock, J. Nanda, S. J. May, Y. Gogotsi and M. W. Barsoum, *Adv. Funct. Mater.*, 2016, **26**, 3118–3127.
96. J. Ling, W. Zhai, W. Feng, B. Shen, J. Zhang and W. G. Zheng, *ACS Appl. Mater. Interfaces*, 2013, **5**, 2677–2684.
97. J. Liang, Y. Wang, Y. Huang, Y. Ma, Z. Liu, J. Cai, C. Zhang, H. Gao and Y. Chen, *Carbon*, 2009, **47**, 922–925.
98. D.-X. Yan, P.-G. Ren, H. Pang, Q. Fu, M.-B. Yang and Z.-M. Li, *J. Mater. Chem.*, 2012, **22**, 18772–18774.
99. B. Shen, Y. Li, W. Zhai and W. Zheng, *ACS Appl. Mater. Interfaces*, 2016, **8**, 8050–8057.
100. H.-B. Zhang, Q. Yan, W.-G. Zheng, Z. He and Z.-Z. Yu, *ACS Appl. Mater. Interfaces*, 2011, **3**, 918–924.
101. H.-B. Zhang, W.-G. Zheng, Q. Yan, Z.-G. Jiang and Z.-Z. Yu, *Carbon*, 2012, **50**, 5117–5125.
102. Z. Chen, C. Xu, C. Ma, W. Ren and H. M. Cheng, *Adv. Mater.*, 2013, **25**, 1296–1300.
103. N. Yousefi, X. Sun, X. Lin, X. Shen, J. Jia, B. Zhang, B. Tang, M. Chan and J. K. Kim, *Adv. Mater.*, 2014, **26**, 5480–5487.
104. M. Arjmand, M. Mahmoodi, G. A. Gelves, S. Park and U. Sundararaj, *Carbon*, 2011, **49**, 3430–3440.
105. M. H. Al-Saleh and U. Sundararaj, *Carbon*, 2009, **47**, 1738–1746.
106. Z. Zeng, H. Jin, M. Chen, W. Li, L. Zhou and Z. Zhang, *Adv. Funct. Mater.*, 2016, **26**, 303–310.
107. M. H. Al-Saleh, W. H. Saadeh and U. Sundararaj, *Carbon*, 2013, **60**, 146–156.
108. Z. Liu, G. Bai, Y. Huang, Y. Ma, F. Du, F. Li, T. Guo and Y. Chen, *Carbon*, 2007, **45**, 821–827.
109. N. Li, Y. Huang, F. Du, X. He, X. Lin, H. Gao, Y. Ma, F. Li, Y. Chen and P. C. Eklund, *Nano Lett.*, 2006, **6**, 1141–1145.
110. N. C. Das, Y. Liu, K. Yang, W. Peng, S. Maiti and H. Wang, *Polym. Eng. Sci.*, 2009, **49**, 1627–1634.

111. Y. Chen, H. B. Zhang, Y. Yang, M. Wang, A. Cao and Z. Z. Yu, *Adv. Funct. Mater.*, 2016, **26**, 447–455.
112. A. Ameli, P. Jung and C. Park, *Carbon*, 2013, **60**, 379–391.
113. S. Kuester, C. Merlini, G. M. Barra, J. C. Ferreira Jr, A. Lucas, A. C. de Souza and B. G. Soares, *Composites, Part B*, 2016, **84**, 236–247.
114. L. Kong, X. Yin, M. Han, X. Yuan, Z. Hou, F. Ye, L. Zhang, L. Cheng, Z. Xu and J. Huang, *Carbon*, 2017, **111**, 94–102.
115. K. Yao, J. Gong, N. Tian, Y. Lin, X. Wen, Z. Jiang, H. Na and T. Tang, *RSC Adv.*, 2015, **5**, 31910–31919.
116. B. B. Rao, P. Yadav, R. Aepuru, H. Panda, S. Ogale and S. Kale, *Phys. Chem. Chem. Phys.*, 2015, **17**, 18353–18363.
117. M. Crespo, N. Méndez, M. González, J. Baselga and J. Pozuelo, *Carbon*, 2014, **74**, 63–72.
118. B. Yuan, C. Bao, X. Qian, L. Song, Q. Tai, K. M. Liew and Y. Hu, *Carbon*, 2014, **75**, 178–189.
119. Y. Chen, Y. Wang, H. Zhang, X. Li, C. Gui and Z. Yu, *Carbon*, 2015, **82**, 67–76.
120. M. Arjmand, A. A. Moud, Y. Li and U. Sundararaj, *RSC Adv.*, 2015, **5**, 56590–56598.
121. F. Fang, Y.-Q. Li, H.-M. Xiao, N. Hu and S.-Y. Fu, *J. Mater. Chem. C*, 2016, **4**, 4193–4203.

CHAPTER 4

Graphite Nanoplatelet–Carbon Nanotube Hybrids for Electrical Conducting Polymer Composites

HANG ZHAO,*[a] DELONG HE[b] AND JINBO BAI*[b]

[a] State Key Laboratory of Photon-Technology in Western China Energy, and Institute of Photonics & Photon-Technology, Northwest University, Xi'an 710069, China; [b] Laboratoire de Mécanique des Sols, Structures et Matériaux, CNRS UMR 8579, Centrale-Supélec, Université Paris-Saclay, 8-10 rue Joliot Curie, Gif-sur-Yvette 91190, France
*Emails: hang.zhao@nwu.edu.cn; jinbo.bai@centralesupelec.fr

4.1 Introduction

Carbon nanotubes (CNTs), with their excellent mechanical, electrical and thermal properties, have received wide attention and in-depth investigation in recent decades. In particular, with characteristics of high aspect ratio structure and ultra-low density, CNTs have been one of the most important structural and functional additives in the composite field. Nevertheless, CNTs are hard to disperse homogeneously in a polymer matrix, due to their relatively strong surface energy.

Rationally designing a special CNT construction could be a useful approach to solving this dispersion issue. The typical hybrid construction whereby CNT arrays are synthesised vertically onto certain substrates can be obtained through a catalytic chemical vapour deposition method. These regular arrays are able to keep CNTs from aggregating and improve the interaction between neighbouring CNTs. Thus, a significant improvement in functional properties

Inorganic Materials Series No. 6
Two-dimensional Inorganic Nanomaterials for Conductive Polymer Nanocomposites
Edited by Chaoying Wan, Xingyi Huang and Chris Bowen

Published by the Royal Society of Chemistry, www.rsc.org

can be obtained in the corresponding composites. In this chapter, we describe an interesting hybrid construction whereby graphite nanoplatelet (GNP)–CNT hybrids (GCHs) were successfully synthesised, with CNT arrays well synthesised onto the GNP substrate. GCHs can provide significant regulation of dielectric and electrical properties of polymer-based composites because of their distinctive features of complete carbon composition, high conductivity and low inherent density. Thus, the regulation between the structural features of GCHs and key parameters in chemical vapour deposition (CVD) synthesis, and the corresponding influence of incorporation of GCHs on the internal conducting network in composites, will be discussed in detail.

4.2 Carbon-based Nanostructures and Their Functional Characteristics

4.2.1 Carbon Nanotubes

In 1991, Iijima reported the typical nano-scale, acicular-like carbon nanotubes (CNTs), with great potential in microelectronics and microscopy applications.[1,2] Since then the many excellent inherent characteristics and potential applications of CNTs have been reported, and they have been considered as one of the most popular materials for such applications.[3,4]

In general, single-walled carbon nanotubes (SWCNTs) and multi-walled carbon nanotubes (MWCNTs) are two main types of CNTs, due to their different radial organisation. SWCNT is a single molecule, which is composed of a graphene-based cylindrical tube. MWCNTs are composed of several concentric graphene-based cylinders, which is usually considered as a meso-scale graphite system.[5] Both SWCNTs and MWCNTs with open ends are composed of carbon atoms, which are bonded in a large number of hexagonal lattices. Because of the intense van der Waals interaction, the closely packed SWCNTs can exhibit an approximate inter-tube distance of 3.15 Å. As for MWCNTs, the interwall space and diameter are ∼3.4 Å and several to hundreds of nanometres, respectively. Typically, the length of both types of CNTs can reach from several micrometres to centimetres. The ultra-high aspect ratio (the ratio of the length in the axial direction to the diameter in the radial direction) endows CNTs with a large number of excellent inherent features.

The various atomic arrangement of SWCNTs are always described by the tube chirality (or helicity) (Figure 4.1). The carbon atom orientation can be classified into three main categories: the armchair structure, the zig-zag structure and the chiral structure according to the nanotube chirality.[6,7] These various atom-arrangement structures can lead to different intrinsic features shown by the SWCNTs. Specifically, all armchair SWCNTs exhibit a metallic characteristic and an empty band gap of 0 eV. Semiconductor-type SWCNTs display a diameter-dependent band gap range of *ca.* 0.18–1.8 eV.[8] However, as there are a certain number of tubes with their own corresponding chiralities in a MWCNT, it is difficult to precisely evaluate the physical features of a single MWCNT.

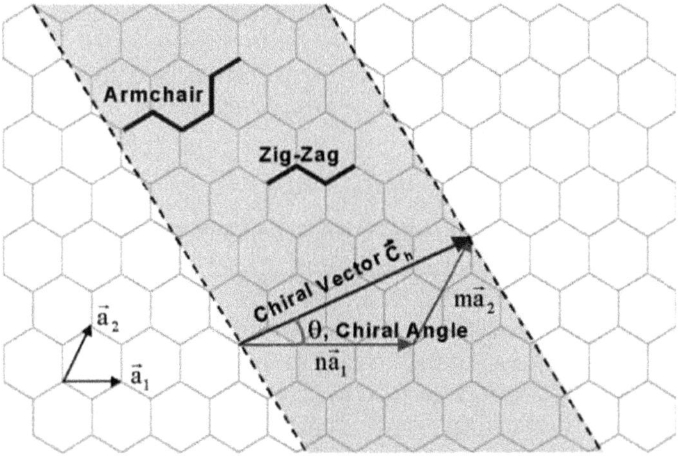

Figure 4.1 Three chirality mode of SWCNTs: armchair, zig-zag and chiral. Reproduced from ref. 7 with permission from Elsevier, Copyright 2001.

Both types of CNTs can be synthesised *via* three main approaches: electric arc discharge, laser ablation and CVD.[7,9] The carbon source, heat resource and catalyst precursor are key issues during the CNT synthesis process.

4.2.1.1 Electric Arc Discharge

In electric arc discharge (EAD), two highly pure graphite sticks are employed as an anode and cathode, and are placed end to end. A stable electric arc is generated by a voltage source under a He atmosphere. The average temperature of the plasma regions during the discharge process between electrodes is as high as 4000 K.[10] At such high temperatures, the carbon atoms generated from the vapourisation of the graphite anode deposit and re-accumulate onto the surface of the cathode and the internal cavity wall, which leads to the formation of nanotubes and other carbon materials. In particular, if the electrodes are doped by a tiny amount of metallic catalyst, SWCNTs can be synthesised.[11,12]

4.2.1.2 Laser Ablation

Laser ablation (LA) is similar to electric arc discharge, in that it is another physical vapourisation procedure at high temperature. In the laser ablation equipment, an introduced laser beam is focused onto the metallic catalyst-modified graphite target. CNTs can only be fabricated under the extreme conditions required to evaporate carbon, near 1200 °C, and under an inert gas atmosphere.[13,14]

Although EAD-synthesised CNTs always display a high degree of graphitisation, a series of subsequent complex purification processes is unavoidable. LA-synthesised CNTs usually exhibit an excellent quality and purity, however, this approach is subject to a low output–input ratio.[15,16] However, for the purpose of endowing composites with the inherent superiorities of CNTs as much as possible, the large-scale production of CNTs becomes a key challenge.

Compared with the previously mentioned problems of electric arc discharge and laser ablation processes, very recently mass production of CNTs with high purity and production efficiency has been realised by a gas-phase synthesis technology. Moreover, the complex purification process can be also cut down.

4.2.1.3 Chemical Vapour Deposition

CVD is an important gas-phase technology, which has been extensively utilised in CNT synthesis in recent decades. During CVD reaction, due to the high reaction temperature, molecules in the inert gas atmosphere are decomposed into active agents that participate in the reaction. These obtained active components then re-aggregate and re-construct, forming the target constructions. Typical CVD reactors can be divided into vertical, horizontal and barrel types, according to the various forms and applications required. Generally, the horizontal type is selected to carry out the floating catalytic CVD process.

The catalytic CVD procedure was first reported in the mid-twentieth century, and carbon tubular structures with a diameter range of 50–100 nm were synthesised by Radushkevich at 600 °C using an iron catalyst.[17] It was found that iron carbide first formed through dissolving carbon in iron, onto which both carbon deposition and graphene formation can take place. Similarly, metallic catalysts, such as iron, cobalt and nickel particles, exhibit an outstanding catalytic activity for CNT synthesis. Other metallic materials can be also employed as catalysts during CNT synthesis, such as copper, gold, silver and platinum.[18,19] In addition, solid organometallocenes have been used as catalyst precursors for CNT growth, such as ferrocene, cobaltocene, nickelocene, *etc.* Compared to traditional single-use metallic catalysts, organometallocenes can greatly speed up the carbon source decomposition.[17] In particular, the catalyst activity is mainly dependent on not only particle size, but also its metallic nature (different valence electron configurations). Based on the 'size-effect', the diameter of the synthesised CNTs is generally positively related to the catalyst size.[20–22]

The carbon source is another critical issue in providing carbon atoms for CNT growth during a catalytic CVD procedure. The carbon sources composed of various hydrocarbons always exist in gas form, such as carbon monoxide, methane, acetylene and ethylene, *etc.*, or in liquid form, which can be evaporated as gas form, such as xylene, benzene, toluene, cyclohexane, *etc.*[23–30] Carbon source decomposition under high temperature takes places both in the gas phase and on the surface of catalyst particles.[15] As reported in some studies, in comparison with using only one type carbon source, utilisation of multi-carbon sources can effectively optimise the catalytic activity of CNT synthesis over a wide temperature range.[31]

Through the CVD approach, CNTs can be grown onto a variety of substrates with different chemical composition and structure. According to the different intrinsic characteristics of substrates, as-prepared CNTs can exhibit various structures and constructions. For instance, perpendicularly aligned large-area CNT arrays can be synthesised on large and flat substrates. Ren *et al.* successfully obtained a large number of CNT arrays on nickel foils and glass at a temperature less than 700 °C, aiming at applications in electron emission.[32] The synthesised CNT forests with high electrical conductivity have been considered as an appropriate substitute for traditional metals in next-generation interconnects in integrated circuits and electrodes of supercapacitors. The area density of SWCNT forests can be improved to 8.3×10^{12} and 1.5×10^{13} tubes cm^{-2} by tuning the CVD parameters.[33,34] In addition to quartz and glass substrates, carbon fibres and polymeric fabrics are also very important and promising substrates for CNT array growth.[35,36] The specific surface area of CNT array attached carbon fibres or semiconductor fabric can be significantly enhanced, which is beneficial to the reinforcement of polymer-based composites; see Figure 4.2. Veedu *et al.*[37] prepared a three-dimensional (3D) SiC microfibre fabric with a grown MWCNT array by hybridisation to construct a filled epoxy-based composite, which exhibited significantly improved mechanical properties in comparison with the bulk fabric/epoxy composite. Gui *et al.*[38] fabricated a typical 3D architecture by using CVD, where the morphologies and orientations of each CNT array can be well arranged and regulated. The electrical and mechanical properties of this carbon-based

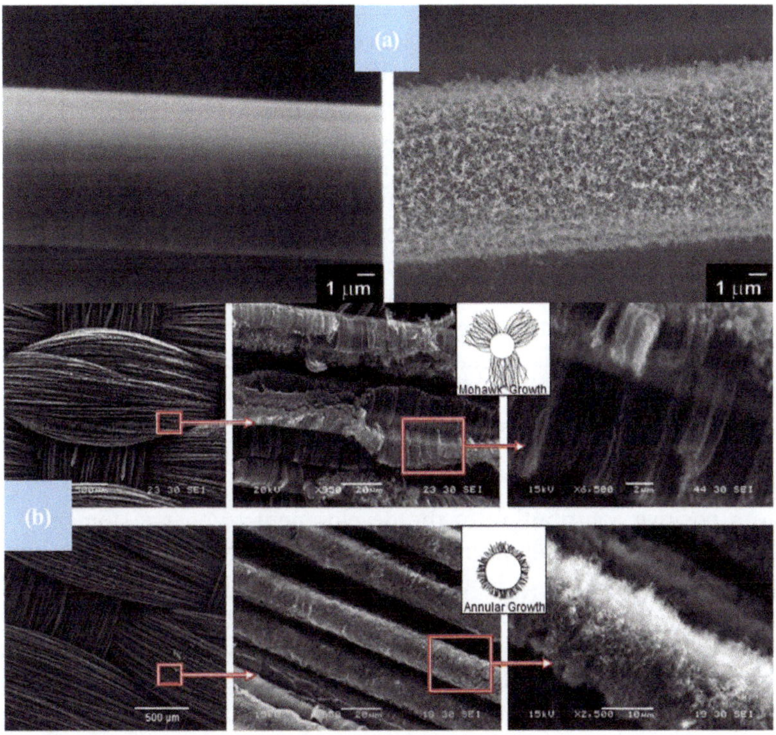

Figure 4.2 SEM images of (a) pristine carbon fibre and carbon fibre-CNTs. Reproduced from ref. 35 with permission from AIP Publishing, Copyright 2002. (b) Blurry construction with long and short CNTs on fibres, respectively, before being dispersed in the polymer matrix. Reproduced from ref. 36 with permission from Elsevier, Copyright 2014.

stereoscopic architecture can be regulated by optimising the structure and patterns of sponge and array forms of the CNTs.

In the last 25 years, a very large number of patents and articles have been proposed focusing on research of the construction, synthesis, inherent features and potential applications of CNTs and CNT-loaded composites. CNTs display not only ultra-high mechanical properties, but also excellent electrical conductivity.

The elastic modulus of a single CNT is characterised by measurement and analysis of the oscillation amplitude through tunnelling electron microscopy (TEM). Deforming the tightly fixed CNTs using the atomic force microscope (AFM) tip and recording the exerted force according to the displacement from the equilibrium position is another useful way to evaluate the mechanical behaviour of single CNTs.[39–41] In addition, the tension strength of CNTs can be characterised by using AFM tips.[42,43] Compared with all the commercial fibres, the much superior elastic modulus and tensile strength of a single MWCNT were calculated as approximately 1 TPa and 100 GPa, respectively.[39,44] As mentioned, their typical size, helicity construction and metallic or semiconductor state endow CNTs with excellent electrical performance. Carried current densities of 10^9 A cm^{-2} and electrical conductivity ranging from 10^2 to 10^6 S cm^{-1} have been precisely measured using a four-probe testing device.[33,45] The conductivity range obtained can be ascribed to the fact that MWCNTs are composed of a certain number of nanotubes with various chirality forms.[46,47] Moreover, the special structure also allows fast phonon transmission along the axial direction, which is responsible for the higher intrinsic thermal conductivity ($\sim$3500 W m^{-1} K^{-1}) than diamond, and most other materials.[48,49] It is also worth noting that the thermal expansion of MWCNTs is isotropic and usually negligible.[3]

Therefore, CNTs are a promising material that can be applied in many fields, such as transparent flexible electrodes, advanced composites, energy storage, microelectronics, sensors, actuators, biotechnology, *etc.*[9] In recent decades, with rapid development of industrialisation, the issue of ever-increasing consumption of energy resources is a problem that urgently requires a solution. Thus, it is crucial to develop novel, efficient and environmentally friendly energy generation, conversion and storage technology. Fortunately, CNTs could make a positive contribution to the development of energy storage. For instance, CNTs can effectively improve the rate capability of $LiCoO_2$ cathodes and graphite anodes and the cycle life of lithium-ion batteries.[50–53] Using CNTs as catalyst support would significantly reduce the Pt usage in the field of fuel cells, which the corresponding cost also reduced.[54,55] It is noteworthy that transparent electrodes composed of SWCNTs have been successfully incorporated into commercial photovoltaic devices. Compared with traditional capacitors, supercapacitors show larger capacitances and a faster charging rate. The CNT-based electrode is able to effectively amplify the capacitance of supercapacitors, which is preferable for applications requiring high storage capacity. CNTs also exhibit great application potential in electron emitters due to their ultra-low threshold electric field. In particular, CNTs are able to endow field-emitting devices with the superior features of more stable emission and a longer lifetime.[56–58] Another practical application for CNTs is flat panel displays. Compared with ordinary liquid crystal displays (LCDs), CNTs are capable of effectively reducing power consumption, improving response rate and brightness, and widening the viewing angle and operating temperature range of advanced displays.[4,9]

With features of high aspect ratio and high bending strength, CNTs make a valuable contribution to the property optimisation of novel nanoprobes, which have been applied to high-resolution imaging, nanolithography, nanoelectrodes, *etc.* For instance, an individual CNT has been applied as a scanning probe for visualising microscopic objects. Two CNTs can be employed as a nanotweezer to precisely tune the position of the nano-scaled structure.[59–61]

Therefore, the distinctive characteristics of chiral atomic structures, high aspect ratio and large specific surface endow CNTs with an excellent overall properties and great potential for application in a number of fields. Through rationally regulating the processing parameters of CVD reactions, CNTs with various morphological constructions can be synthesised successfully. The investigation of CNT-loaded polymer-based composites is another important issue, which will be discussed in detail in subsequent sections.

4.2.2 Graphite Nanoplatelets

Compared with the human-made synthetic CNTs, graphite is a typical stacked carbon structure of a certain number of individual graphene layers, which exists naturally on earth with abundant reserves. Graphite nanoplatelets (GNPs) are composed of dozens of stacked graphene layers, exhibiting an interlayer distance of *ca.* 0.34 nm;[62] see Figures 4.3 and 4.4. The structural feature of GNPs is in between graphene and graphite. Carbon atoms in-plane are covalently bonded, showing the properties of individual graphene. However, along the thickness direction, the stacked graphene layers are bonded through weak non-covalent van der Waals interactions, which exhibit the features of natural graphite. Moreover, GNPs possess a nano-scale thickness and a micro-scale in-plane diameter. GNPs also display an inherent large aspect ratio in the radial direction and a giant specific surface, which is theoretically as large as 2630–2965 $m^2 g^{-1}$.[63] The morphological size and aspect ratio of GNPs can be precisely regulated by different fabrication approaches, such as heat treatment, ultrasonication, oxidation, microwave irradiation, intercalation *etc.*[64–68] Two-dimensional (2D) structured GNPs with large aspect ratio and specific surface can be effectively utilised in the optimisation of structural stability and functional properties in the field of composites.

Nanoclay is also a natural 2D multilayered material, which has been widely employed in various fields. However, GNPs exhibit a lower density and much more superior electrical and thermal conductivity than clay. From the perspective of the microconstruction of GNPs, electronically three sp^2 orbitals are formed by the hybridisation between 2p orbitals and an s orbital. Then, three nearest-neighbour carbon atoms in the honeycomb lattice,

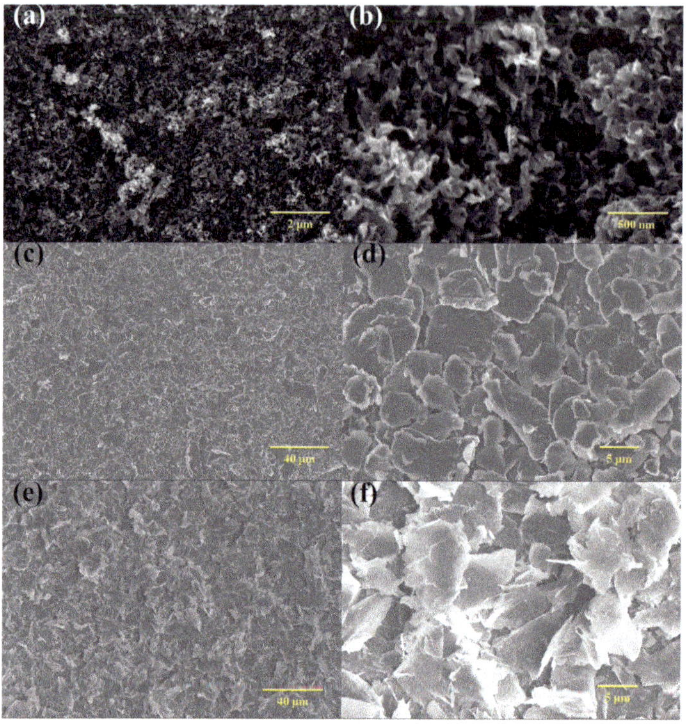

Figure 4.3 SEM micrographs of GNPs with different dimensions. Reproduced from ref. 66 with permission from IOP Science, Copyright 2019.

together construct the localised and energy-stable σ-bonds. In addition, the $2p_z$ orbital is occupied by a residual single-valence electron. The fast migration of $2p_z$ electrons is mainly responsible for the outstanding in-plane electrical conductivity of graphene (~6000 S cm^{-1}).[69–72] Thus the electric current is more easily conducted in-plane in graphene rather than across layers. This structural feature also holds for phonon propagation, which is responsible for the highly anisotropic acoustic and thermal properties of GNPs. It is noteworthy that in contrast to the complex process and large energy consumption required for CNT synthesis, highly pure GNPs can be mass produced from abundant natural graphite resources by relatively facile processes.[69] Therefore, GNPs as a low-cost substitute for CNTs have been applied in different fields, particularly that of composites.

Graphene with a typical mono-layered structure and outstanding intrinsic features was first discovered in 2004, receiving much attention from all over the world. The 2010 Nobel Prize in Physics was awarded for the preparation of graphene and can be considered as a surprising milestone, which further drives an unprecedented upsurge in the graphene-related investigations. Nevertheless, there is still a gap between the theoretical superiority of graphene and its practical applications in polymer-based composites. On one

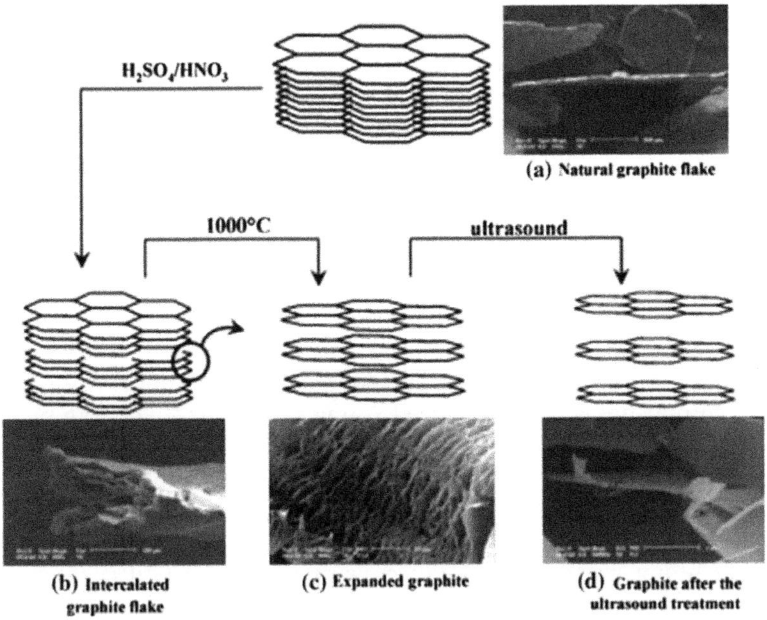

Figure 4.4 Schematic illustration of the procedures for graphite architecture modification after various treatments.
Reproduced/Adapted from ref. 62 with permission from John Wiley & Sons, Copyright 2010.

hand, the issue of mass production of well-separated individual graphene has not been completely solved, which is necessary for the practical fabrication of composites. On the other hand, it is unavoidable that well-exfoliated graphene re-aggregates and re-stacks during dispersal in the polymer matrix. In fact, when the stacked structure is composed of more than ten graphene layers, the interlayer coupling effect gives rise to a transformation from graphene-dominant to graphite-dominant behaviour. Thus, to preserve both the individual separation and inherent superiorities of graphene in matrices is another critical challenge. In fact, combining the superiorities of graphene (high aspect ratio, excellent electrical and thermal properties) and graphite (low cost and mass production), GNPs can be extensively used in composites as an optimal substitution of graphene.

4.2.2.1 Rheological Behaviour

The melt viscosity of GNP-filled composites is much higher than that of the graphite-filled counterpart, which is directly dependent on both the aspect ratio and exfoliation degree of the loaded GNPs.[73,74] However, thanks to its graphite-like self-lubricating properties, GNP-loaded composites usually display a very low melt viscosity and excellent processability, compared with other nanocarbon material-filled composites at certain volume fractions.[75]

4.2.2.2 Mechanical Properties

When loading nanoparticles with high specific surface areas into a polymer matrix, a strong interfacial interaction between two phases benefits the load transfer effect within composites. Different from CNTs, GNPs with a large aspect ratio could endow composites with a 2D area reinforcement. The practical reinforcement effect is strongly dependent on the dispersing state and morphological size of the selected GNPs. Compared with carbon black or CNTs, GNPs with their 2D structure can more easily establish interactions with the polymer matrix and restrict the movement of macromolecular chains. The strengthening interface formed between GNPs and the matrix is beneficial for improving the mechanical properties of the composites. In addition, the rigid backbone constructed by loading a number of GNPs also endows composites with an enhanced Young's modulus.[67] For example, with a loading as low as 5 wt% GNPs into epoxy, the obtained composite showed more than 10% improvement in its overall mechanical properties compared with the bulk counterpart.[76] However, some abnormal reinforcing effects were reported as well; for instance, the mechanical strength first increases at a critical filling fraction then reduces with increasing GNPs loading.[77–79] This decrease could be explained by the stress concentration and weak interfaces due to poor dispersion. Therefore, it is necessary to rationally optimise not only the GNPs dispersion in matrices, but also the effective intercalation of macromolecular chains into the interlayered structure.[80]

4.2.2.3 Electrical Properties

Generally, GNP-loaded composites usually show larger percolation threshold values than their CNTs-loaded counterparts, due to their relatively low interlayer electrical conductivity. The stable electrical conductivity level obtained by GNP-based conductive networks is kept around 10^{-3} S cm^{-1}.[81] Many smart sensors are designed based on the accurate response of the electrical conductivity of a material to environmental conditions, such as temperature, humidity, pressure and strain. Yu *et al.*[82] found that at a critical temperature the positive temperature coefficient effect can be transferred to a negative one in GNP/carboxymethyl polyvinyl alcohol composites. This is mainly ascribed to the thermal expansion of the matrix not only damaging the conducting network, but also speeding up the thermal disturbance of the conducting charges. Chen *et al.*[83] designed a GNP-loaded silicone elastomer-based composite with a low percolation threshold of 1.36 vol%, which exhibits an ultrasensitive positive piezo-resistive response under small pressures (0.3–0.7 MPa).

4.2.2.4 Dielectric Performance

The electrical energy storage of a material can be promoted by improving its overall dielectric permittivity and dielectric strength. Different from insulating composites, the dielectric permittivity of conductive filler-loaded composites is significantly enhanced when the conductive phase fraction is close to the percolation threshold, according to percolation theory. GNP with its large aspect ratio and electrical features can endow a composite with enhanced overall permittivity at a relatively low loading fraction. In fact, the improvement in permittivity of a composite is dependent on the degree of exfoliation of the selected GNP. Xu *et al.*[84] obtained a relatively low percolation threshold of 1 wt% from a GNP/poly (methyl methacrylate) (PMMA) composite, which is much lower than that of the traditional graphite-filled counterpart. Compared with a majority of polymers, poly(vinylidenfluoride) (PVDF), with its electronegative composition, exhibits a surprisingly high permittivity (~10) at room temperature. Thus, PVDF and its copolymers have been widely selected as the polymer matrix in recent investigations concerning dielectric energy storage composites. Li *et al.*[85] prepared a GNP/PVDF composite by combining solution mixing and compression moulding, showing a high dielectric permittivity of 173 and a low dielectric loss of 0.65. Optimal performance was obtained at the GNP loading close to the threshold (2.4 wt%). He *et al.*[86] prepared a series of acid-intercalation exfoliated GNP/PVDF composites, from which a dielectric permittivity as high as 2700 (measured at 100 Hz) could be obtained near the threshold (1.01 vol%), which is 270 times higher than that of neat PVDF. When the GNP loading fraction is further increased, composites have the potential to be applied in the field of microwave absorption and electromagnetic interference shielding.[87,88]

Generally, the majority of functional properties of composites can be tuned by regulating the thickness of the GNPs and their dispersion in the matrix. Thus, for the purpose of further optimising the overall performance of GNP/polymer composites, several potential strategies have been studied, such as chemical functionalisation to the GNP surface, regulation of GNP orientation, utilisation of multi-type carbon materials as functional fillers, *etc.*

4.2.3 Carbon-based Micro–Nano Hybridisation Constructions

The surface functionalisation approach can improve not only the dispersion of CNTs in matrices, but also the interfacial interaction of CNTs with matrices. However, chemical functionalisation reactions are usually accompanied by powerful separating and dispersing processes, where CNTs are probably broken into tiny fragments. This breaking of the CNT structure inevitably influences the overall performance of CNT-loaded composites, particularly, their mechanical and electrical behaviour. Moreover, defects in graphitic structures can usually be formed during both chemical functionalisation and heteroatom-doping processes. The bond hybridisation transfers from the sp^2 to sp^3 mode, which gives rise to a subsequent transformation from initial oxygenated sites to local carrier-scattering sites. In this transformation the carrier mobility is reduced and the π conjugation is also disrupted. In particular, the π conjugation determines the mobility of electrons and phonons, which further influences the electrical and thermal conduction of CNTs, respectively.[89] In fact, the large quantity of concentrated acids or strong oxidants consumed during the functionalisation and doping processes also aggravates environmental pollution.

In addition to the approaches of optimising the surface compositions and processing modes of CNTs, recently the dispersion of CNTs in a matrix has been shown to be effectively improved by CNT hybridisation. This type of hybridisation is constructed by vertically synthesising CNT arrays onto certain micro-scaled substrates. In our previous studies, several insulating micro-scale materials, such as micro-SiC and microspherical Al_2O_3, were utilised as substrates to participate in the construction of CNT array hybridisations by a floating-catalyst CVD method.[31,91–95] Figure 4.5 shows the SEM morphology of the as-synthesised SiC-CNT and Al_2O_3-CNT hybrids. The influence of the CVD experimental parameters and substrate type on the morphological features of the CNT array hybrids was investigated in detail.

On one hand, the aligned CNT array greatly enlarges its interfacial surface with the polymer matrix, which is of benefit for improving the structural strength of the composites. Ci *et al.*[91] found that the Young's modulus of epoxy was enhanced by 23.5% through adding a certain amount of SiC-CNT hybrids into the system. Li *et al.*[96] compared the influence of the CNT array morphological features of Al_2O_3-CNT hybrids on the mechanical behaviour of epoxy-based composites. The elastic modulus and tensile strength of the composites could be regulated by modifying the aspect ratio of the

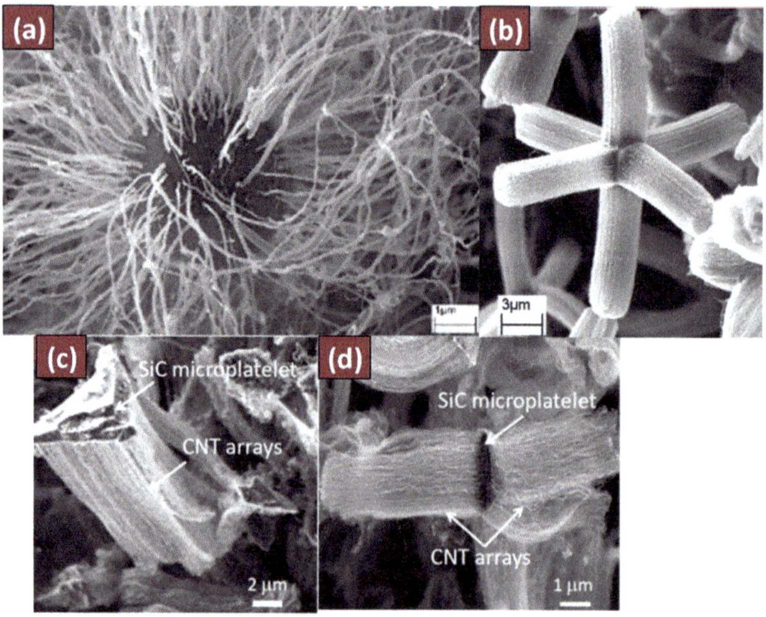

Figure 4.5 SEM images of microstructures for Al_2O_3-MWCNT hybridisation: (a) the urchin-like Al_2O_3-MWCNTs hybrid and (b) the six-branch constructed Al_2O_3-MWCNTs hybrid. Adapted from ref. 31 with permission from Elsevier, Copyright 2010. SEM images of SiC-CNT hybridisation: (c) 'mono-direction' and (d) 'dual-direction'. Adapted from ref. 90 with permission from American Chemical Society, Copyright 2014.

synthesised CNTs. In addition to mechanical strengthening, the thermal properties of the composites can also be optimised by these hybridisations. Bozlar *et al.*[97] obtained an epoxy-based composite with significantly improved thermal conductivity through addition of a very small quantity of Al_2O_3-CNT hybrids. This was explained by the contact thermal resistance between CNTs being reduced by the aligned CNT array structure.

On the other hand, a hybrid architecture has great potential for avoiding aggregation of CNTs, thus effectively improving both the filler–matrix interfacial area and polarisation. Yuan *et al.*[98] prepared a SiC-CNT hybrid/PVDF composite that exhibits both a high permittivity (>800 at 100 Hz) and a low percolation threshold (~1.47 vol%). Li *et al.*[95] found that the as-synthesised monodirectional SiC-CNT hybrid endowed the PVDF-based composite with a lower percolation threshold, in comparison with the dual-directional SiC-CNT hybrid-filled counterpart. In fact, due to the particular construction of the CNT hybrids, an internal conducting network can be easily formed, meaning that the detecting signal for structural failures can be collected and sent back more efficiently. Li *et al.*[96,99,100] found that the *in situ* sensing ability of fabric-reinforced epoxy-based composites is strongly determined by the aspect ratio of the CNT array, the substrate type and synthesised hybrid construction. Thus, the well-designed CNT hybrid exhibited potential for promising application in *in situ* non-destructive testing in the aerospace sector.

4.2.4 The Conductive Network in Composites

Due to their distinctly different electrical behaviour, polymer-based composites are divided into pre-percolative and post-percolative composites, respectively, according to the percolation threshold. Before the threshold, composites with a typical non-conductive behaviour exhibit potential for application in dielectric energy storage. Due to the formation of an electrically conducting network, post-percolative composites usually show a typical conducting behaviour. Before discussing the dielectric and electrical conductive functional properties of CNT hybrid-loaded composites, some of the theoretical background is introduced in the following sections.

4.2.4.1 Pre-percolation: Dielectric Composites

The dielectric property is an important physical feature of various materials. The practical application of capacitor energy storage is established on the basis of the dielectric property of a composite in the pre-percolation region. As illustrated in eqn (4.1), the complex dielectric permittivity is a frequency-dependent physical quantity.

$$\varepsilon^*(\omega) = \varepsilon'(\omega) - j\varepsilon''(\omega) \tag{4.1}$$

where ω is the angular frequency of alternating electrical field, and $\varepsilon'(\omega)$ and $\varepsilon''(\omega)$ are related to the dielectric permittivity and loss, respectively. Moreover, compared with the permittivity in vacuum ($\varepsilon_0 = 8.854$ pF m^{-1}),

$\varepsilon'(\omega)$ is generally denoted as ε_r for assessing the electrical energy storage ability of materials. The loss tangent ($\tan\delta(\omega)$), defined as the ratio of $\varepsilon''(\omega)$ and $\varepsilon'(\omega)$, is generally employed to evaluate the frequency-dependent heat energy dissipation.

$$\tan\delta(\omega) = \varepsilon''(\omega)/\varepsilon'(\omega) \tag{4.2}$$

When an external electric field is applied to a dielectric material, the positive/negative charges in the filler, matrix and interface separate from the balanced state. The formation of electric dipole moments can be defined as polarisation. In particular, as shown in Figure 4.6, the polarisation of dielectrics mainly includes electron polarisation, ionic polarisation, dipole polarisation and interfacial polymerisation.[101]

1. *Electron polarisation.* Generally, an atom, composed of both an electron cloud shell with negative charges and an atomic nucleus possessing positive charges, coincidently exhibits electrically neutral features. However, the electron cloud shell deforms under an applied external electrical field, leading to the formation of small electric dipole moments, namely, electron polarisation. This polarisation mode can be observed at an ultra-high frequency range (10^{14}–10^{16} Hz), and occurs in any material.
2. *Ionic polarisation.* The ionic polarisation is generally observed in ionic crystals composed of equal numbers of cations and anions, such as NaBr, KCl and NaCl. When applying an external electrical field, the cations and anions of the ionic crystal migrate along the electrical field. The initially formed chained arrangement is broken, giving rise to the generation of an electric dipole moment. Ionic polarisation occurs at a high frequency range of approximately 10^{12} Hz.
3. *Dipole polarisation.* There is an inherent permanent electric dipole in some molecules, for instance, HCl. Under thermal excitation, the permanent electric dipoles orientate randomly. However, these permanent dipoles are inclined to arrange themselves under an applied external electric field. Electric dipole moments along the electric field are also formed. This polarisation triggered by dipole orientation occurs at a relatively low frequency range ($\sim 10^6$ Hz). Compared with electron and ionic polarisation, dipole polarisation displays a stronger temperature dependence.[102]
4. *Interfacial polarisation.* Charges accumulate on the interface between heterogeneous materials or different regions of a homogeneous material. These charges at the interface move along the direction of the applied external electric field, leading to interfacial polarisation.[103,104] Moreover, various mobile charges exist within composites, such as electrons, holes and ion vacancies, which make an additional contribution to the interfacial polarisation. In fact, a number of opposite charges are attracted to electrodes by charges accumulating at the interface. Under a constant electrical field, the permittivity of a material is related to the input of electric charges. Thus, increasing charges caused by interfacial polarisation lead to enhancement of the permittivity. Due to the large amount

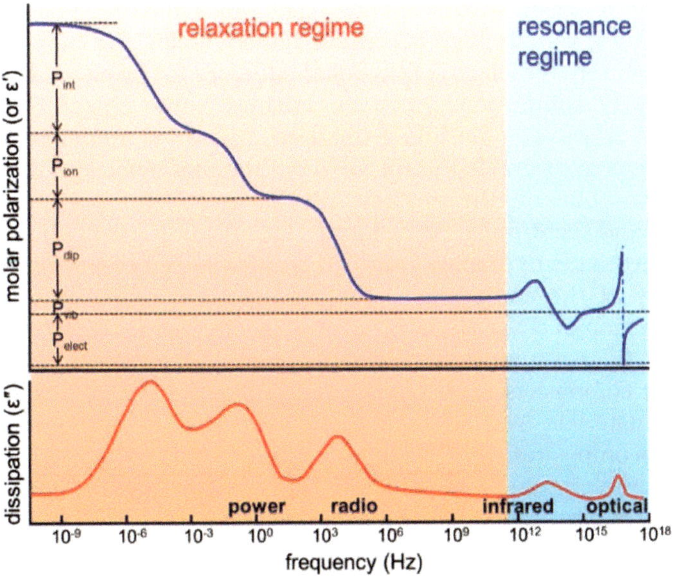

Figure 4.6 Imaginary and real part of the dielectric permittivity as a function of frequency.
Reproduced from ref. 101 with permission from American Chemical Society, Copyright 2014.

of interface that exists in a heterogeneous composite, the interfacial polarisation makes a considerable contribution to determining the dielectric permittivity of the multiphase system. Thus, several approaches focused on augmentation of interfacial polarisation have been studied in recent years, such as adjusting the interfacial interaction between fillers and matrix, enhancing the effective interfacial area, *etc.* In particular, the widely reported methods for surface functionalisation, improvement of dispersion and amelioration of aggregation are all based on this strategy. Interfacial polarisation is usually characterised in a low frequency region, ranging from 10^{-2} to 10^{2} Hz.

The dielectric film capacitor is a promising candidate for next-generation advanced energy conversion and storage. Compared with other energy storage devices, the dielectric film capacitor exhibits a much larger power density. However, the relatively low energy density limits its practical application. The energy density of a dielectric (U_e) is mainly determined by its intrinsic dielectric permittivity (ε_r) and an applied external electrical field (E).

$$U_e = \frac{1}{2}\varepsilon_r\varepsilon_0 E^2 \tag{4.3}$$

According to eqn (4.3), synergistically improving both the dielectric permittivity and dielectric strength is the main strategy and challenge for the effective optimisation of the energy density of a dielectric composite. In recent decades, various high-ε_r ceramic nanoparticles and nanowires have been widely used as dielectric functional additives for increasing the overall permittivity of composites. However, obvious improvements are usually obtained at relatively large filling fractions, where the amount of filler aggregation and undesirable interfacial and structural defects would increase rapidly. Consequently, the intrinsic superiorities of the polymer, such as flexibility and continuity, deteriorate significantly. Another frequently adopted method is to prepare conducting filler-loaded composites, and the permittivity of the composites is significantly enhanced when the filler loading fraction is close to the percolation threshold. However, addition of conductive filler results in a severe reduction of the dielectric breakdown strength of composites, which is fatal for the practical application of these dielectric materials. Therefore, improving the dielectric strength of percolative nanocomposites needs further effort in the long term.

4.2.4.2 Percolation Behaviour

Percolation theory was originally proposed as a classical mathematical tool, concentrating on investigating the overall influence of the degree of interconnection of each component on a huge and disordered system.[105] In order to establish a connection between microstructures and the functional performance of heterogeneous materials, the percolation threshold was then successfully utilised to provide a mechanistic explanation for various physical phenomena.

As shown in Scheme 4.1, at low loading fractions, conductive fillers are separated from each other by the polymer matrix, and the insulating matrix mainly dominates the electrical behaviour of the composites. With augmented filling, adjacent single conductive units connect to each other, forming local conductive clusters. As the number of clusters increases, they

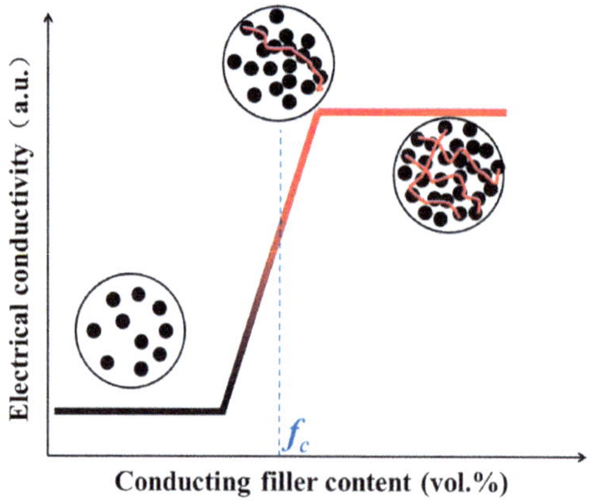

Scheme 4.1 Schematic diagram of the non-linear variation in the properties of composites near the f_c. The geometric phase transition of conducting fillers is shown in the insets.

are connected to each other. At a critical conductive filler loading, a 3D conducing network is successfully built, where the electrical behaviour of the system is transferred from insulating mode to conducting mode. Typically, this critical filling fraction is the percolation threshold f_c, which can be theoretically evaluated by the following classical percolation power laws.[105,106]

1. When the conducting phase fraction (f_{filler}) is very close to, but lower than, f_c, the conductivity of the composite (σ_c) can be estimated by the following law:

$$\sigma_c \propto (f_c - f_{\text{filler}})^{-s} \tag{4.4}$$

   where s is the critical exponent, which is dependent on the loading filler dimension, for instance, $s = 1.1$–1.3 or 0.7–1.0 for fillers with 2D or 3D constructions, respectively.
2. When $f_{\text{filler}} > f_c$, both continuous percolating paths and completed conductive networks throughout the polymer matrix are established. σ_c of the composite is determined by the following equation:

$$\sigma_c \propto (f_{\text{filler}} - f_c)^t \tag{4.5}$$

   where t is also a critical exponent related to the filler dimension, typically, $t = 1.1$–1.3 for 2D fillers; $t = 1.6$–2.0 for 3D fillers.
3. In conditions where f_{filler} is extremely close to f_c, namely, $|f_{\text{filler}} - f_c| \to 0$, σ_c of the composite is described as:

$$\sigma \propto \omega^u \tag{4.6}$$

   where $\omega = 2\pi\nu$, ν is the frequency and $u = t/(s+t)$ is the relevant critical exponent.

Besides the σ_c value of a percolative composite showing a significant increase of several orders of magnitudes, the corresponding permittivity is also greatly enhanced when $f_{\text{filler}} \sim f_c$. The relevant law is presented as follows.

$$\varepsilon'_c \propto (f_c - f_{\text{filler}})^{-s'} \quad \text{for } f_{\text{filler}} < f_c \tag{4.7}$$

where s' is a critical exponent of approximately 1. Moreover, this is not always observed in practical continuous systems.[106]

In fact, the f_c value of a composite is directly determined by the intrinsic electrical features and physical dimensions of the conductive filler and filler–matrix interfacial interaction. Moreover, both the filler dispersion and distribution states in the matrix play important roles in determining the f_c value of a composite as well. In comparison with conductive fillers with a spherical shape, loading ellipsoidal fillers is a useful method to reduce the f_c value of the composites. This is explained by the interconnection between neighbouring ellipsoidal fillers being built more easily than that between spherical fillers. A number of simulation and experimental investigations have shown that the f_c values of composites are negatively correlated with the aspect ratio of the loading fillers.[107,108]

4.2.4.3 Post-percolation: Conducting Composites

After formation of the internal conducting network, how the interaction between conductive fillers influences the conducting network in the structural and overall electrical behaviour of a material will be discussed in detail in this section.

As a main electrical interconnection mode, ohmic contacts form from the direct physical contact between overlapped conductive fillers, such as point-to-point connections between CNTs and the interlayer connections of GNPs. Since there is a large ohmic contact resistance between single CNTs, CNTs always exhibit a much lower test electrical conductivity than an individual CNT ($\sim 10^4$–10^7 S m^{-1}).[109] It is reported that the ohmic contact resistance values between metallic/metallic-type or semiconducting/semiconducting-type SWCNTs are characterised as 100–400 kΩ in experimental studies, which is consistent with the theoretical range (100 kΩ–3.4 MΩ).[110,111] When conducting fillers are dispersed in polymer matrices, ohmic contact also occurs in local aggregates. Moreover, the probability of ohmic contact is augmented by increasing the conductive filler content when $f_{\text{filler}} > f_c$.

Compared with ohmic contact, another mechanism, that of the tunnelling effect, is usually considered as dominant in the formation of conducting networks in heterogeneous conducting filler/insulating matrix composites. It is noteworthy that the tunnelling effect means that filler particles are connected to each other electrically rather than geometrically. Well-dispersed conducting fillers are separated by the insulating matrix in the composite, which forms a number of microcapacitors. When an external electric field is applied to the composite, a local electric field is generated in each microcapacitor. If the sandwiched insulating thin layer cannot endure the growing local electric field, a short circuit takes place through the microcapacitors. This gives rise to a large leakage current, which is defined as the tunnelling effect.[112,113] The overall electrical behaviour of a composite is dependent on the contest between the Coulomb blockade effect and the tunnelling effect occurring at adjacent conducting fillers. According to the tunnelling model, the tunnelling conductivity between neighbouring fillers (σ_{tun}) is described as:[114]

$$\sigma_{\text{tun}} \propto \exp\left(-\frac{l-2b}{d}\right) \tag{4.8}$$

where l is the centre-to-centre distance between conducting units, and b and d are the radius of the conducting unit and the characteristic tunnelling space, respectively. It was found that the maximum tunnelling distance of a CNT/polymer composite was calculated to be about 1.8 nm.[109] The parameter $(l-2b)$ indicates the real spacing distance between the edges of neighbouring particles, being dependent on the concentration of incorporated filler. When $f_{\text{filler}} \rightarrow f_c$ the value of $(l-2b)$ reduces dramatically, leading to an exponentially enhanced overall electrical conductivity for the composite. In addition, with ever-increasing filler loading content, the sandwiched insulating matrix layer becomes ultra-thin $(l \rightarrow 2b)$. Besides the tunnelling current, a large number of quasi-ohmic contacts take place in the inner composite, where the conducting network is built with a steady construction. Simmons *et al.*[115] provided a method for estimating the tunnelling resistance value (R_{tun}) between two neighbouring CNTs approximately as:

$$R_{tun} = \frac{V}{AJ} = \frac{h^2 d}{Ae^2\sqrt{2m\lambda}} \exp\left(\frac{4\pi d}{h}\sqrt{2m\lambda}\right) \quad (4.9)$$

where J is the tunnelling current density, V the electrical potential difference, e the quantum of electricity, m the mass of an electron, d the distance between CNTs, h Planck's constant, λ the barrier height and A the cross-sectional area of the tunnel. Based on the theoretical simulation, Hu *et al.*[116] obtained a negative correlation between σ_{tun} and the distance between CNTs for different λ values.

However, the critical exponents in the percolation power law do not exhibit excellent universality, which can be ascribed to the fact that only the tunnelling between adjacent filler units with the closest distance is taken into account.[117] For this reason, several abnormally large t values of about 5–10 have been reported in some percolative composites.[105,118]

This section offers an introduction on the microstructure, general synthesis strategy and inherent characteristics of CNTs and GNPs. Then, the dielectric and conducting properties and promising applications of several CNT- or GNP-filled polymer-based composites are reviewed. Thanks to their intrinsic features, CNTs and GNPs could endow polymer-based composites with various outstanding functional properties. Nevertheless, both challenges and opportunities exist for the further improvement of properties of these carbon-based composites. We will discuss these issues in detail in the following sections, in order to establish the relationship between synthesising procedure and hybrid construction, and to regulate and improve the electrical functional properties of CNT hybrid-filled composites.

4.3 Synthesis and Characterisation of CNT-based Hybrids

4.3.1 Introduction

In recent decades, CNTs have been frequently used as a crucial nanofiller in the domain of composites. As mentioned previously, CNTs are notable for their extraordinary intrinsic morphological features and excellent functional characteristics. By loading CNTs into polymer matrices, the various properties of the composites, such as dielectric permittivity, mechanical strength and thermal and electrical conductivities, can be effectively improved.[9,48,50,119–121]

However, it has been reported in many studies that undesirable aggregations are easily generated in polymer matrices due to the high aspect ratio of CNTs. The effective interface between CNTs and the polymer matrix is significantly decreased with agglomeration of CNTs in the composites. Inhomogeneous dispersion, poor distribution and related mechanical weak interfaces are all ascribed to these CNT aggregations in some extent, which are harmful to the mechanical strengthening of the composites. In addition, with regard to electrical performance, CNT aggregation gives rise to a relatively high contact resistance, a weakened interfacial polarisation and an extremely reduced breakdown strength. In fact, all these problems are critical to the dielectric and electrical conducting properties of the composites.[89]

In recent years, it has been confirmed that the synthesis of CNTs on micro-scale substrates is beneficial to release the CNT aggregation and optimise the multifunctional characteristics of composites. Unlike other hybridisation constructs using semi-conductive materials as substrates, complete carbon-composed GNP–CNT hybrids (GCHs) have been synthesised. The GCHs exhibit a number of superior intrinsic characteristics, such as coupling construction and low intrinsic density.[31,92–93,95,122]

In this section, synthesis using the CVD method is presented first. As decomposition of the carbon source progresses, a large number of CNTs are synthesised and aligned vertically in array onto the surface of GNP substrates, from which the GCH is formed. Second, in order to establish the relationship between construction of the obtained GCHs and CVD synthesis, several important parameters during CVD synthesis, such as temperature, reaction time and gas composition, are investigated. In fact, the synthesis conditions are strongly responsible for the final construction of the as-synthesised products. Furthermore, the relevant characterisations provide a better understanding of the morphological construction and inherent features of GCHs. Based on these structural characteristics of GCHs, improvement in the functional properties in various polymer-based composites will be discussed in detail in the following sections.

4.3.2 The Synthesis of GCHs

The principal materials used in the synthesis are summarised in Table 4.1.

As schematically illustrated in Scheme 4.2, the GCHs can be synthesised in a horizontal quartz tube by the CVD process. The as-received GNPs were homogeneously deposited onto a quartz plate surface, which was subsequently placed into the reactor. Then, the CVD system was heated to the selected temperature (from 550 to 850 °C) under an atmosphere of Ar–H_2 mixture.

Table 4.1 Information on the main materials used in the experiment.

Material	Molecular formula	Physical state	Purity	Notes
Xylene	C_8H_{10}	Liquid	98.5% +	Assay, isomers plus ethylbenzene, Alfa Aesar
Ferrocene	$Fe(C_5H_5)_2$	Solid	99%	Alfa Aesar
Argon	Ar	Gas		Carrier gas
Hydrogen	H_2	Gas		Carrier gas
Acetylene	C_2H_2	Gas		Carbon source
GNP particles		Solid		Hundreds of nm in thickness; <5 μm in diameter, Xiamen K-Nano Co.

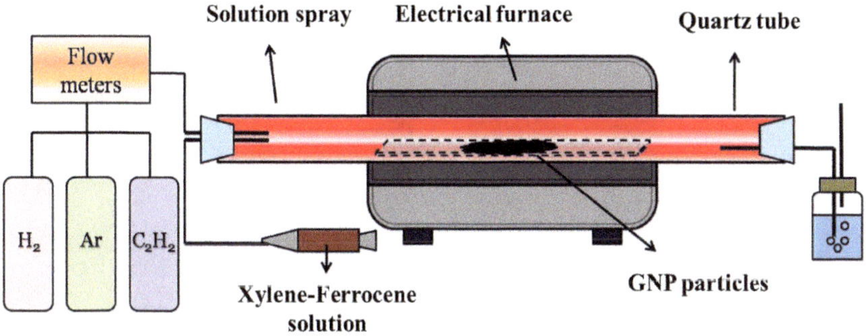

Scheme 4.2 Schematic of the CVD system for synthesising GCH particles. Reproduced from ref. 119 with permission from American Chemical Society, Copyright 2015.

The mixing gas velocity was precisely regulated by an electronic multichannel flow meter. The ferrocene-dissolved xylene suspension was uniformly injected into the reaction system by a precise controlling device. The vaporised suspension was transferred to the reaction region by the carrier gas (Ar–H_2 mixture). Moreover, acetylene was also added into the system to participate in the synthesis of GCHs. After the fixed synthesis duration, the furnace was cooled down to room temperature under the carrier gas atmosphere. Notably, the total gas velocity during the synthesis process was maintained at 1 L min^{-1}.

4.3.3 Regulation of Parameters in CVD Synthesis of GNP–CNT Hybrid Constructions

In fact, two main reactions are taking place in the GCH synthesis process: thermal decomposition of the carbon source and decomposition and re-aggregation of the catalyst precursor. Both these reactions are strongly influenced by the key parameters during synthesis, which play determining roles in both the morphological characteristics of the grown CNT arrays and the organisation of the synthesised hybrid. Thus, the relationships between key conditions in CVD, such as reaction temperature, hydrogen ratio and synthesis time, and the construction of the final hybrids are studied in detail.

4.3.3.1 Temperature

For the purpose of assessing the influence of reaction temperature on the construction of GCHs, the relevant CVD procedures were performed at various temperatures. The hydrogen ratio and synthesis duration were fixed as $H_2:Ar = 0.3:0.7$ L min^{-1} and 10 min, respectively. Figure 4.7a–b, c–d, e–f and g–h show the SEM morphological images of various GCHs synthesised at 550 °C, 650 °C, 750 °C and 850 °C, respectively. It can be clearly seen that when the CVD process was carried out at a low temperature (~550 °C), a relatively small number of CNTs were synthesised onto the GNP surfaces, with a wide distribution of diameter. Moreover, the multilayered stacked structure of the substrate is well

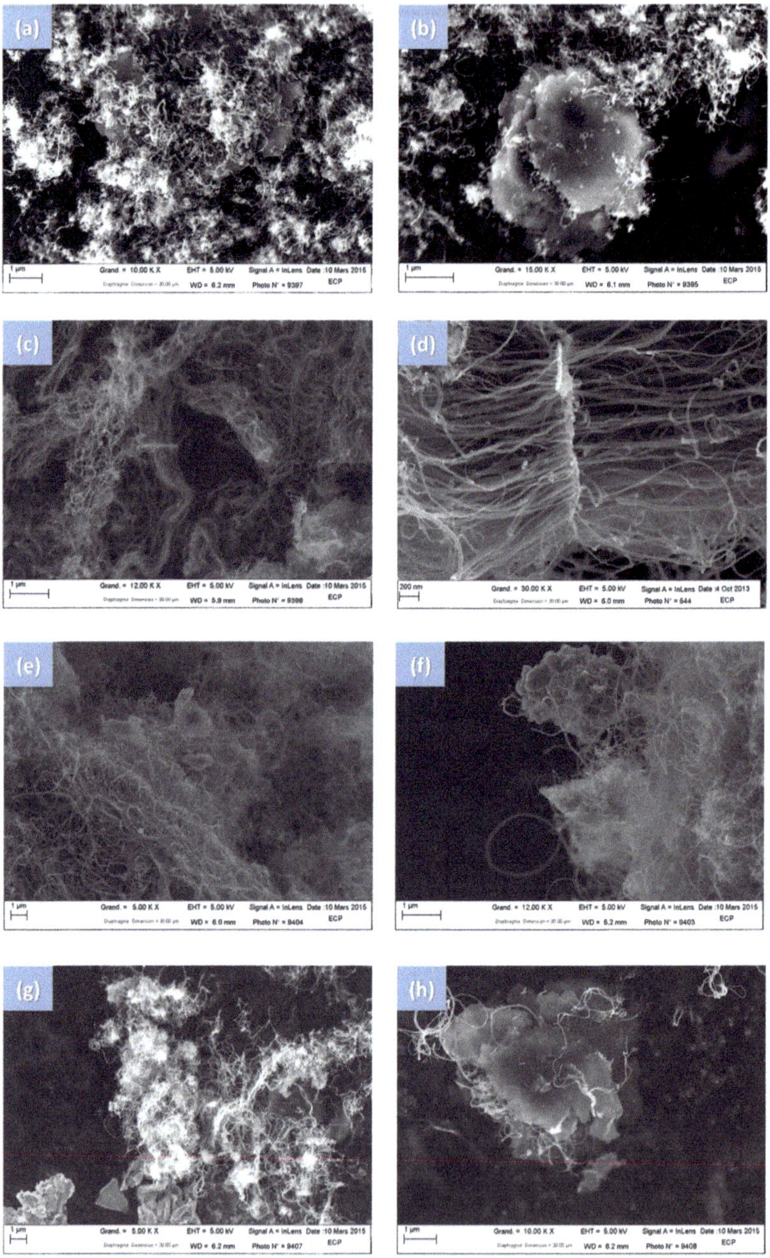

Figure 4.7 SEM images of GCH microstructures at four CVD synthesis temperatures: (a, b) 550 °C, (c, d) 650 °C, (e, f) 750 °C, (g, h) 850 °C. (d) Reproduced from ref. 119 with permission from American Chemical Society, Copyright 2015.

preserved during the reaction, which is the same as for the as-received GNPs. When the synthesis temperature is increased to 650 °C, a large number of CNT arrays are successfully synthesised and aligned onto the surfaces of individual GNP substrates. Thus, complete GCH constructions are formed. When the synthesis temperature is above 750 °C, the average length of the synthesised CNT array displays an obvious increase. However, it is observed that the area number density of synthesised CNT arrays reduces at higher temperature ranges, particularly for the product synthesised at 850 °C.

Rather than in a gas atmosphere, the catalyst precursor is more easily decomposed on the surface of the GNP substrate under a low-temperature environment. The activation energy of the decomposition reaction is reduced due to the interaction between the substrate and the reactants. Thus, a positive environment is generated for CNT nucleation.[31] When the synthesis temperature reaches 550 °C, the stacked GNPs are difficult to separate into individual ones due to insufficient energy in the reactor. Thus, the aforementioned sparsely distributed CNTs are synthesised onto surfaces of the multilayered GNPs. However, the amount of decomposed xylene is limited by the high decomposition temperature of xylene. The growth of CNT arrays is therefore inhibited due to the low concentration of effective carbon species[31,92,93,122] The as-received multilayered GNPs can be exfoliated into a large number of separated GNP units with the reaction temperature increasing to 650 °C. These flexible large-area platforms are able to promote not only catalyst particle deposition, but also formation of 'dual-directional' GCH construction. In fact, accompanied by the contribution of the catalyst, after experiencing complex processes such as dissolution, diffusion and precipitation of the carbon source, CNT arrays can be finally synthesised. With elevating of the reaction temperature, the reaction rates for dissolution and diffusion are boosted as well. When the synthesis temperature is elevated to over 750 °C, the concentration of carbon in the system is extremely high. Under these conditions, the dissolving effect becomes the dominant reaction during synthesis. Thus, the carbon atoms aggregate and cover the surface of the GNP substrates in the form of amorphous carbon, which terminates the CNT growth.[17]

4.3.3.2 Hydrogen Ratio

Hydrogen, as one of the carrier gases, is significant for tuning the construction of the CVD products and determining the decomposition of carbon sources and catalyst precursors.[123] Figure 4.8 illustrates the SEM morphology of GCH constructions synthesised with various ratios (0% to 30%) of hydrogen to total gas flow rate, namely the H_2/Ar ranges from 0 to 0.3/0.7 L min^{-1}. The reaction temperature and synthesis duration were fixed as 650 °C and 10 min, respectively. As demonstrated in Figure 4.8(a, b), a number of abnormal 'sticks' were synthesised onto GNP substrates *via* the CVD process without H_2 injection, with diameters (beyond 100 nm) that are significantly larger than ordinary CNTs. This abnormal construction can be further assessed in detail from the relevant TEM morphology (see Figure 4.9). Compared with CNTs,

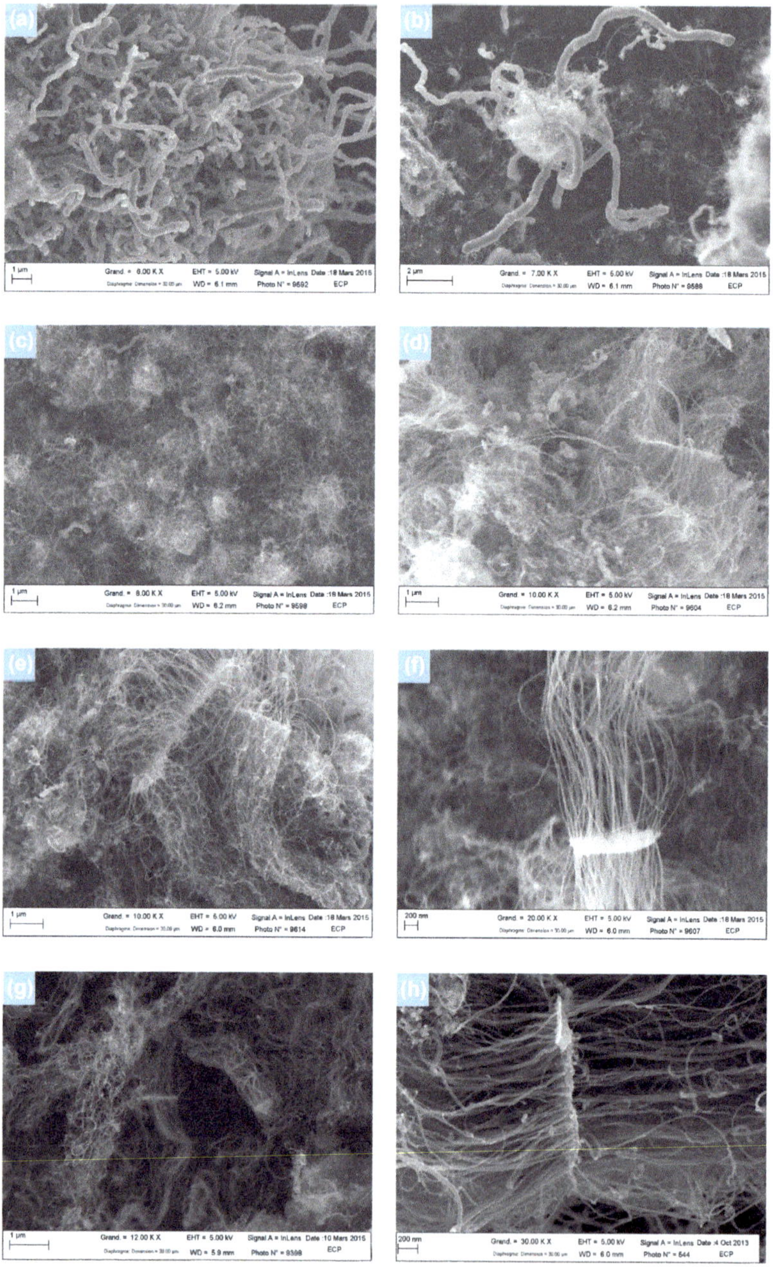

Figure 4.8 SEM micrographs of GCH hybridisation synthesised by floating CCVD method under different ratios of hydrogen to total carrier atmosphere: (a, b) 0%, (c, d) 10%, (e, f) 20%, (g, h) 30%.
(h) Reproduced from ref. 119 with permission from American Chemical Society, Copyright 2015.

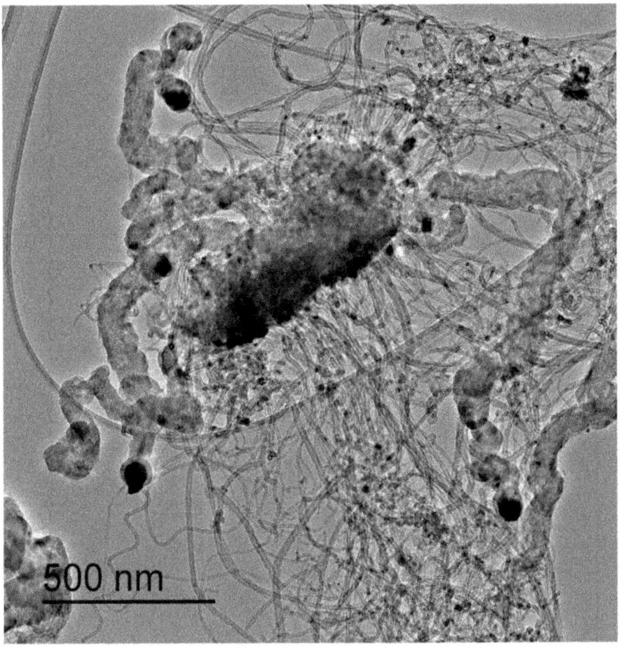

Figure 4.9 TEM images of the 'stick' consisting of non-crystalline carbon.

there is no characteristic tube structure observed in these 'sticks'. Thus, the 'sticks' may be mainly composed of amorphous carbon. In fact, during the synthesising process, sensitive iron catalysts can be effectively protected by H_2 by being wrapped in accumulating amorphous carbon. Thus, the utilisation of H_2 is beneficial to prolong the active lifetime of the catalyst.[124,125] Of note is that the amorphous carbon sticks coexist with CNTs in Figure 4.8(c, d), but are hardly observed in Figure 4.8(e–h), which indicates that the enhanced hydrogen ratio is able to inhibit the formation of amorphous carbon sticks. Moreover, when the hydrogen ratio increases, the average length of the CNTs shows a gradual reduction; however, the corresponding mean diameter still remains in a stable range. Therefore, the aspect ratio of the synthesised CNT arrays decreases with enhancing hydrogen ratios.

4.3.3.3 Reaction Duration

Figure 4.10 presents the evolution process in the morphology of GCHs obtained from various synthesis duration times (5–40 min). The temperature and hydrogen ratio employed in the synthesis were fixed at 650 °C and 30%, respectively. It is observed that the 'dual-directional' GCH constructions are achieved from CVD synthesis with all the selected reaction durations. Figure 4.11 shows the relationship between the mean length of the as-synthesised CNTs and the reaction time. It is seen that the length of the as-synthesised CNT array exhibits a quasi-linear improvement with prolonging the reaction duration, with the corresponding slope by linear fitting being 0.22.

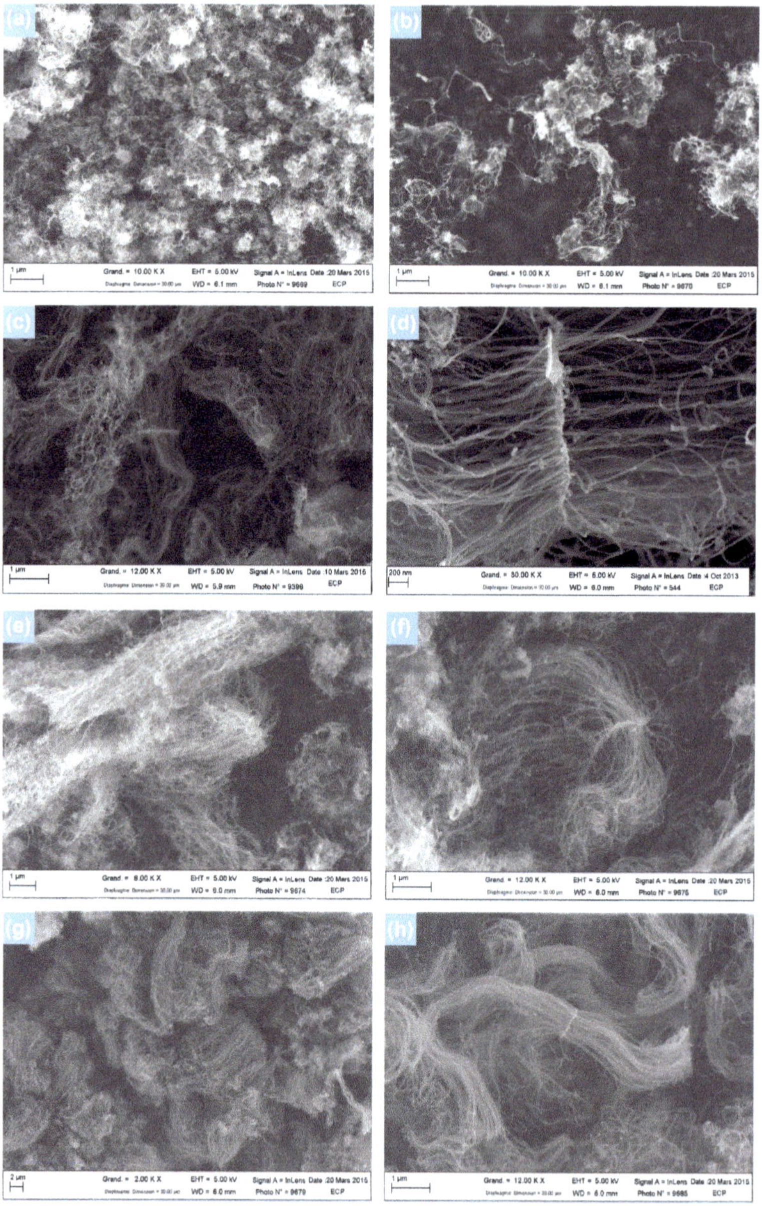

Figure 4.10 SEM images of GCH architecture generated by the floating CCVD method with varied synthesis times: (a, b) 5 min, (c, d) 10 min, (e, f) 20 min, (g, h) 40 min.
(d) Adapted from ref. 119 with permission from American Chemical Society, Copyright 2015.

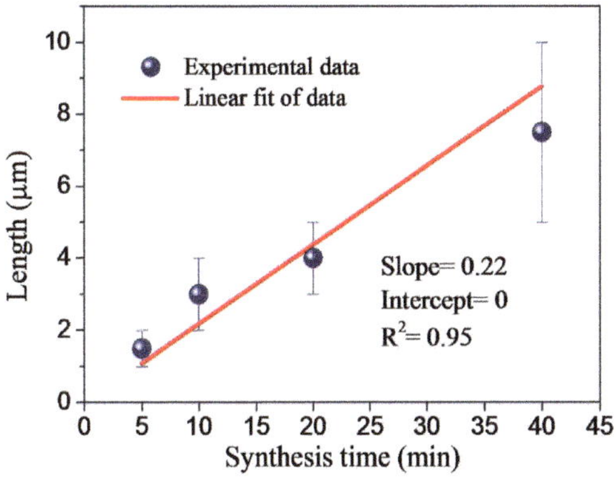

Figure 4.11 The dependence between the mean length of grown CNTs and variation in synthesis duration.

4.3.4 Comprehensive Analysis of GNP–CNT Hybrids

As mentioned, the electrical behaviour of a composite is highly related to the aspect ratio of the loaded CNTs. For overall consideration of the relationship between the morphological characteristics of GCHs and the key parameters of synthesis, a typical GCH construction synthesised under conditions of 650 °C, 30% hydrogen ratio, and 10 min, was selected as the functional filler for composites in the following discussion. Since the selected GCH construction shows a moderate mean length and aspect ratio for the CNT array, this is beneficial to ameliorate CNT entanglement in the polymer matrix. Therefore, several characterisations and the corresponding analyses based on GCH construction were performed, with the purpose of figuring out how this construction is capable of endowing related composites with significant improvements in various functional properties.

4.3.4.1 Composition of GCHs and By-products From CVD Synthesis

The catalyst is an indispensable component in floating-catalyst CVD synthesis. In order to figure out what kind of residual by-products can be formed during CVD synthesis and whether and how these by-products can influence functional properties of composites, relevant investigations need to be carried out. During the synthesis, the catalyst precursor ferrocene ($Fe(C_5H_5)_2$) was first dissolved in xylene solvent. Then, the suspension was injected into the reacting area in the furnace for the purpose of efficiently synthesising the carbon hybrid construction. In our previous results, a series of complex chemical reactions have been proved to occur in the furnace.[93,94,126] The following formula expresses the general reaction of the catalyst precursor:

$$Fe(C_5H_5)_2 + H_2 \Rightarrow Fe + 2C_5H_6 \tag{4.10}$$

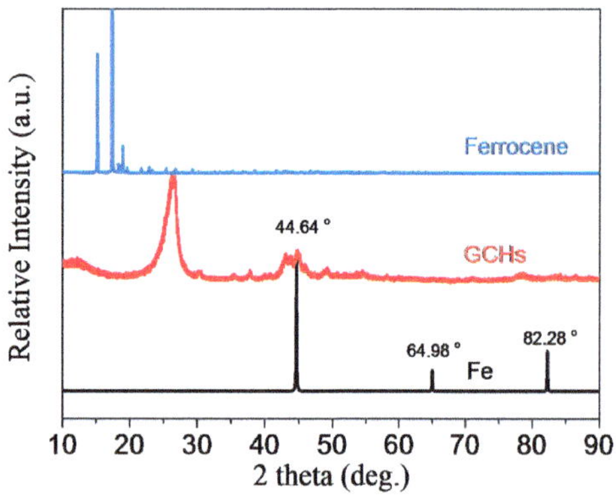

Figure 4.12 X-ray diffraction patterns of ferrocene and GCHs.

As the whole synthesis process is carried out under an Ar atmosphere, iron is identified to be the main residual product of the catalyst precursor. Figure 4.12 exhibits the X-ray diffraction (XRD) patterns of ferrocene, as-synthesised GCHs and iron.[127,128] From the XRD patterns, the characteristic peak of ferrocene is hardly observable in the GCH spectrum, which indicates that there is no residue of the catalyst precursor in the synthesised products. Moreover, three characteristic peaks at 44.64°, 64.98° and 82.28° are found in the XRD pattern of iron. The characteristic peak at 44.64° in the XRD pattern of the GCH confirms that iron exists in the GCHs as the catalyst by-product.

As shown in Figure 4.13(a), the quantity of iron can be estimated by thermogravimetric analysis (TGA). The mass fraction of iron (iron oxides) is about 5 wt% in the whole production process. The amount of residual iron in the GC-filled composites (discussed in the following sections) is quite small, less than 0.1 vol%. Figure 4.13(b) presents the TEM image of the by-product particles. It is observed that their average length and diameter are about 20 nm and 8 nm, respectively. Notably, most iron particles can be observed in the inner tube of the as-synthesised CNTs, which makes a negligible contribution to building the conducting network in the polymer matrix. Thus, it is concluded that the influence of the catalyst by-product to the electrical properties of the composites can be ignored.

Besides the tiny concentration of metallic catalyst, the CNT/GNP mass ratio is another important parameter to assess the composition of the GCH construction quantitatively. Moreover, the CNT/GNP mass ratio plays an important role in influencing the internal conducting network in the host polymer and the overall properties of the composites. Although TGA measurements are frequently used in analysing the composition concentration of other ordinary CNT-based hybridisation constructs,[98,129] it is hard to distinguish as-synthesised CNTs from GNP through loss of mass, because of the complete carbon composition. Only a single heat flow characteristic peak at 586 °C can be observed over the whole temperature range during differential

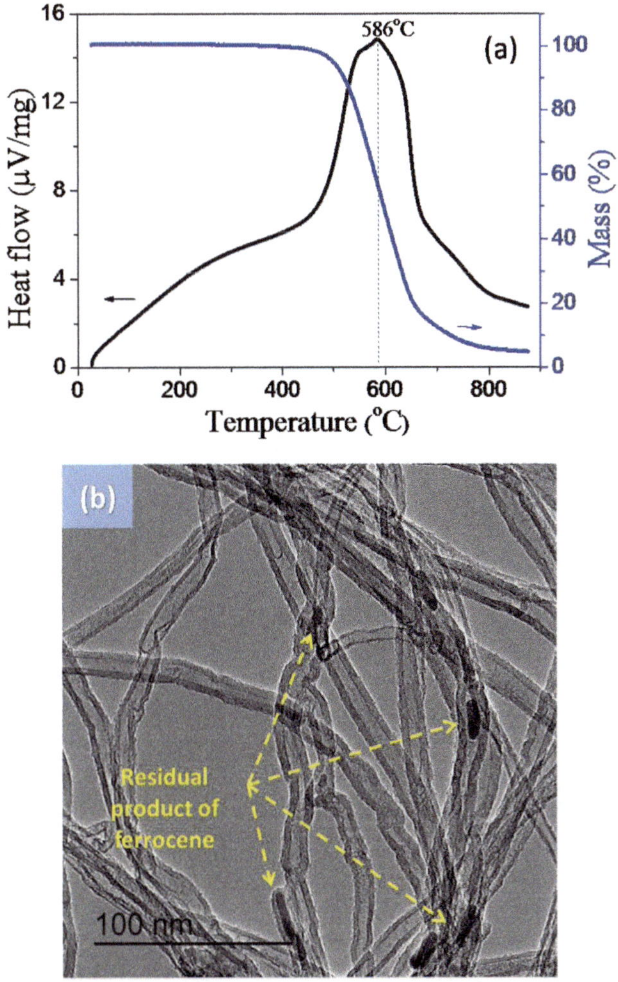

Figure 4.13 (a) TGA and DSC curves of the classic GCH structure, and (b) TEM morphology of the by-product of the catalyst.

scanning calorimetry (DSC)/TGA testing, which indicates the integrated thermal decomposition of both CNT arrays and GNP substrates.

4.3.4.2 Structural Characterisation

The CNT structure can be generally confirmed by the characteristic structure of the carbon walls with axial and internal cavities. High-resolution TEM (HRTEM) characterisation of synthesised GCHs and CNTs is presented in Figure 4.14. From the TEM image of an individual CNT, the typical multi-wall structure and inner cavity can be found, which corresponds to previous research.[31]

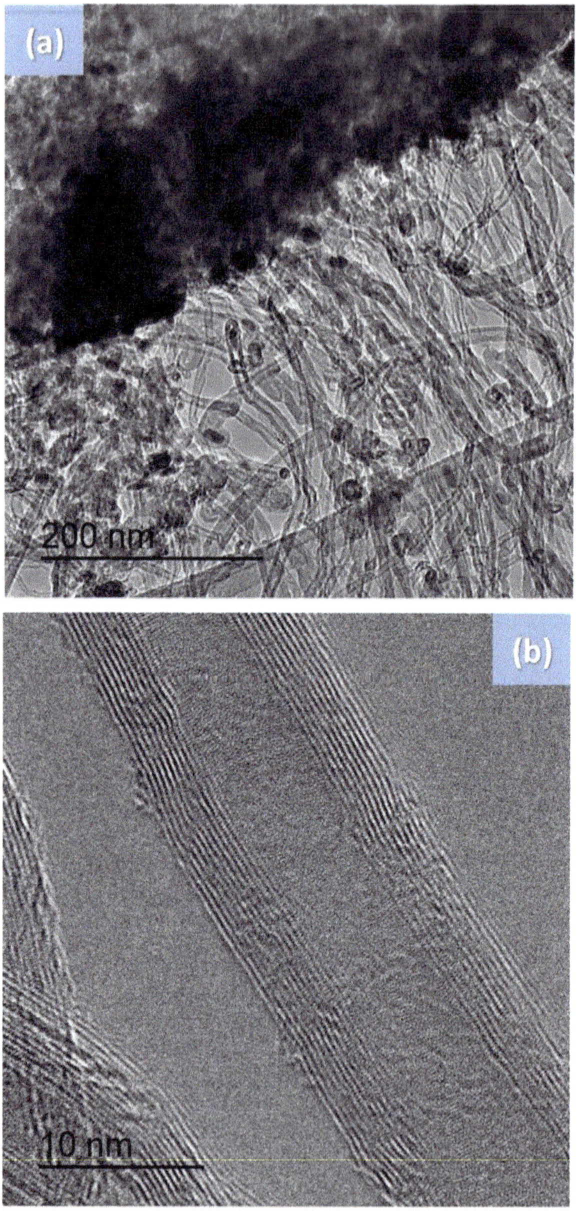

Figure 4.14 (a) TEM morphology of part of the typical GCH hybridisation. (b) Detailed construction of the synthesised CNTs exhibited by a high-magnification TEM image.
(b) Reproduced from ref. 136 with permission from the Royal Society of Chemistry.

Furthermore, in order to estimate the structure of the CNTs, the useful approach of Raman spectroscopy is applied. Two main vibrational modes, G-band (~1580 cm^{-1}) and D-band (~1330 cm^{-1}) are shown in the Raman spectra of carbon materials. The G-band represents the graphite mode, being related to the ordered sp^2 hybridised carbon structure. The D-band represents the defect mode, being associated with local defects and structural imperfections. In order to assess the structural integrity of CNTs, the relative intensity ratio of the D- and G-bands (I_D/I_G) is defined. Figure 4.15(a) exhibits the Raman spectrum of as-received GNPs, as-synthesised GCHs and CNTs synthesised on a quartz substrate through the same CVD procedure.

In addition to the results from HRTEM morphology, the positions of several characteristic peaks in the Raman spectrum further confirm that CNT structures have been successfully synthesised onto the GNP surface as well.[130,131] As illustrated in Figure 4.15(b), the measured I_D/I_G values of the as-synthesised GCHs and CNTs are 1.06 and 1.38, respectively, implying that

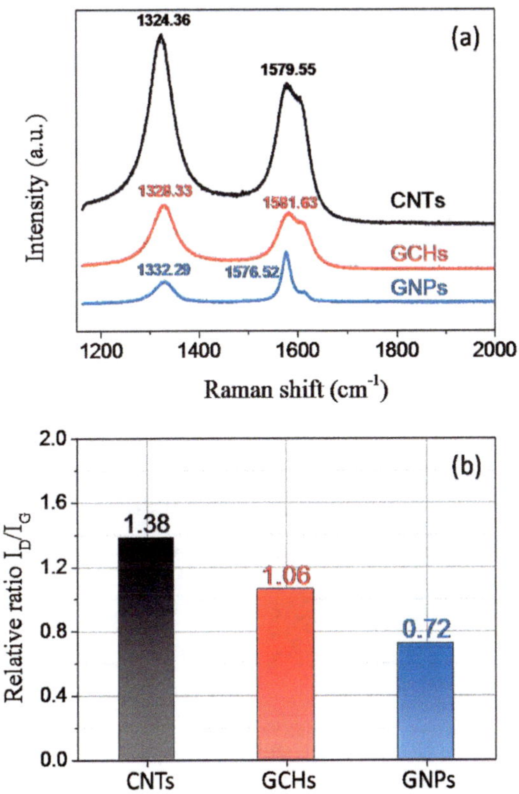

Figure 4.15 The characterisation of CNTs (black), GCHs (red) and GNPs (bule). (a) Raman spectrum and (b) the calculated I_D/I_G values of CNTs, GCHs and GNPs.

some structural defects do exist in the carbon crystal lattice. In fact, the structural integrity of synthesised CNTs is further optimised by regulating the CVD process, which is also one important factor in our future studies.

4.3.4.3 *Thermal Exfoliating Effect*

As mentioned, the influence of thermal exfoliation is important for determining the construction of GCHs. A GNP substrate composed of multi-natural flake graphite sheets is demonstrated in Figure 4.16(a). It can be seen that the average thickness is approximately hundreds of nanometres and the diameter is about 1–4 μm. The 1D–2D coupling structures of synthesised GCHs are presented in Figure 4.16(b, c). The vertically aligned CNTs have a length of 2–4 μm and a diameter of 15–20 nm. A relatively complete hybrid structure can be obtained by TEM characterisation, which is demonstrated in Figure 4.16(d). It is seen that the CNT array–GNP substrate interfacial connection is relatively stable.

It is noted that almost all of the CNT array clusters are perpendicularly synthesised onto the surface of the independent GNP substrate with a

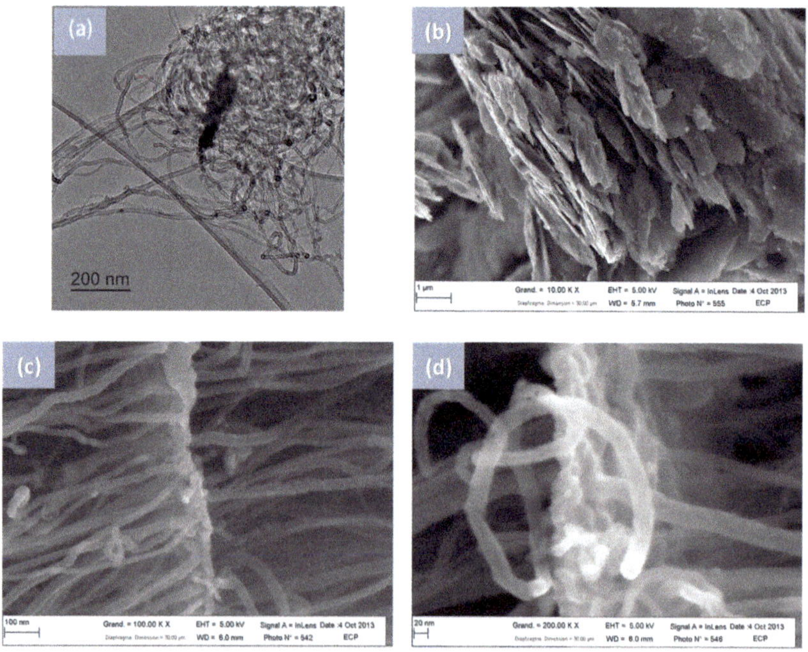

Figure 4.16 SEM morphological images of (a) the multilayer stacked structure of pristine GNPs; (b, c) the architecture of GCH particles in which CNTs are grown on the surface of the substrate of individual GNP layers; (d) TEM image of the GCH hybridisation microstructure. (a–c) Reproduced from ref. 119 with permission from American Chemical Society, Copyright 2015. (d) Reproduced from ref. 136 with permission from the Royal Society of Chemistry.

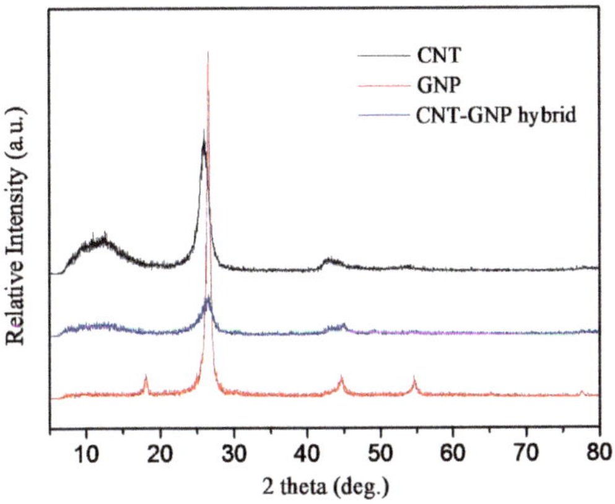

Figure 4.17 The XRD patterns of stacked GNPs, as-received CNTs and as-synthesised GCH particles.
Reproduced from ref. 119 with permission from American Chemical Society, Copyright 2015.

thickness below 20 nm, rather than onto the surface of the much thicker stacked GNP aggregations. In fact, the required high temperature is important for the GCH synthesis. On one hand, the high temperature is able to provide sufficient energy to meet the requirements for carrying out the relevant chemical reactions, while, on the other hand, it can drive the thermal exfoliation effect to exfoliate the as-received GNP stacked construction into a large number of much thinner, independent GNP substrates.

As shown in Figure 4.17, the XRD patterns of as-received CNTs, GNP aggregations and as-synthesised GCHs further identify the thermal exfoliating influence of GNPs. It can be seen that the characteristic peak at 26.46° sharply reduces from GNPs to GCHs, which indicates that the interlayered distance in the individual GNP substrate is constant. However, the degree of exfoliation of GNP aggregation is improved. In contrast with the sharp peaks in the GNP pattern, the characteristic peaks in the GCH pattern at 44° and 54.3°, with a gentle slope, are in accordance with characteristic peaks of CNTs. Thus, the XRD patterns again verify the GCH construction.

Therefore, the thermal exfoliating effect is responsible for effectively improving the specific surface area of GCHs and the filler–matrix interface area in composites. The relevant improvement in structural and functional properties of various composites is discussed in detail in the following sections.

4.3.5 Section Summary

With a floating-catalyst CVD (CCVD) process, well-organised GNP–CNT hybrid structures were synthesised. The CNTs were perpendicularly arrayed on surfaces of individual GNP substrates with no pre-treatment. The final GCH constructions are closely dependent on key parameters of the synthesising process, such as synthesis temperature, hydrogen ratio and reaction duration time.

In particular, a low synthesis temperature cannot provide the reaction with sufficient energy for exfoliating stacked multilayered GNPs into individual ones. Moreover, xylene decomposition is limited by low synthesis temperatures, which also results in inhibition of CNT growth by the low effective carbon species concentration. However, at high temperature, CNT growth can be terminated by the amorphous carbon formed and accumulated on the surface of the GNP substrate. Thus, in order to achieve appropriate complete construction GCHs, the temperature during the CVD reaction is set as 650 °C. Moreover, hydrogen is also able to influence the construction of the hybridisation directly during the synthetic process. Although the growing hydrogen fraction in a mixed atmosphere has little regulation of the diameter of the synthesised CNTs, it is able to reduce their length and aspect ratio gradually. Besides, it is observed that the average length of CNT arrays is quasi-linearly dependent on the reaction duration. According to linear fitting, the slope of the linear relationship equals 0.22. The typical GCH construction synthesised under conditions of 650 °C, 30% and 10 min was selected as the functional filler of various composites. Notably, the mass ratio between CNTs and GNP in this type of GCH is approximately 1, as obtained from the TGA measurement.

As the synthesis progresses, the ferrocene catalyst is totally transformed into a tiny amount of iron (oxide) particles, which exist in the inner tube of CNTs. Thus, these residual by-products make little difference to the overall electrical properties of the composites. A significant multi-wall and internal cavity structure are observed from the as-synthesised CNTs, which provides an important structural foundation for the subsequent discussion. In addition, the mono-layered GNP substrate is achieved through the thermal exfoliating effect. Therefore, this typical GCHs construction exhibits an extremely large specific surface area, which is critical for the improvement of the functional properties of polymer-based composites.

4.4 The Dielectric Behaviour of GNP–CNT Hybrid/PVDF Composites

4.4.1 Introduction

In recent years, power dissipation, high flexibility and miniaturisation have been the main developing trends in designing and fabricating next-generation advanced dielectric materials with high energy density.[106,132–135] Due to the intrinsic electronegativity of fluorine atoms and a large electrical dipole moment, poly(vinylidene fluoride) (PVDF) exhibits promising dielectric and piezoelectric features. Based on these characteristics, applications based on PVDF and its copolymers have been widely investigated, such as piezoelectric sensors, microactuators and, particularly, as a host matrix in composite film capacitors.

Compared with other thermoplastic polymers, neat PVDF displays a relatively high permittivity ($\varepsilon_r \sim 10$). In order to make use of this inherent superiority and further improve the ε_r of PVDF-based composites, several ordinary methods have been reported for years, such as loading high-ε_r

nanoceramics or electrically conductive fillers into composite system. However, these traditional strategies can result in the irreversible loss of flexibility and insulating properties, which is detrimental to the processing and practical applications of dielectric composites.[106,132]

For electrical energy storage materials based on CNT-filled composites, the effective filler–matrix interfacial area and interaction can be weakened by CNT aggregation. Moreover, this undesirable aggregation can become the trigger point for dielectric breakdown, which limits the energy density of the composite to a very low level. Therefore, effective improvement of the dispersion of the CNT in the polymer matrix is still an important challenge that needs further investigation.

As mentioned in previous sections, the way in which the surface is modified has proven to be an effective method to ameliorate CNT dispersion in the matrix. However, the compromise between the intrinsic electrical properties of the conducting filler and dispersion of CNTs requires further intensive study. Moreover, the dispersion of nanofillers is also determined by the processing method used. Compared with the solution-mixing method frequently used in the laboratory, the melt-mixing process with the feature of being solvent-pollution-free is able to provide a stable dispersion for industrial mass production of composites with multiphase compositions. Through the melt-mixing process, the molecule-level interaction between nanofillers and the matrix is achieved under high temperature and a strong shear force.

In this section, we present typical CNT array hybridisation/PVDF composites showing a relatively low f_c and improved dielectric performance. The as-synthesised GCH constructions with vertically aligned CNT arrays were prepared as previously described. Different from ordinary insulating substrate–CNT hybrids, such as Al_2O_3-CNTs and SiC-CNTs, the electrically conducting GNP substrate is able to endow GCHs with some distinctive advantages, such as low intrinsic density, all-carbon composition and completely conducting coupling structure. It is worth noting that each individual GNP acts as a micro-scale wire concentrator making all the grown CNTs electrically conducting. The melt-mixing process that improves both the filler dispersion and interfacial interaction was carried out. Compared with the GNP/CNT/PVDF ternary counterpart, the GCH/PVDF binary composite exhibits a significantly lower f_c value and more stable thermal properties. The dielectric and percolation behaviour of GCH/PVDF composites was analysed in depth. The implication is that the strategy of constructing CNT hybrids is feasible for use for reducing the f_c value of polymer-based composites.

4.4.2 The Preparation of PVDF-based Composites

The CNT/GNP mass ratio of the as-synthesised GCHs was controlled as 1 by regulating the CVD synthesis parameters. The GCH/PVDF binary and GNP/CNT/PVDF ternary nanocomposites with various loading fractions were prepared though melt-blending by utilising a twin-screw micro-compounder (Micro 5 cm^3 Twin Screw Compounder, DSM) at a temperature beyond

200 °C. The length and diameter of the selected CNTs are about 5–15 μm and 20–40 nm, respectively. Both a relatively long duration and low shearing rate are important to preserve complete construction of the hybrid and good dispersion during the blending process. Finally, well-dispersed samples were obtained through injection moulding.

4.4.3 Structural Characteristics and Functional Properties of PVDF-based Composites

4.4.3.1 *PVDF-based Binary and Ternary Composites*

To evaluate the influence of the structural superiority of GCHs on the dielectric properties of PVDF-based composites, GNP/CNT/PVDF ternary composites were fabricated as well. For the composition of GCHs, the loading ratio between GNPs and CNTs was precisely regulated as 1 : 1. Scheme 4.3 schematically illustrates the difference in composition and structure between the two types of composites. The detailed compositions of the binary and ternary composites are displayed in Table 4.2. The designations B-*n* and T-*n* are used to represent the binary and ternary composites filled with *n* wt% conducting phase, respectively.

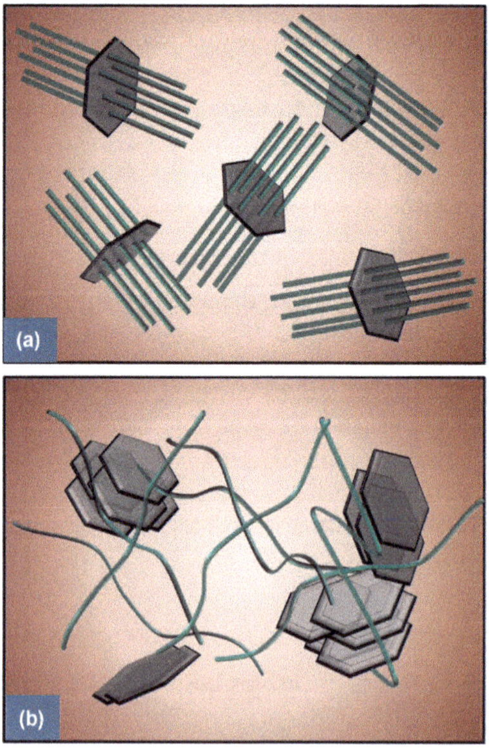

Scheme 4.3 Schematic of the structural difference between (a) binary and (b) ternary composites.

Table 4.2 The concentration of every component in the GCH/PVDF binary composites and the GNP/CNT/PVDF ternary composites. Reproduced from ref. 136 with permission from the Royal Society of Chemistry.

	GCHs	GNPs		CNTs		Total conductive phase	
	(wt%)	(wt%)	(vol%)	(wt%)	(vol%)	(wt%)	(vol%)
GCH/PVDF binary composites							
B-1.7	1.7					1.7	1.42
B-4	4					4	3.38
B-6	6					6	5.09
B-6.5	6.5					6.5	5.52
B-7	7					7	5.95
B-8	8					8	6.81
B-8.5	8.5					8.5	7.24
B-9	9					9	7.67
B-10	10					10	8.54
GNP/CNT/PVDF ternary composites							
T-2		1	0.79	1	0.89	2	1.69
T-4		2	1.59	2	1.79	4	3.38
T-6		3	2.40	3	2.70	6	5.09
T-6.5		3.25	2.60	3.25	2.92	6.5	5.52
T-8		4	3.20	4	3.61	8	6.81
T-10		8	4.02	8	4.52	10	8.54

4.4.3.2 SEM Morphologies of PVDF-based Binary and Ternary Composites

Figure 4.18(a) presents an image of the morphology of GNPs, which served not only as the substrate for GCHs, but also the conducting filler in the ternary composites. Two-dimensional multilayered constructions with a diameter of 1–4 μm and thickness of hundreds of nm are observed. The large area of the GNPs provides an ideal platform for catalyst deposition and CNT array growth. However, the bulk CNTs exist in the form of entanglement; see Figure 4.18(b).

Figure 4.18(c) demonstrates the SEM microscopy of a fractured cross-section of the GCH/PVDF composite, where both GNP and grown CNT arrays are tightly embedded in the PVDF matrix. This indicates that the GCH–PVDF interfacial adhesion is relatively strong. Compared with the CNT-filled system, both the greatly enlarged interfacial area and strong interfacial adhesion between GCHs and the matrix are able to make an important structural contribution to the superior overall properties obtained from the GCH/PVDF composites. SEM microscopy of the fracture section of the ternary composite is also shown in Figure 4.18(d), where the independent GNPs and CNTs are also dispersed and well embedded in PVDF.

4.4.3.3 XRD Patterns

Figure 4.19 presents the typical XRD patterns of various conducting fillers, neat PVDF and binary and ternary composites. It can be seen that the

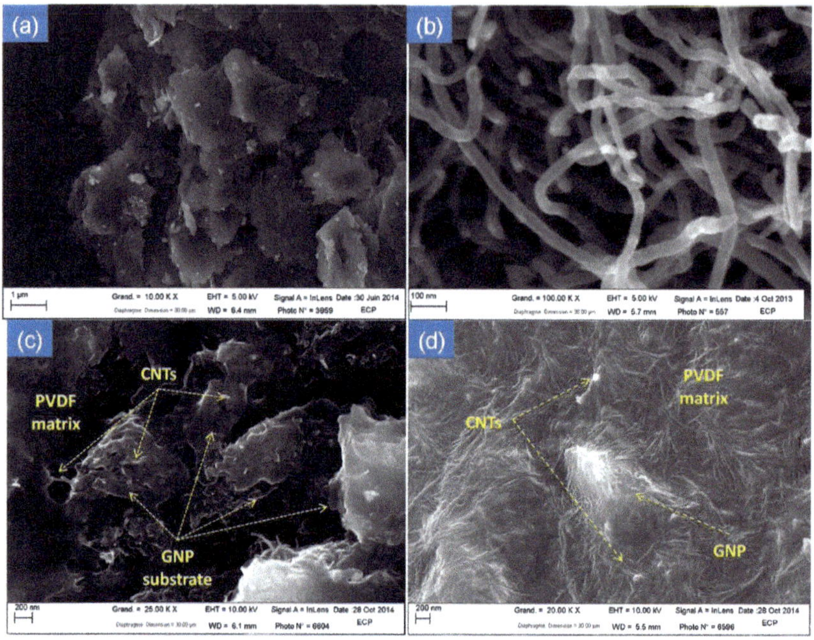

Figure 4.18 SEM morphology of (a) as-obtained GNPs and (b) as-obtained curly CNTs. The SEM images of (c) the fractured cross-sections of GCH/PVDF binary composites and (d) GNP/CNT/PVDF ternary composites. (a, c, d) Reproduced from ref. 136 with permission from the Royal Society of Chemistry.

characteristic peak around 26.46° in the XRD pattern of GCHs displays a significant decrease in intensity and a slight left-shift in position compared with that of GNPs. This variation could be ascribed to the thermal exfoliation of multilayered GNPs at high temperature during the CVD process, which was discussed in Section 4.3.4.3. Moreover, there is a characteristic peak (26.4°) shown in XRD pattern of the ternary composite whose position is between the same peak in GNP curve (26.6°) and the CNT curve (25.9°). This implies that both GNPs and CNTs are well distributed in the composite.

Three characteristic XRD peaks (17.6°, 18.3° and 19.9°) are found from the patterns of neat PVDF polymer and binary and ternary composites, which correspond to the α-crystalline phase of PVDF. It is noteworthy that there is an obvious characteristic XRD peak at 20.26° in the XRD pattern of the GCH/PVDF composite, which represents the β-crystalline phase of PVDF. Thus, the crystalline type and relative crystalline composition of PVDF can be influenced by the incorporation of filler. Notably, incorporation of GCHs could make a contribution to the formation of the β-crystalline phase of PVDF, further influencing the overall properties of the PVDF-based composite.

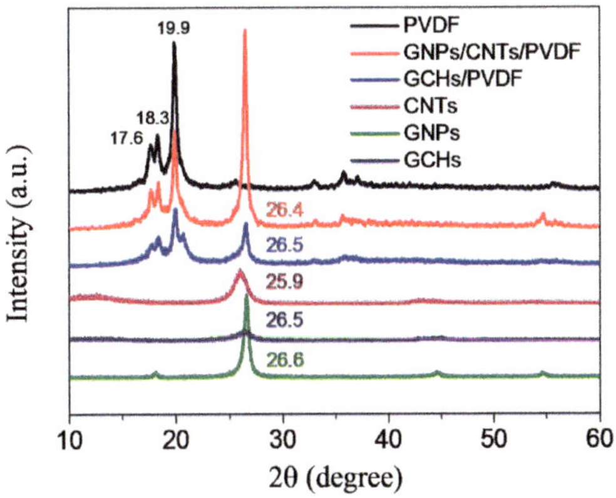

Figure 4.19 The XRD patterns of as-obtained CNTs and GNPs, as-synthesised GCHs, neat PVDF, binary and ternary composites.
Reproduced from ref. 136 with permission from the Royal Society of Chemistry.

4.4.3.4 The Influence of Frequency on Dielectric Behaviours in Binary and Ternary Composites

Figure 4.20 illustrates the frequency-dependent dielectric behaviour of binary and ternary composites with various carbon-loading content (measured by Solartron 1260, 10 Hz to 10 MHz). In the low frequency range ($<10^4$ Hz), when f_{GCHs} <7 wt%, the ε_r of binary composite shows frequency-independent behaviour. However, it is observed that ε_r values of the composite reduce with increase in the frequency (when the frequency $>10^4$ Hz). Moreover, the reduction rate accelerates with enhanced GCH loading (see Figure 4.20(a)). As shown in Figure 4.20(b), when f_{GCHs} is low, the loss tangent values of the composite are always kept at a low level (~0.2) throughout the whole testing frequency range. Meanwhile, the corresponding dielectric relaxation occurring in neat PVDF could be observed. In addition, when loading relatively high fractions of f_{GCHs}, the composites present sharply enhanced loss tangent values, which is attractive in electromagnetic shielding applications.

As demonstrated in Figure 4.20(c), when loading low GCH fractions, the alternating current (AC) conductivity (σ_{AC}) of the composites is enhanced linearly with increasing frequency, while, with increasing f_{GCHs}, the σ_{AC} values exhibit both a significant improvement of several orders of magnitudes and a typical frequency-independent behaviour in the low frequency range.

In comparison with their binary counterparts, the ternary composites exhibit a distinct dielectric behaviour, particularly at high filler loadings (f> 6 wt%) (see Figure 4.20(d–f)). In addition, over the whole of the measured frequency range, all ternary composites filled with various loadings of GNPs and CNTs show an electrically insulating behaviour.

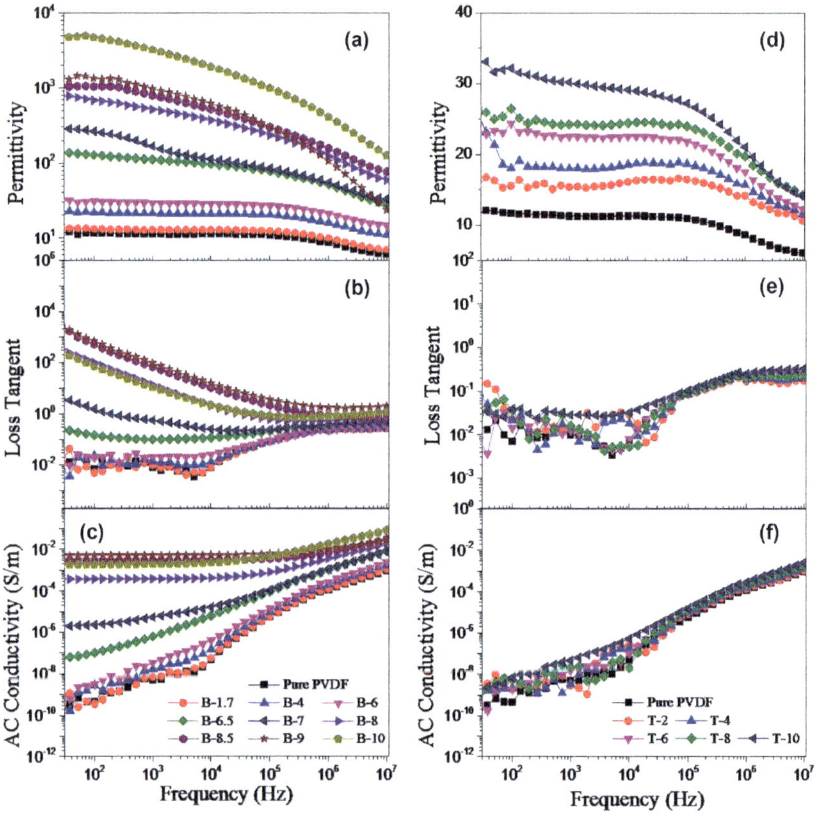

Figure 4.20 The dielectric permittivity (a, d), loss tangent (b, e) and AC conductivity (c, f) as a function of frequency for binary and ternary composites at room temperature, respectively.
Reproduced from ref. 136 with permission from the Royal Society of Chemistry.

4.4.3.5 The Percolation Threshold of Binary Composites

Figure 4.21 establishes a relationship between σ_{AC} of GCH/PVDF binary composites and the GCH loading at room temperature. It is observed that the σ_{AC} of the binary composite first grows gradually and subsequently increases sharply with increasing GCH loading. As mentioned before, the jump in the development of the electrical property takes place around the percolation threshold, implying that an internal conductive network is established. Based on percolation theory, the f_c of the binary system can be assessed from σ_{AC} values obtained from the composites filled with various GCH loading using the following power laws:

$$\sigma_{AC}(f_{GCHs}) \propto (f_c - f_{GCHs})^{-s'} \quad \text{for } f_{GCHs} < f_c \tag{4.11}$$

$$\sigma_{AC}(f_{GCHs}) \propto (f_{GCHs} - f_c)^{t} \quad \text{for } f_{GCHs} > f_c \tag{4.12}$$

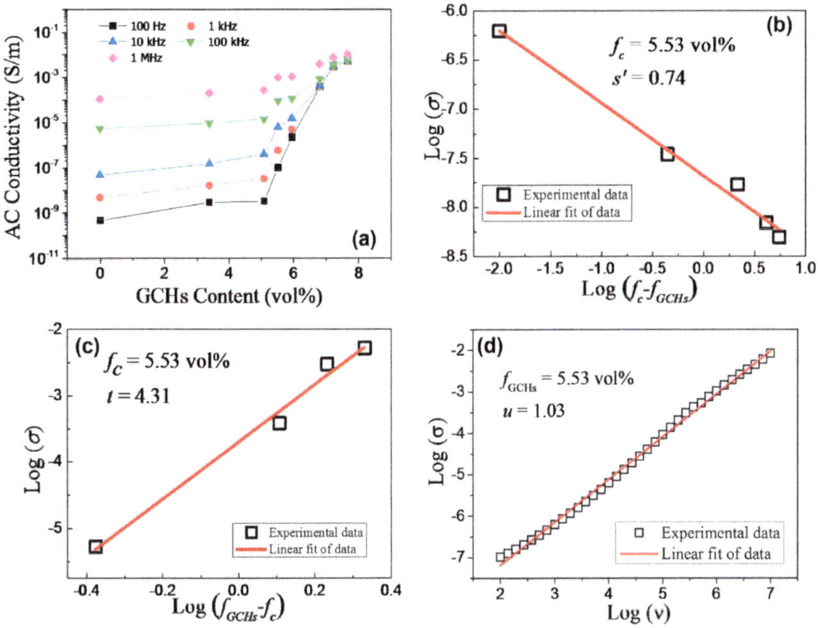

Figure 4.21 (a) AC conductivity (1 kHz frequency) of the GCH/PVDF binary composites filled with various volume fractions of GCHs. (b, c) The AC conductivity best matches for eqn (4.11) and (4.12). (d) The experimental AC conductivity is linearly fitted according to eqn (4.13). Reproduced from ref. 136 with permission from the Royal Society of Chemistry.

where s' and t represent the critical exponent in the insulating and conducting regions, respectively. Based on eqn (4.11) and (4.12), an equivalent f_c (~5.53 vol%) is achieved from the optimal linear fitting based on σ_{AC} in both the insulating and conducting regions; see Figure 4.21(b, c). The critical exponent s' is calculated as 0.74, which is close to the universal range ($s' \approx 0.8–1$). However, the calculated t is about 4.31, being larger than the universal values ($t \approx 1.6–2$). The deviation of the achieved s' and t might be ascribed to the irregular GCH hybridisation structure. Moreover, it is assumed that the tunnelling effect only takes place between nearest adjacent conducting units, which could be the main reason for the non-universal critical exponents of the percolation power laws.[105]

When $f_{GCHs} \rightarrow f_c$, the threshold power law can be presented as:

$$\sigma \propto \omega^u, \quad \varepsilon \propto \omega^{u-1} \tag{4.13}$$

where $\omega = 2\pi\nu$, and ν and u are the frequency and critical exponent, respectively. The experimental u value obtained from the sample B-6.5 is 1.03, however, the theoretical value is calculated as: $u = t/(s' + t) = 4.31/(0.74 + 4.31) = 0.85$. These two values are considered to be in correspondence. The theoretical fitting based on the experimental results is shown in Figure 4.21(d).

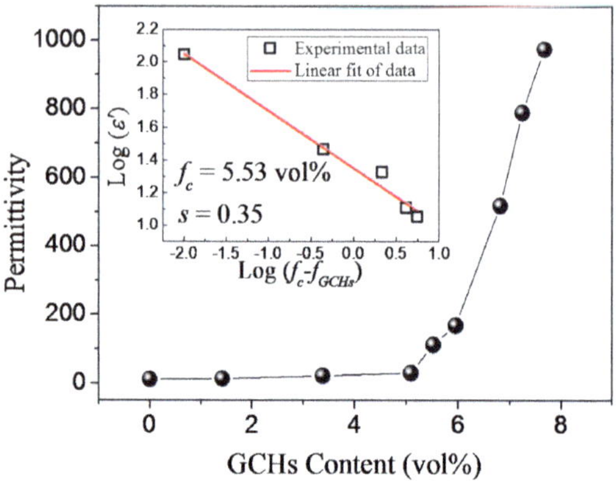

Figure 4.22 The dielectric permittivity (1 kHz) of the GCH/PVDF binary composites filled with different volume fractions of GCHs at room temperature. The best fit of the permittivity to eqn (4.14) is shown in the inset. Reproduced from ref. 136 with permission from the Royal Society of Chemistry.

When $f_{GCHs} \rightarrow f_c$, not only the σ_{AC} values, but also the ε_r of the binary composite is greatly enhanced.[105] Figure 4.22 illustrates the relationship between ε_r and f_{GCHs} measured at 1 kHz, which is evaluated by the following equation:

$$\varepsilon'(f_{GCHs}) \propto (f_c - f_{GCHs})^{-s} \quad \text{for } f_{GCHs} < f_c \tag{4.14}$$

From the log–log linear fitting values of f_c =5.53 vol% and $s = 0.35$ are deduced (see inset of Figure 4.22).

4.4.3.6 *Improved Overall Dielectric Permittivity of Composites*

The influence on the dielectric property caused by incorporating fillers is evaluated by the relative ε_r enhancement: $\varepsilon/\varepsilon_{PVDF}$. In general, when the $\varepsilon/\varepsilon_{PVDF}$ of ceramic/polymer composites increases to 10 at 1 kHz, the corresponding loading fraction will be as high as >40 vol%. At such a loading amount, the intrinsic flexibility of the host polymer is destroyed. However, as demonstrated in Figure 4.23(a), compared with neat PVDF, a surprising $\varepsilon/\varepsilon_{PVDF}$ value as high as 400 is obtained from the binary composite filled with 8.5 vol% GCHs. Furthermore, the $\varepsilon/\varepsilon_{PVDF}$ value achieved from the CNT/PVDF counterpart with a similar loading fraction fabricated through melt-mixing is only about 8.[134] Therefore, with their highly conducting 'hub' structure, GCHs as a functional filler are not only able to alleviate CNT entanglement and aggregation, but also accelerate the formation of the internal conductive network in the composite.

Moreover, a low loss tangent is necessary for the practical application of dielectric composites. In particular, a loss tangent below 0.1 is usually an acceptable range in the industrial field. However, high loss tangent resulting

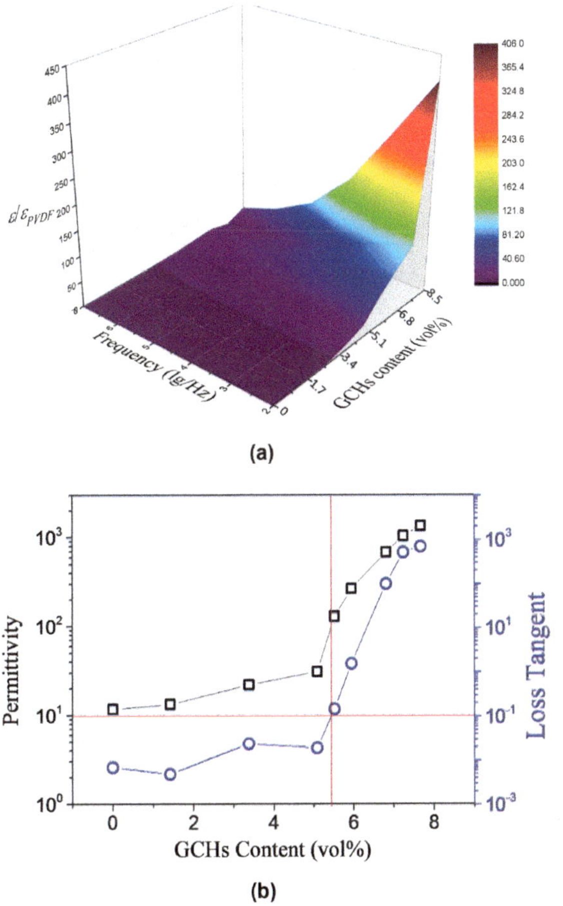

Figure 4.23 (a) Dependence of $\varepsilon/\varepsilon_{PVDF}$ values on GCH volume fractions and frequency. (b) Loss tangent and relative permittivity at 1 kHz of neat PVDF and GCH/PVDF composites.

from enhancement in ε_r becomes an on-going and difficult issue for the practical application of percolative composites. For the purpose of comparing the overall dielectric behaviour of GCH/PVDF composites, Figure 4.23(b) displays the relationship between ε_r, loss tangent and the corresponding GCH loading fractions, where the base loss tangent of 0.1 is shown as a horizontal red line. In addition, the vertical red line in Figure 4.23(b) indicates the relationship between the obtained and corresponding f_{GCHs}. It is observed that the maximum ε_r of the binary composite of about 10^5 can be obtained below the threshold. Such a large improvement in ε_r makes GCH-filled composites promising for relevant high-ε_r applications.

It is noteworthy that GCH/PVDF binary composites exhibit a very low f_c value compared with their ternary counterparts. Moreover, this is only 50% of the value for CNT/PVDF binary composites ($f_c = 10.4$ vol%).[136]

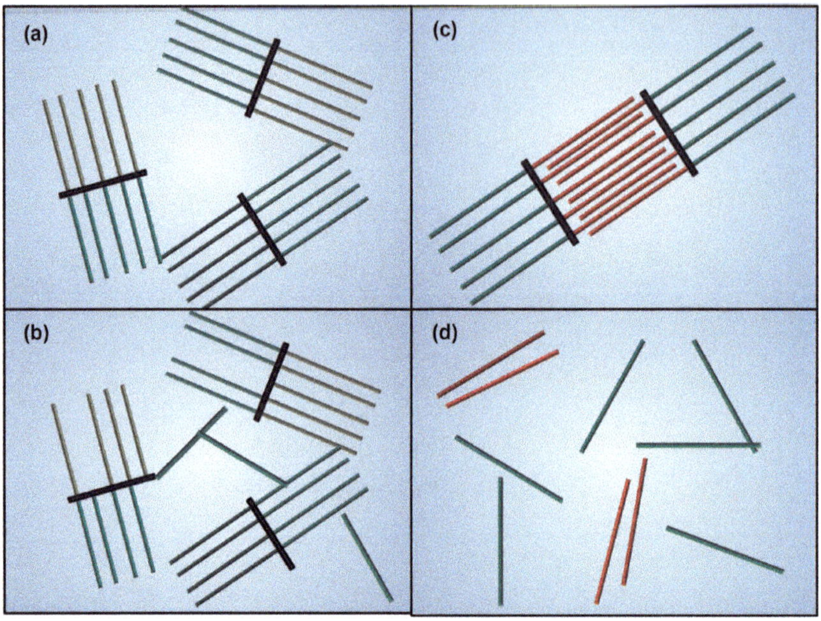

Scheme 4.4 Schematics of four potential electrical interconnections existing in GCH/PVDF composites: the electrical interconnection between (a) GCHs–GCHs and (b) GCHs–CNTs; the microcapacitor construction formed by (c) GCHs–GCHs and (d) CNTs–CNTs. The green sticks, black sticks and pairs of red sticks represent the synthesised CNTs, GNP substrate and microcapacitor construction, respectively. Reproduced from ref. 136 with permission from the Royal Society of Chemistry.

Scheme 4.4 demonstrates various probable types of electrical interaction existing in the GCH/PVDF binary composite. As mentioned, the GCH construction, with its relatively large effective space, works as a 'wire concentrator', making every single CNT 'tubular conductor' electrically connected. When an electrical connection takes place between the single CNT of GCH-A and another single CNT of the neighbouring GCH-B, every CNT of both these GCHs will be electrically connected; see Scheme 4.4(a). Nevertheless, a certain number of CNTs would inevitably be detached from the GNP substrate during the high-power dispersing process. However, it is possible that these detached CNTs might serve as 'bridges' in making an electrical connection between neighbouring GCHs; see Scheme 4.4(b). Therefore, the increased possibility of the occurrence of both electrical tunnelling and ohmic contact is mainly responsible for the low f_c obtained in GCH/PVDF binary composites.

The outstanding dielectric properties achieved can be ascribed to the following features. First, besides optimised dispersion, the melt-blending procedure is able to form a particular molecular-level interaction between the π-electron clouds of the GCHs and the electronegative PVDF macromolecular chains. Compared with the frequently adopted solution-blending approach, melt-blending usually makes composites that exhibit larger f_c values. Nevertheless, the method of melt-blending is still a reliable, easy processing and low cost method for mass production at industrial scale. Second, the ameliorated CNT aggregation makes a large contribution to enlargement of the interfacial area and improvement in the interfacial polarisation between fillers and matrix. Thus, the strengthened Maxwell–Wagner–Sillars (MWS) effect gives rise to enhanced overall permittivity.

In addition, the significant ε_r enhancement occurring near the f_c is ascribed to the typical microcapacitor model. In fact, a series of microcapacitors composed of the filler–matrix–filler (conducting electrode–insulating dielectrics–conducting electrode) can be formed by the inner conductive fillers loaded in the composite. When filler loading is close to f_c, the distance between two conductive fillers (two electrodes) reduces, leading to an increase in the local capacitance of each microcapacitor. Integrating the enhanced local capacitance of a large number of microcapacitors, the overall capacitance increases significantly, giving rise to an effective improvement in the overall permittivity of the composites. Owing to the conducting characteristics of the GNP substrate and the grown CNTs, a complex microcapacitor system can be formed in GCH-loaded composites. This is different to the CNT-filled composites with random CNT orientation, in which mono-directionally aligned CNT arrays of neighbouring GCHs can insert into to each other and form a spatial microcapacitor network; see Scheme 4.4(c). In addition, another part of the local microcapacitor construction is composed of detached CNTs, which are randomly dispersed in the polymer matrix; see Scheme 4.4(d). Both these two microcapacitor models are responsible for the ε_r enhancement occurring in GCH/PVDF binary composites.

Besides the aforementioned strengthened interfacial polarisation and the complex microcapacitor network, the crystalline phase composition in PVDF strongly influences the overall dielectric properties of the composites as well. In general, the α- and β-phase are two important crystalline forms in semi-crystalline PVDF. In fact, the α-phase is a nonpolar crystalline structure with a *TGTG′* (*T-trans*, *G-gauche*+, *G′-gauche-*) dihedral conformation arrangement in a centrosymmetric unit cell. However, the total *TTTT* conformation arrangement gives rise to a polar non-centrosymmetric β-phase crystalline PVDF. Owing to its polar non-centrosymmetric structure, the β-phase possesses an intense dipolar movement and displays an excellent dielectric property, in comparison with the other main crystalline phases of PVDF.[137–140]

The XRD patterns of binary and ternary composites with various filler loadings are summarised in Figure 4.24. As previously mentioned, the characteristic peak at 20.26° in the XRD pattern represents the β-phase of PVDF. It is worth noting that this characteristic peak can be observed for each GCH/PVDF composite with different f_{GCHs}, but is absent from any GNP/CNT/PVDF counterpart.

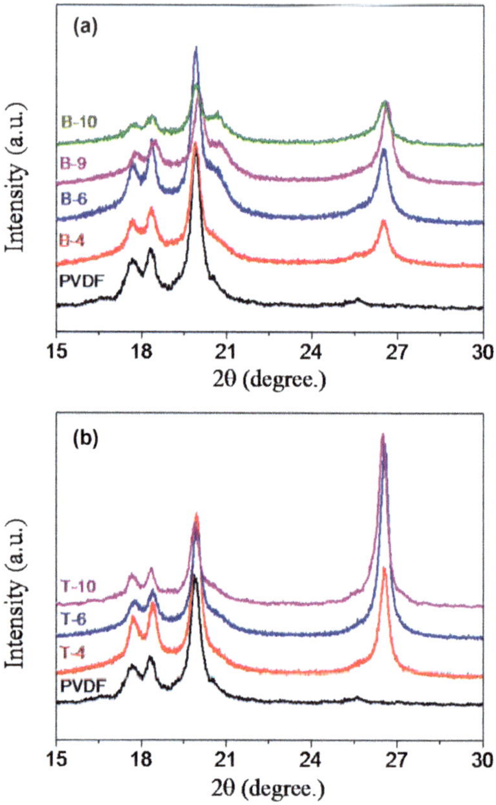

Figure 4.24 XRD patterns of (a) neat PVDF and binary composites: B-4, B-6, B-9 and B-10; (b) neat PVDF polymer and ternary composites: T-4, T-6 and T-10.
Reproduced from ref. 136 with permission from the Royal Society of Chemistry.

Fourier-transform infrared (FTIR) spectroscopy was employed to further verify the crystalline structures and phase composition in neat PVDF polymer and the relevant composites. Figure 4.25(a) shows the obtained FTIR spectra, in which the annotated transmittance bands 766 cm^{-1} and 840 cm^{-1} represent the α- and β-phase of PVDF, respectively. Theoretically, the relative fraction of β-phase PVDF ($F(\beta)$) is calculated by:[139]

$$F(\beta) = \frac{A_\beta}{(K_\beta/K_\alpha)A_\alpha + A_\beta} \tag{4.15}$$

where A_α and A_β are the area of the characteristic peaks at 766 cm^{-1} and 840 cm^{-1}, respectively, and K_α and K_β are the absorption coefficients at the respective wavenumber, with values of 6.1×10^4 and $7.7\times10^4\, cm^2\, mol^{-1}$, respectively.

The corresponding $F(\beta)$ of neat PVDF, and typical binary and ternary composites were calculated according to eqn (4.15) and are shown in Figure 4.25(b). It is observed that the calculated $F(\beta)$ values of binary composites B4, B6 and B10 are greater than those of the ternary counterparts T4, T6 and T10, respectively. It is noteworthy that, compared with neat PVDF, the $F(\beta)$ value of the B-10 composite are enhanced by as much as 150%. This unexpected result is mainly ascribed to the stretching deformation being generally able to align PVDF macromolecular chains into crystallised a construction, and promote PVDF crystalline transformation from the *α-phase to the* β-phase. In fact, a strong torque is added to GCHs during the molten injection process, which prompts GCHs to re-orientate along the injecting direction. Compared with the CNT-entanglement and GNP-stacked construction in ternary composites, re-orientated GCHs in binary

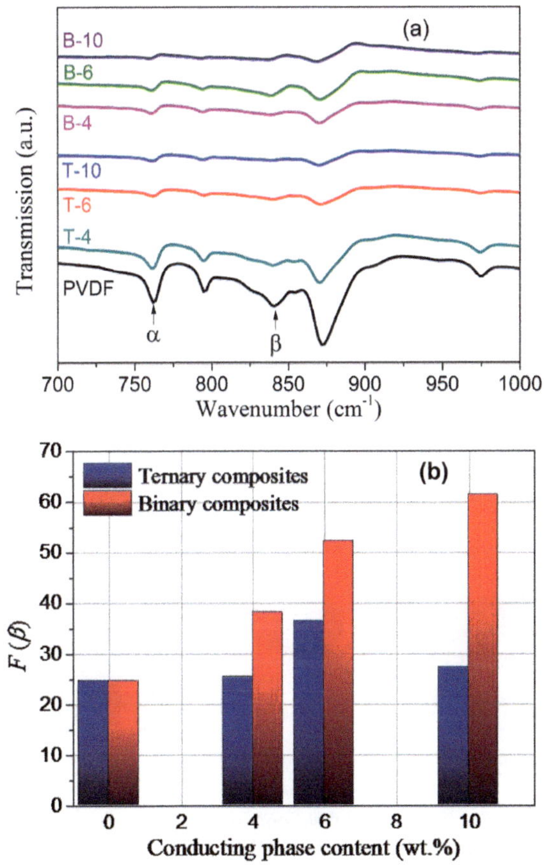

Figure 4.25 (a) FTIR spectra and (b) proportion of β-phase of neat PVDF, binary and ternary composites with different content of fillers, respectively. Reproduced from ref. 136 with permission from the Royal Society of Chemistry.

composites make an important contribution to the alignment of PVDF macromolecular chains into a crystallised construction along the aligning CNT arrays. This is responsible for not only the formation of β-phase PVDF, but also the enhancement of the dielectric property in the binary composite.

4.4.3.7 Temperature-dependent Dielectric Properties

In order to assess the thermal stability of various composites, the temperature-dependent dielectric behaviours are investigated. Figure 4.26(a, b) presents the relationship between the permittivity of the samples (B-6.5 and T-6.5), the testing temperature (−40 °C to ∼150 °C) and testing frequency (1 kHz to 10 MHz). It is observed that the ε_r values of the binary composite are much greater than that of the ternary counterpart. Under the testing temperature range, the ε_r values of the composites are negatively related to the testing frequency, which is similar to the condition at room temperature.

Moreover, the ε_r values of the composites are enhanced with increasing ambient temperature. This temperature-dependent behaviour can be explained by the fact that the semi-crystallised PVDF displays its inherent T_g and

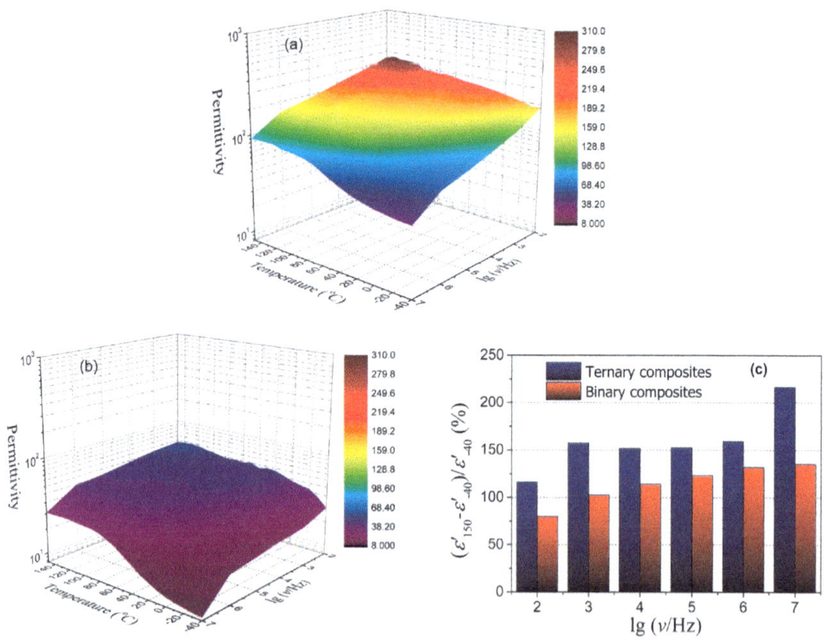

Figure 4.26 3D surface plot of relative permittivity as a function of frequency and temperature for (a) GCH/PVDF binary composite B-6.5 and (b) GNP/CNT/PVDF ternary composite T-6.5. (c) Comparison of the variation percentages of permittivity of the binary and ternary composites with frequency at two testing boundary temperatures: −40 °C and 150 °C. Reproduced from ref. 136 with permission from the Royal Society of Chemistry.

T_m at about −20 °C and ∼160 °C, respectively. At low temperature, the movement of the PVDF macromolecular chains is limited by the crystalline construction, and it is difficult to make timely responses to the applied electrical field. However, with increasing temperature, the movement of PVDF macromolecular chains, particularly the chains in the crystalline region, becomes gradually more significant.[141–143] Furthermore, the ever-increasing polarisation leads to enhancement of the permittivity at a high temperature range.

As shown in Figure 4.26(c), in order to quantitatively evaluate the thermal stability of the binary and ternary composites, the relative ε_r variation $((\varepsilon'_{150} - \varepsilon'_{-40})/\varepsilon'_{-40})$ was calculated *via* the ε_r values measured at −40 °C (ε'_{-40}) and 150 °C (ε'_{150}). Moreover, the binary composites show a more significant frequency-independent behaviour than the ternary composites, throughout the whole testing frequency range. The better thermal stability obtained in GCH/PVDF binary composites is mainly ascribed to the strengthened MWS effect, which is derived from a strong GCH–PVDF interfacial interaction.[132,136]

4.4.4 Section Summary

In conclusion, GCH/PVDF composites with enhanced dielectric properties were fabricated through melt-blending and injection moulding. The high-power melt-mixing procedure endows GCHs with a homogeneous dispersion. The typical molecular-levelled fillers–matrix interaction can also be obtained through this pollution-free mechanical blending approach. The electrical behaviour of the GCH/PVDF binary composites changed from the semiconducting to conducting state with increasing GCH loading. Compared with ternary composites, a low f_c (5.53 vol%) in the binary composites was achieved based on classical percolation theory. The enhanced dielectric properties obtained in the binary composites are mainly ascribed to both the internal microcapacitor network composed by GCHs, and the GCH-induced crystal phase transition in PVDF. Furthermore, compared with loading GNPs and CNTs as co-fillers, filling with GCHs was capable of endowing PVDF-based composites with better thermal stability.

4.5 Flexible GCH/PDMS Piezo-resistive Composites

4.5.1 Introduction

The piezo-resistive (PR) effect is a classical electrical function property of conducting composites. PR material is usually deformed under an external pressure or field, where the deformation and applied external stimuli can be detected and exhibited as observable electrical resistance variation.[144–147] Due to their ability to quantitatively evaluate the applied external load, PR materials have frequently been investigated and show a wide range of applications, such as in health monitoring, artificial skin, strain/stress sensors and wearable electronics.[83,148–151] Typically, the repeatable PR behaviour is usually achieved with a conducing filler fraction beyond the f_c, where the internal conductive network formed can be partly damaged by external stimuli and then recover to its original construction when the loading is released.

Multiposition recognition, high sensitivity and multiconditional response are the main areas of research in the investigation of PR materials, and in which improving the PR sensitivity of the composite is always an important issue. It is implied that even a tiny stimulus can be detected by a significant resistance change. To effectively enhance the PR sensitivity of composites, retaining the flexibility of the composite and construction of an internal conductive network are the two main strategies. In fact, many studies have concentrated on retaining the overall flexibility through reducing the f_c of the composites.[152] The construction of the internal conductive network is determined by the morphological features, loading concentration and dispersion state of the incorporated conducting units. Several studies have achieved excellent PR responses in zero-dimensional (0D) carbon-based or metallic-based conducting units loaded into composites.[148,153] However, the construction of the conducting network in these systems usually requires a large loading content, which is fatal for retaining the flexibility of the composites. In recent decades, with its outstanding inherent features, 2D-structured graphene has frequently been utilised as a competitive conductive unit in polymer-based composites.[154,155] However, as previously mentioned, the large surface energy needed and complicated exfoliating and dispersing processes make use of mono-layered graphene in composites difficult on a massive scale. A high PR sensitivity and a low f_c were obtained in our previous study, where an alkyl-functionalised graphene/silicone rubber composite was fabricated.[156] However, mass production is limited by the large consumption of chemical solvents in this strategy. Compared with graphene, 2D-structured GNPs have been shown to endow different composites with outstanding functional properties in recent research.[157–159] In addition to 0D and 2D carbon-based structures, 1D structured CNTs have also been used for improving the electrical conductivity of composites. However, it is found that CNTs with high aspect ratio usually make opposite contributions to the f_c reduction and enhancement of PR sensitivity. First, the f_c value of a composite can be reduced by using CNTs with higher aspect ratio, which has been reported in many studies. Moreover, it is facile for CNTs with a low aspect ratio to orientate along the deformation direction, which gives rise to the corresponding composite exhibiting a high PR sensitivity to external stimuli.

To overcome these problems, as-synthesised GCHs with distinct morphological construction and extraordinary inherent features are employed as conductive units in constructing the internal conducting network. These conducting 'micro-hub' structures are able to accelerate the interconnection between them, which benefits the formation of the conducting network.

Owing to its intrinsic merits of outstanding thermal stability, elasticity, biocompatibility, ease of processing and low cost, poly(dimethylsilicone) (PDMS) is widely used as the host elastomer for PR composites in various applications.[160]

In this section, the influence of GCHs incorporation on the PR performance of PDMS-based composites is investigated in detail. The relevant physical mechanism is also discussed in detail. Compared with PDMS-based composites loaded with an equivalent concentration of CNTs or GNPs, the addition of GCHs is capable of endowing the PDMS composite with an ultra-low f_c and a highly sensitive positive PR behaviour. The obtained temperature-dependent PR behaviour offers GCH/PDMS composites with the possibility of being applied as a temperature sensor. Furthermore, the experiment of finger-motion recognition indicates that the GCH/PDMS composite with high PR sensitivity could potentially be applied as an advanced next-generation artificial skin.

4.5.2 The Preparation of PDMS-based Piezo-resistive Composites

The GCHs were synthesised through the CVD method discussed in Section 4.2. A hydroxyl-terminated PDMS (50 000 mPa s, Sigma-Aldrich) was selected as the host polymer. The process of dispersing GCHs with various loading fractions into PDMS was undertaken in a three-roll mill (EXAKT 80, Germany) at room temperature. The moderate mechanical blending provide GCHs with homogeneous dispersion in PDMS and strong interfacial adhesion with PDMS. The cured composites were cut into samples of a certain shape and size, in preparation for subsequent characterisation. The general procedure for preparing the GCH/PDMS composites is illustrated in Scheme 4.5.

4.5.3 Structural Characteristics and Functional Properties of Carbon Nanomaterial-loaded Composites

4.5.3.1 Morphological Characterisation

Figure 4.27(a) presents the SEM image of a cross-section of a GCH/PDMS composite, where the GCHs are observed to be homogeneously dispersed in

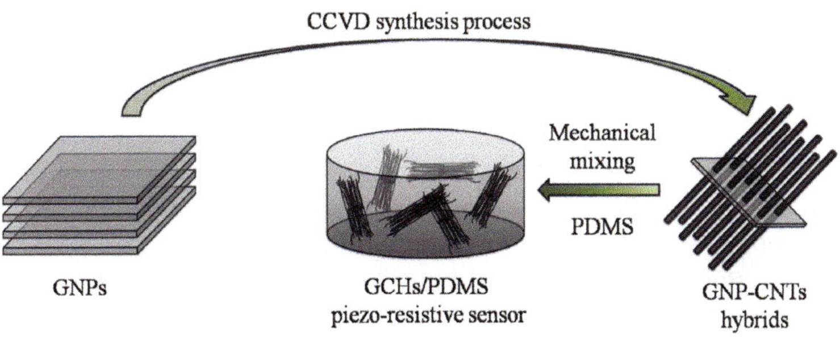

Scheme 4.5 Schematic of the GCH synthesis and GCH/PDMS composites fabrication process.
Reproduced from ref. 119 with permission from American Chemical Society, Copyright 2015.

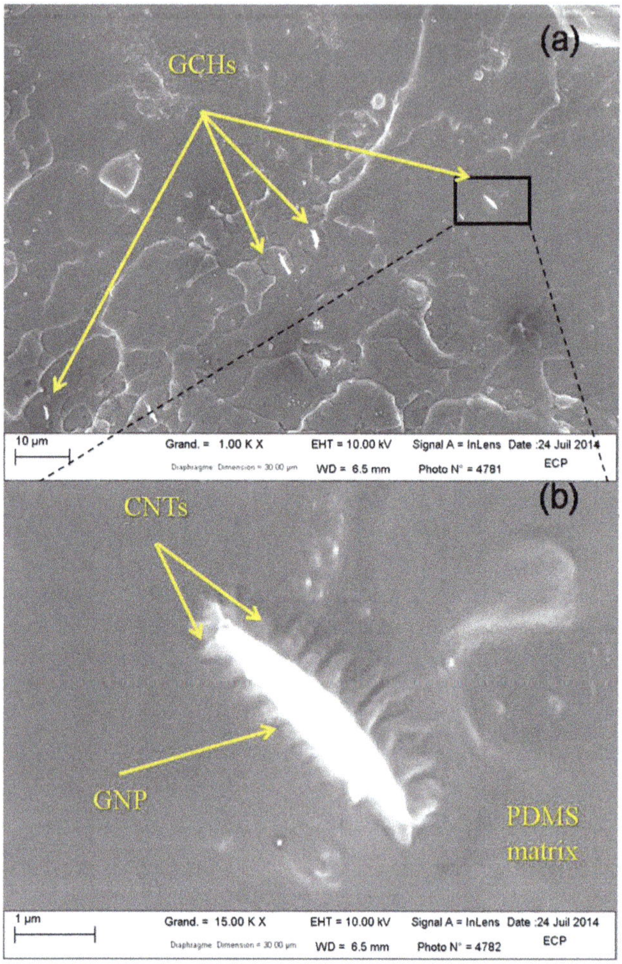

Figure 4.27 (a) The SEM morphology of the fractured cross-sections of the GCH/PDMS composite. (b) High-magnification image showing good compatibility of GCH particles and the matrix.
Reproduced from ref. 119 with permission from American Chemical Society, Copyright 2015.

the PDMS matrix without obvious aggregation. The three-dimensional (3D) coupled construction is able to significantly expand the effective contact area with the polymer matrix, which gives rise to an enlarged interface and enhanced interfacial interaction. Moreover, the insulating PDMS matrix can insert into the space between individual CNTs in the synthesised CNT array. As shown in Figure 4.27(b), the typical conducting 'hub' units are well embedded in the PDMS matrix without observable damage. Thus, the moderate mechanical mixing procedure has potential to be scaled up for industrial production of GCH/PDMS composites.

4.5.3.2 Dielectric Behaviour

In order to identify the f_c value and find the appropriate f_{GCHs} range corresponding to high PR sensitivity of the system, the dielectric behaviour of the GCH/PDMS composites was measured. As shown in Figure 4.28, the ε_r values of the composites are positively related to f_{GCHs}. It is noteworthy that compared with neat PDMS (ε_r ~3.3 at 10^3 Hz), adding as little as 1.5 vol% GCHs makes the overall ε_r value of the composite increase by 700 times. When $f_{GCHs} > f_c$, a significant frequency-dependent ε_r variation is observed, which can be explained by the newly formed internal conductive networks being sensitive to the applied alternating electric field. Moreover, the polarisation in the composite is not able to catch up with the alternation of the applied electric field in time.

Since the occurrence of dielectric percolation is a direct indication of the formation of the conductive network, the electrical properties of GCH/PDMS composites were evaluated. The f_{GCHs}-dependent variation in σ_{AC} values of the GCH/PDMS composites is illustrated in Figure 4.29, in which a significant electrical transition from an insulating state (σ_{AC}~10^{-9} S m^{-1}) to a conductive state (σ_{AC} ~10^{-2} S m^{-1}) with increasing f_{GCHs} is observed. Moreover, when f_{GCHs} increased from 0.5 to 0.9 vol%, a sharp increase in σ_{AC} beyond four orders of magnitude was obtained in the GCH/PDMS composites. This may correspond to the on-going construction of the internal conductive network.

The f_c of the GCH/PDMS composite was obtained by theoretical fitting according to the power law of eqn (4.12). As presented in the inset in Figure 4.29, based on experimental σ_{AC} values, the optimal fitting result offers an ultra-low $f_c = 0.64$ vol%. The critical exponent $t = 1.76$ is in accordance with the universal values ($t \approx 1.6$–2) for the 3D-conducting unit-

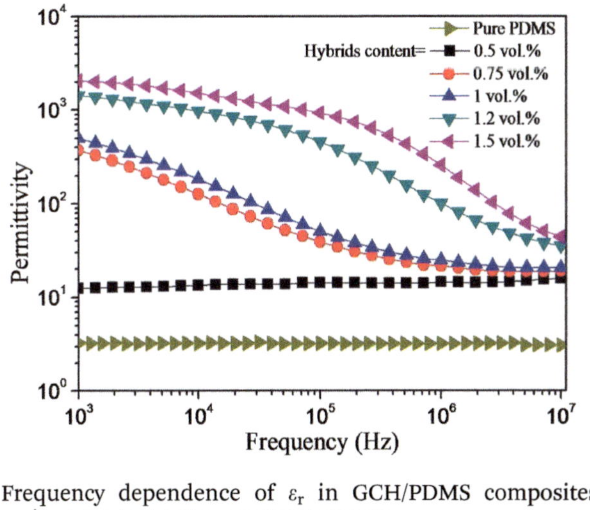

Figure 4.28 Frequency dependence of ε_r in GCH/PDMS composites filled with various contents at room temperature.

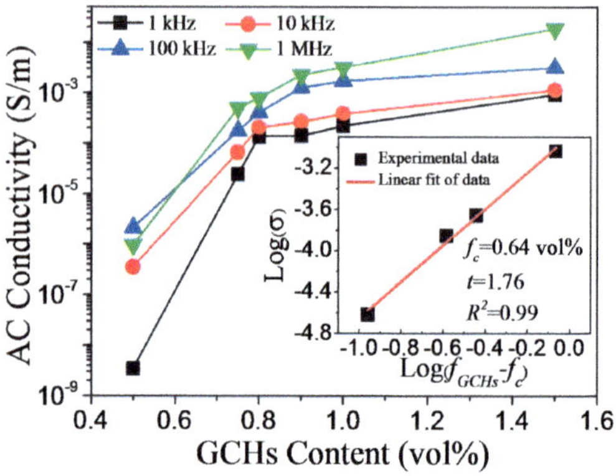

Figure 4.29 Dependence of AC conductivity on GCH content of the GCH/PDMS composites with different test frequency. The inset presents the best fit of linear fit of log(σ) *vs.* log($f_{GCHs} - f_c$).
Reproduced from ref. 119 with permission from American Chemical Society, Copyright 2015.

filled composites. Notably, the obtained f_c of the GCH/PDMS composite is extremely low in comparison with the majority of other carbon nanomaterial/PDMS composites, to our knowledge. This is mainly attributed to the conducting 3D 'hub' structure, which leads to the efficient construction of the internal conductive network.

4.5.3.3 *Piezo-resistive Property*

Real-time piezo-resistive behaviour was evaluated through a universal testing machine (Instron 5544) and a source meter (Keithley 2400). Repeated loading–unloading compression stress (0 to 0.32 MPa) was applied to surface-metallised composite samples, where the volume electrical resistance was measured in real time and recorded in the system. In order to investigate accurately the influence of f_{GCHs} on the PR property of the composites, more than three parallel samples for each composition were characterised. Figure 4.30 illustrates the relationship between the resistance variation (R/R_0) and the applied pressure. All the GCH/PDMS composites exhibit a typical positive PR behaviour. It is noteworthy that when f_{GCHs} is greater than f_c, more significant PR responses are usually obtained from composites filled with f_{GCHs} that is close to f_c. However, the composites filled with a further increase in f_{GCHs} display a relatively small resistance variation under an external load. In fact, when the internal conducting network is just constructed, namely $f_{GCHs} \sim f_c$, the tunnelling effect is mainly responsible for the electrical interconnection. However, when $f_{GCHs} > f_c$, a stable internal conducting network has been completely constructed, which is not prone to influence from external stimuli. In this stage, the ohmic contact between neighbouring GCHs is the dominant electrical interconnection.

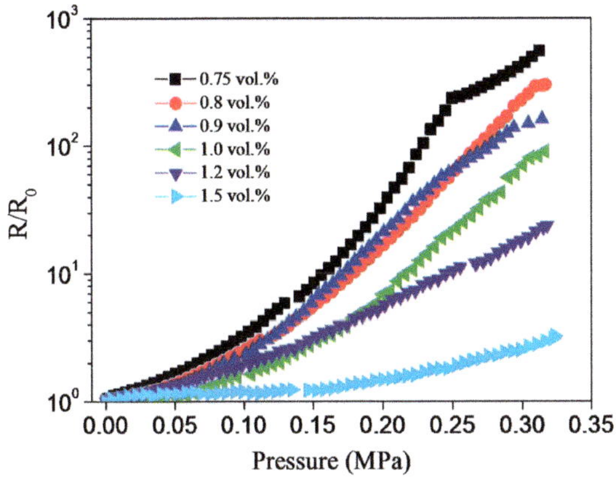

Figure 4.30 The typical PR curves of GCH/PDMS composites with different GCH content under uniaxial compression.
Reproduced from ref. 119 with permission from American Chemical Society, Copyright 2015.

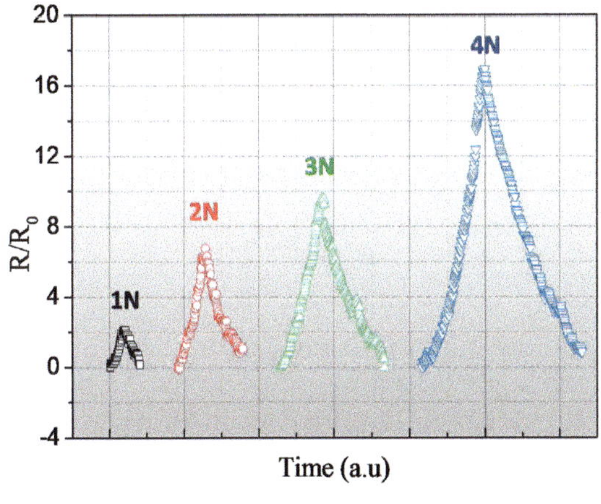

Figure 4.31 At a very low charge, the relative resistance variation of 0.8 vol% GCH/PDMS composites shows excellent piezo-resistive sensitivity.

As previously mentioned, sensitivity is a basic characteristic to evaluate the piezo-resistive behaviour of composites. As shown in Figure 4.31, a high pressure sensitivity of 0.16 kPa^{-1} was obtained from the composite filled with 0.8 vol% GCHs under a compression load as small as 1 N. Moreover, resistance variation can be recognised and detected not only for a tiny loading, but also for a slight difference in loading. Thus, both high pressure sensitivity and quantitative precise loading recognition are able to further promote the practical applications of GCH/PDMS composites in flexible sensing.

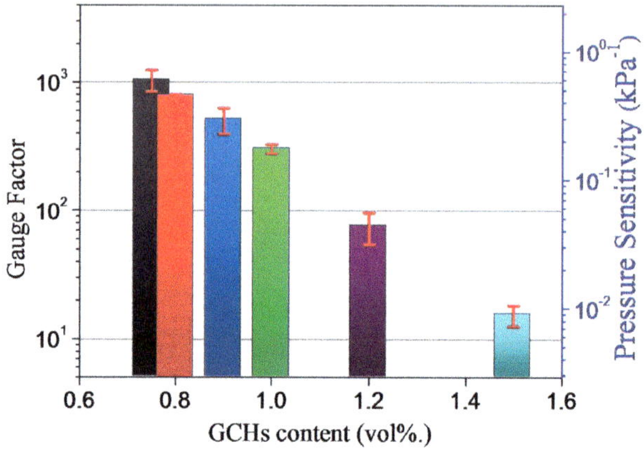

Figure 4.32 Piezo-resistive strain-sensitivity (GF) and pressure-sensitivity (PS) values of GCH/PDMS composites with different GCH fractions. Reproduced from ref. 119 with permission from American Chemical Society, Copyright 2015.

In general, the strain sensitivity and pressure sensitivity of a PR material are assessed quantitatively using two important parameters: gauge factor (GF) and pressure sensitivity (PS), respectively.[158]

$$GF = \frac{\Delta R/R_0}{\Delta l/l_0} = \frac{\Delta R/R_0}{\varepsilon} \tag{4.16}$$

$$PS = \frac{\Delta R/R_0}{p} \tag{4.17}$$

where R_0 and l_0 are the initial resistance and the original thickness of a PR material before the applied loading, respectively. ΔR and Δl are the resistance and thickness variation, respectively, under a certain applied pressure (p) or strain (ε). Figure 4.32 demonstrates the relationship between PR sensitivities (GF and PS values) and fGCHs.

Similarly to the variation in trend of σ_{AC}, both GF and PS values exhibit a negative dependency on increasing f_{GCHs}. Notably, excellent PR sensitivities (GF ~10^3, PS ~0.6 kPa^{-1}) are obtained from the composite filled with 0.75 vol% GCHs under an applied pressure of 0.32 Mpa, which is sufficiently competitive in comparison with other carbon nanomaterial/PDMS counterparts described in recent reports; see Table 4.3.

As illustrated in Scheme 4.6, the obtained high PR sensitivity in GCH/PDMS composites could originate from several sources. First, the typical conducting 'hub' structure is able to accelerate the electrical interconnection between the hubs and improve the efficiency of constructing the internal conducting network. Then, the achieved ultra-low f_c is capable of keeping the composites as flexible as possible. The low mechanical modulus means that the composites exhibit a significant strain even under a very small stress.

Table 4.3 Comparison of the PR sensitivity with other reported carbon material/PDMS composites under similar applied pressures. Reproduced from ref. 119 with permission from American Chemical Society, Copyright 2015.

Type (specific)	f_c (vol%)	PS[a] (kPa^{-1})	GF	Conditions	Ref.
GNP–PDMS	0.9	0.067		0.3 MPa	83
CB–PDMS	—	– 0.001		0.3 MPa	161
CB–PDMS		<0.001		0.3 MPa	162
CNT–PDMS	—		1.25	Strain ~0.4	163
CNT film–PDMS	—	~0.33		3 kPa	15
Carboxyl-MWCNT–PDMS	—	0.013		3 MPa	164
CNT–PDMS	—		~2–3	Strain ~100%	165
P3HT–CNT–PDMS	—	0.0017		0.12 MPa	166
CNT–CB–PDMS	—				167
(AR of CNTs = 117)				0.3 MPa	
(AR of CNTs = 437)		0.01		0.3 MPa	
		0.01			
Alkyl-graphene–PDMS	0.63	<0.1		1 MPa	156
GCHs–PDMS	0.56	0.6	~1000	0.32 MPa	

[a]The pressure or strain sensitivities were calculated and estimated from the curves in the relevant references.

Considering its extraordinary properties, the composite filled with 0.8 vol% GCH was selected to further investigate the piezo-resistive mechanism and to discuss the application potential as a flexible piezo-resistive sensor; see Figure 4.33(a). According to the various increasing rates, the pressure-dependent resistance variation is divided into three stages. These stages for PR behaviour also correspond to the different stages of development during the construction of the internal conducting network, which are shown schematically Figure 4.33(b). As illustrated in the insets in Figure 4.33(a), the PR behaviour of each stage was theoretically fitted based on the modified formulae for tunnelling theory.

The fitting models employed were derived from the electrical resistance model, which is as follows:[83]

$$\frac{R}{R_0} = \left(\frac{Ns}{N_0 s_0}\right) \exp\left[\frac{4\pi}{h}\sqrt{2m\varphi}(s - s_0)\right] \tag{4.18}$$

$$s = s_0\left[1 + \left(\frac{|l - l_0|}{l_0}\right)\right] \tag{4.19}$$

$$p = E\left[\left(\frac{|l - l_0|}{l_0}\right)\right] \tag{4.20}$$

where R, R_0, N, N_0, s, and s_0 represent the total resistance of the composite, the number of conductive paths and the minimum distance between conductive fillers under an applied pressure (p) and initial state ($p = 0$ MPa), respectively. l and l_0 are the practical thickness of the sample under an applied pressure (p) and the original thickness, respectively. h, m and φ are

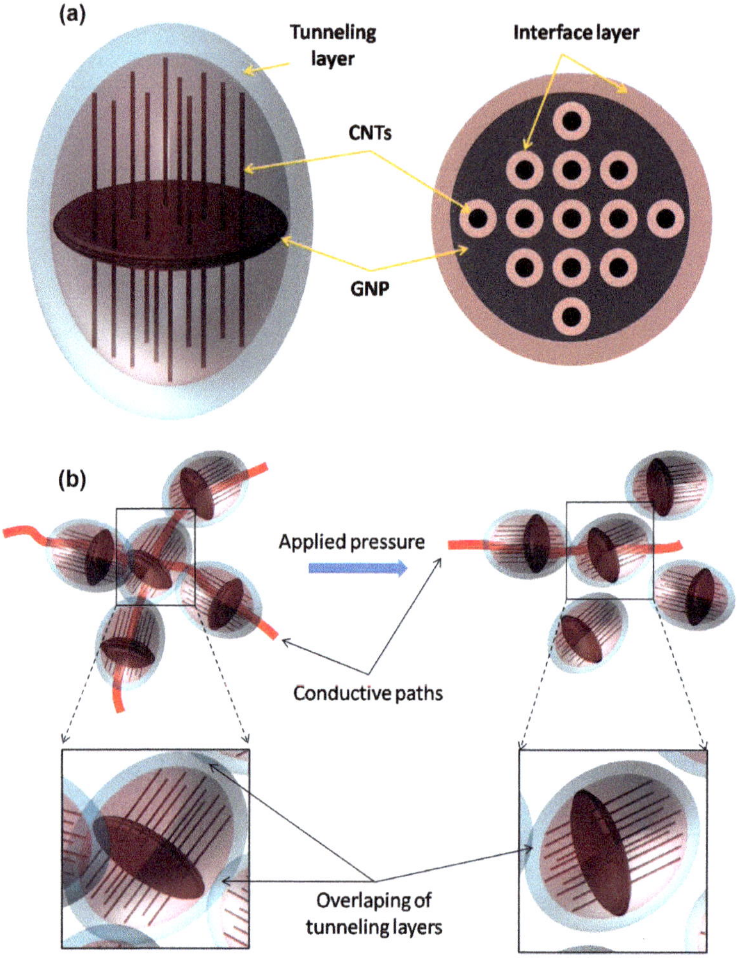

Scheme 4.6 Schematic of (a) enlarged interfacial area and enhanced effective conductive space in the GCH construction. (b) The 'hub' structure is able to build the interconnections with neighbouring GCHs *via* direct contact or the tunnelling effect. The formed interconnection can be easily influenced by applied pressure, resulting in a transformation from the conducting 'hub' to an 'isolated island'.

Plank's constant, electron mass and potential barrier height between neighbouring fillers, respectively. E is the elastic modulus of the composite.

In the first stage, the internal conducting network is constructed through the tunnelling interconnection between incorporated fillers, where a slight electrical resistance variation can be detected over a very small applied pressure range (0–0.04 Mpa). According to the conducting mechanism, this first stage is defined as the tunnelling state. Since the applied pressure is too

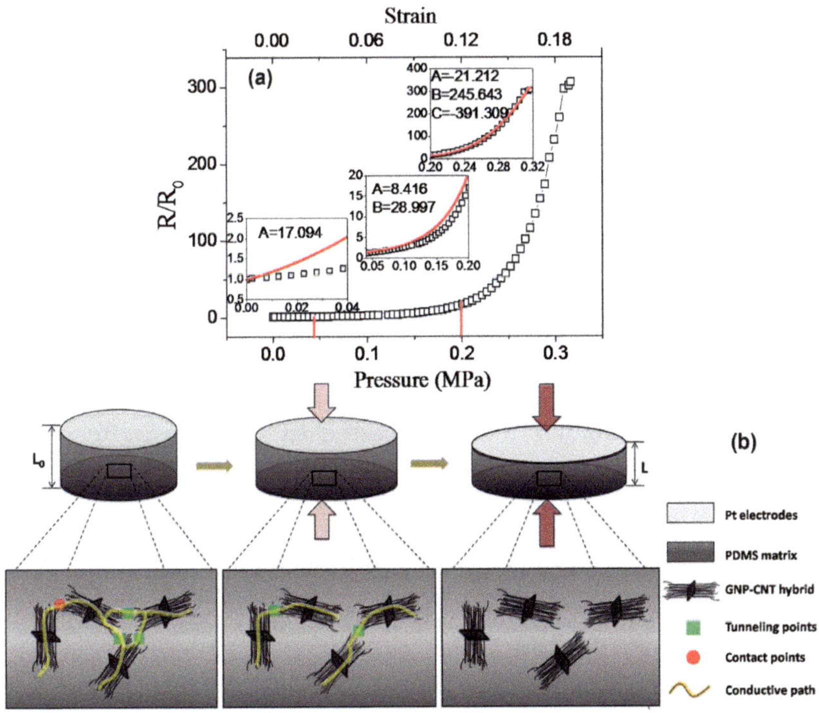

Figure 4.33 (a) The resistance variation under applied loading of GCH/PDMS composites filled with 0.8 vol% GCHs. The inset images display the three-stage theoretical fitting based on experimental data. (b) Schematic illustration of the relationship between the internal conductive network construction and the external applied pressure in the three stages.
Reproduced from ref. 119 with permission from American Chemical Society, Copyright 2015.

small to make GCHs re-orientate along the strain direction, the correlation of the PR behaviour of the composite in the tunnelling state is as follows:

$$N = \frac{N_0}{\exp[A_1(|l - l_0|/l_0)]} \tag{4.21}$$

$$R/R_0 = (1 + p)\exp(A \times p) \tag{4.22}$$

where A_1 is a constant, $A = \left[\frac{4\pi}{h}\sqrt{2m\varphi s} + A_1\right]/E$.

The second stage is the transitional state, in which the applied pressure grows to 0.04–0.2 MPa. At the transitional state, GCH is able to move around and orientate along the direction of deformation of the flexible matrix. With increasing deformation, some original electrical interconnections may be loosened and other parts might be damaged. The loose interconnections still maintain electrical conduction between neighbouring conducting units *via* the tunnelling effect. Nevertheless, the damaged interconnections result in certain conducting paths being broken and several discontinuous points exist in the conducting network. Thus, a moderate resistance variation of the composite can be detected. The relevant pressure-dependent R/R_0 variation is fitted by equations as follows:[83]

$$N = \frac{N_0}{\exp\left[A_2(|l - l_0|/l_0) + B(|l - l_0|/l_0)^2\right]} \tag{4.23}$$

$$R/R_0 = (1 + p)\exp(A\times p + B\times p^2) \tag{4.24}$$

where A_2 and B are constants in fitting, $A = \left[\frac{4\pi}{h}\sqrt{2m\varphi s} + A_2\right]/E$.

With further enhancing the applied p, most GCHs would re-orientate along the deformation. Moreover, the overall extent of orientation reaches a high level, where the minimum distance between GCHs is beyond the extreme tunnelling distance. GCHs are kept apart from each other by the insulating PDMS matrix. Thus, the third stage is defined as the discontinuous state. In fact, the initial conducting network is constructed through interconnection between GCHs. The pressure-derived re-orientation is able to induce whole-network reconstruction of even destruction to some extent. At the discontinuous state, the initial active GCH hubs have become a number of 'isolated island' distributed in the PDMS matrix. In consequence, the GCH/PDMS composites are extremely sensitive to external stimuli. The modified functions employed for fitting the PR behaviour in this stage are given by:

$$N = \frac{N_0}{\exp\left[A_3(|l - l_0|/l_0) + B(|l - l_0|/l_0)^2 + C(|l - l_0|/l_0)^3\right]} \tag{4.25}$$

$$R/R_0 = (1 + p)\exp(A\times p + B\times p^2 + C\times p^3) \tag{4.26}$$

where A_3, B and C are constants, $A = \left[\frac{4\pi}{h}\sqrt{2m\varphi s} + A_3\right]/E$.

The different electrical resistance variation of the tunnelling, transitional and discontinuous states were carefully evaluated through corresponding exponential fittings based on eqn (4.22), (4.24) and (4.26), respectively. The relevant fitting parameters are summarised in Table 4.4.

In addition to high PR sensitivity, the quick response of resistance variation with changing external loading and stable resistance variation in cyclic loading–unloading operation are also determining characteristics for evaluating the application possibilities of PR composites. A typical PR test with multicycle

Table 4.4 The obtained fitting coefficients and R^2 for the three stages of pressure-dependent resistance variation. Reproduced from ref. 119 with permission from American Chemical Society, Copyright 2015.

	Initial state	Transitional state	Discontinuous state
A	17.094	8.416	−21.212
B	—	28.997	245.643
C	—	—	−391.309
R^2	0.966	0.99	0.998

loading–unloading was carried out using the composite sample filled with 0.8 vol% GCHs, in order to evaluate the mechanical–electrical synchrony in real time. As demonstrated in Figure 4.34(a), there is little delay and interference in the real-time mechanical–electrical feedback. Notably, it can be seen that the PR behaviour in the first cycle is different from that in subsequent cycles, which could be explained by the Mullins effect.[168] From the second cycle, the obtained resistance variation in each cycle is stable and significant ($R/R_0 > 90$). Figure 4.34(b) exhibits the PR response stability of the composite over 100 loading–unloading cycles, with a maximum applied p of 0.32 MPa. Although there was a slight drop in maximum R/R_0 and a small rise in unloading resistance as the cyclic test progressed, the resistance variation still remained well beyond 70. This result is ascribed to the continuous multicycled loading–unloading operation being capable of gradually making having a permanent impact on the construction of the conducting network. Typically, this impact on the GCH-constructed conducing network may be determined by following three

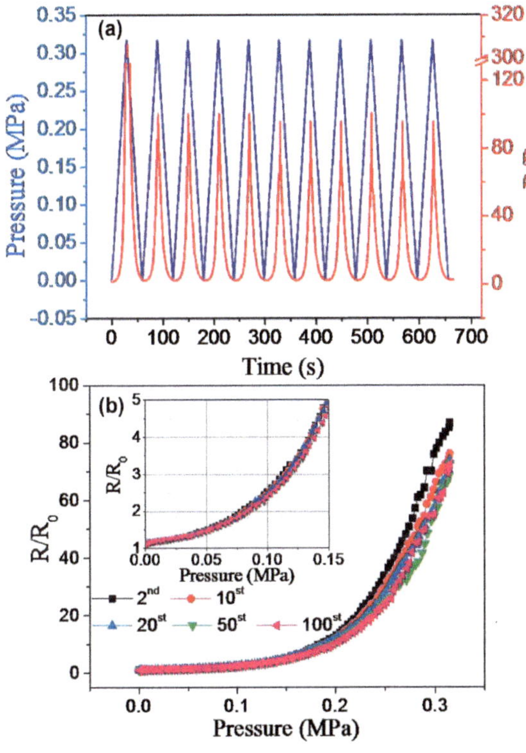

Figure 4.34 (a) The relative resistance variation (R/R_0) and external applied pressure *vs.* time under a cyclic loading–unloading operation. (b) The real-time R/R_0 under 100 cycles. The inset exhibits the corresponding PR response in 100 loading–unloading cycles under a small pressure ranging from 0 to 0.15 MPa.
Reproduced from ref. 119 with permission from American Chemical Society, Copyright 2015.

aspects: (1) the coordination mismatch between the breakdown–reconstruction of the internal conductive network and the loading–unloading of the external applied pressure; (2) the difficulty in attaining a steady state for the GCH–PDMS interfacial interaction; (3) the original GCH construction may be damaged by interfacial friction during the continuous orientation process. Moreover, the small rise in unloading resistance as cycles proceed implies transformation of the conducting networks as well. It is noteworthy that an extraordinary consistency was achieved in the mechanical–electrical response of the composites, even under tiny pressures (<0.1 MPa) (see inset of Figure 4.34(b)).

In addition, to sense and to recognise temperature variation is an important functionality of human skin, which prevents injury in an ambient environment.[144] Thus, electrical response to temperature plays an important role in the design and fabrication of next-generation advanced flexible multicondition sensors. As mentioned, due to the fact that a thermal expansion effect occurring in the polymer matrix can influence the internal conducting network of the composite, a number of conducting filler/polymer composites display an overall resistance variation with changing ambient temperature, namely the PTC or NTC effect. However, compared with other polymers, PDMS, with its cross-linked structure, shows inherently high thermal stability. Figure 4.35

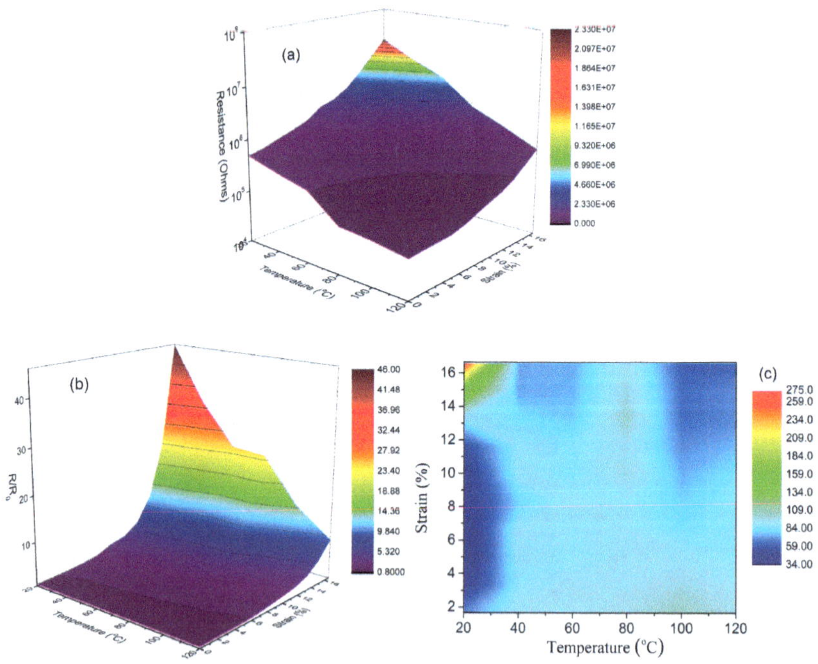

Figure 4.35 Temperature-dependent PR response of GCH/PDMS composites. (a) Resistance variation. (b) Relative resistance enhancement as a function of temperature and applied strain. (c) The gauge factor distribution in the temperature–strain matrix.

presents the temperature-affected PR response of GCH/PDMS composites ($f_{GCHs} \sim 0.8$ vol%) under a wide temperature range (20 °C to ~120 °C).

As show in Figure 4.35(a), similarly to the condition at room temperature, an ordinary positive PR response was obtained throughout all of the measured temperatures. The electrical resistance of the composites reduced remarkably at high testing temperatures, which is attributed to the charge movement being accelerated by the thermal agitation effect. In fact, the charge movement can be further intensified due to the thermally excited free volume expansion of PDMS macromolecular chains. Both these reasons are responsible for the reconstruction of the internal conducting network.[143]

Figure 4.35(b) displays the dependency between testing temperature, R/R_0 and applied strain. Under a certain strain, the relative resistance variation R/R_0 at high temperature range is obviously larger than that at low temperature range. For instance, the maximum R/R_0 obtained at 20 °C is about 45, which is 4.5 times larger than that achieved at 120 °C ($R/R_0 \sim 10$). In fact, owing to the previously mentioned NTC effect in GCH/PDMS composites, the internal conducting network is more stable and less sensitive to external stimuli at high temperature than at low temperature, which is responsible for the observed significant reduction in R/R_0 values.

Gauge factor is a critical parameter to evaluate the strain sensitivity, and its value is influenced by multiple factors, such as intrinsic electrical characteristics, morphological aspect ratio and loading fraction of the conductive fillers, mechanical flexibility of the polymer matrix and the filler–matrix interfacial adhesion, *etc.* Figure 4.35(c) presents a GF value distribution of GCH/PDMS composites in a typical matrix composed of testing temperatures *vs.* applied strains. Notably, when the applied strain increased from 0 to a maximum (~0.17), the GF value of the composite was enhanced from 40 to 270 at room temperature. This trend is easy to comprehend according to the previously discussed theoretical model. In addition, the GF values at high temperature reduced with growing applied strain, which could be attributed to the comprehensive effect of both thermal expansion and strain-actuated macromolecular orientation of the elastic PDMS network.

In order to imitate the surprising features of human skin, a simple flexible finger sensor based on the GCH/PDMS composite was prepared to carry out a finger-motion recognition experiment. As illustrated in Scheme 4.7, the as-prepared film sensor was fixed onto the index finger covered by a stretchable glove. A tip-metalised conducting wire was employed to build an electrical connection between the film sensor and the source meter.

The straight position of the index finger is defined as the initial state, from which the measured resistance is set as the baseline; see Figure 4.36(a). As shown in Figure 4.36(b–d), three simple gestures of the index finger: 2nd joint bending (motion A), 3rd joint bending (motion B) and grasping (motion C) were selected to estimate the mechanical–electrical response in the finger sensor. The tensile strains of the three gestures were 16.9%, 20.5% and 28.9%, respectively. Figure 4.36(e) presents the real-time resistance variation and holding duration in the three gesture changes. The R/R_0 value

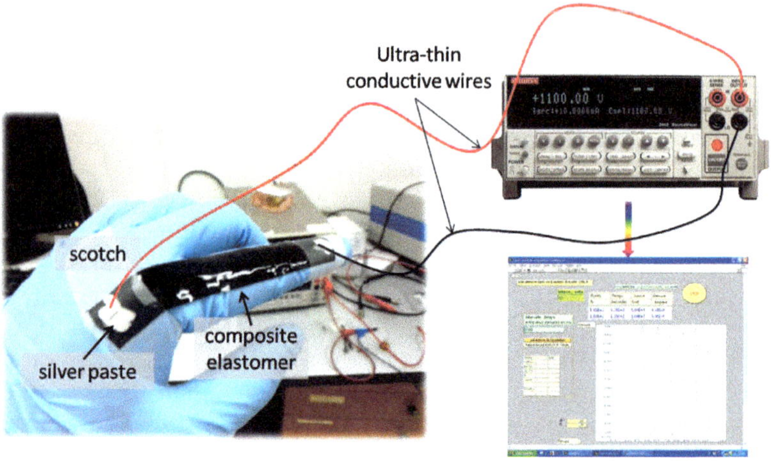

Scheme 4.7 Set-up for the motion-evaluating system using a typical GCH/PDMS composite as a film finger sensor.

obtained in gesture A, B and C equals 0.75, 0.6, 0.13, respectively, which is in accordance with the relative values in strain. Moreover, the observed durations of the resistance variation were also consistent with the duration of holding the gesture. It is worth noting that due to the GCHs in the film preferring to re-orientate along the tensile direction, this leads to the overall resistance of the film being reduced with increasing tensile deformation. The tunnelling effect can easily occur in this high anisotropy system. In accord with the results mentioned, GCHs are able to endow the composite thin film sensor with an ultra-high responding sensitivity and a precise recognition ability. This strategy of constructing an internal conducting network using superior units provides a reasaonable method for designing and fabricating PR sensor applied to next-generation flexible electronics.

4.5.4 CNT/PDMS and GNP/PDMS Composites

The GNP/PDMS and CNT/PDMS composites were fabricated for the purpose of highlighting the advantages of GCHs. The selected CNTs exhibit morphological features similar to those discussed in Section 4.3. As illustrated in Figure 4.37, no matter which are used, CNTs or GNPs are observed to have a good dispersion in the matrix and a relatively close interfacial adhesion with PDMS.

Figure 4.38 shows the comparison of σ_{AC} values among PDMS-based composites loaded with CNTs, GNPs and GCHs, respectively. The general f_c ranges of CNT/PDMS composites ($f_c \sim 1.0$ vol%) and GNP/PDMS composites ($f_c > 1.5$ vol%) can be roughly estimated according to the given σ_{AC} values, which are larger than for GCH-loaded systems. This is because of the highly efficient interconnection among GCHs. As shown in Figure 4.39, in order to ensure that the internal conducting network is constructed completely, the PR response of the composite filled with 1.2 vol% CNTs was

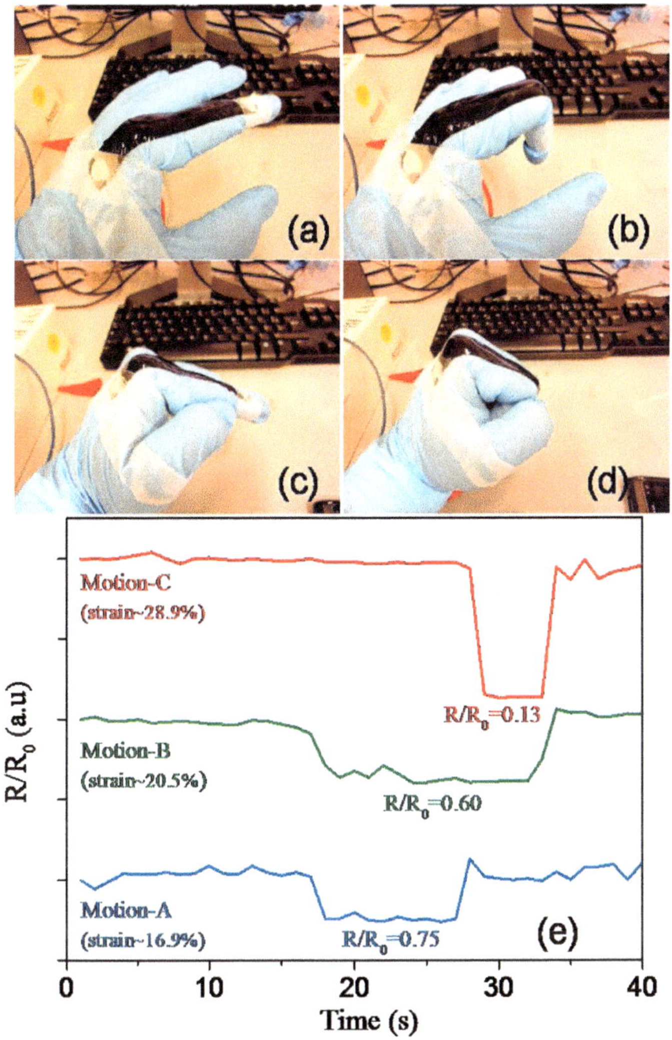

Figure 4.36 The GCH/PDMS film finger sensor. (a) Photograph of the relaxed state of the index finger. (b–d) Photographs of three different gestures (A, B and C) corresponding to the 2nd joint bending, the 3rd joint bending and grasping, respectively. (e) Relative resistance variation for each independent motion.
Reproduced from ref. 119 with permission from American Chemical Society, Copyright 2015.

investigated. The maximum resistance variation of the CNT-loaded counterpart was obtained as 75 under the applied pressure of 0.32 MPa, which is lower than that of the aforementioned GCH/PDMS composites. Considering the same condition of the filler fraction being larger than, but very close to, the f_c value for CNT/PDMS and GCH/PDMS composites, GCHs are truly able to endow the composite with a superior PR response.

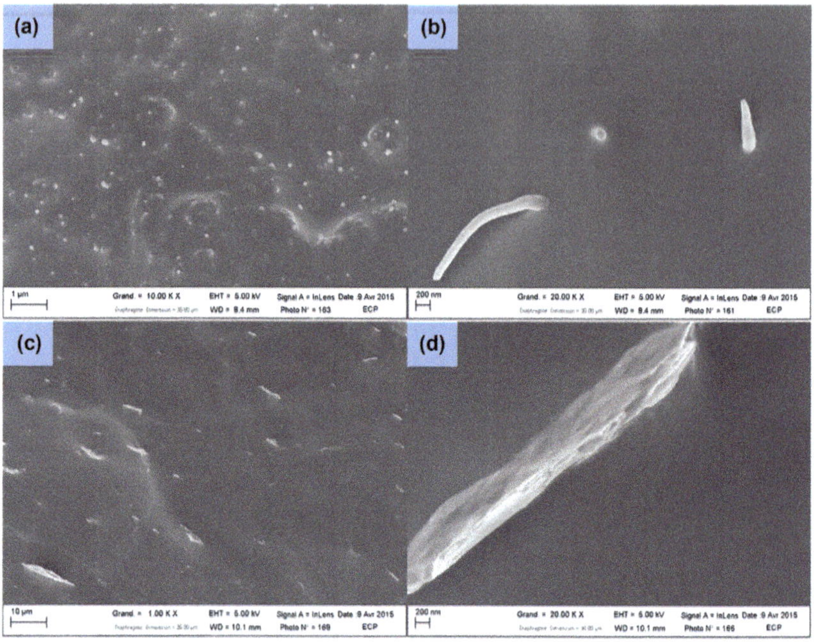

Figure 4.37 The SEM images of the fractured cross-section of (a) CNT/PDMS composite and (c) its high-magnification image, and (b) GNP/PDMS composite and (d) its high-magnification image.

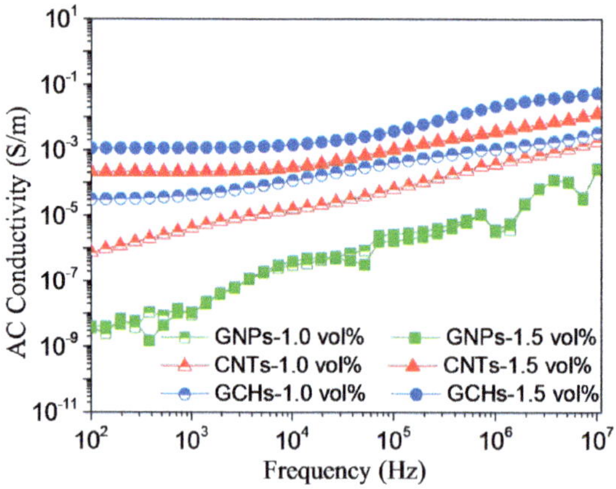

Figure 4.38 The AC conductivity of PDMS-based composites filled with 1.0 vol% and 1.5 vol% GNPs, CNTs and GCHs, respectively.

4.5.5 Section Summary

In conclusion, flexible GCH/PDMS composites exhibiting an outstanding piezo-resistive response were fabricated through mechanical blending. The highly conductive GCHs with 3D construction can be well dispersed in the

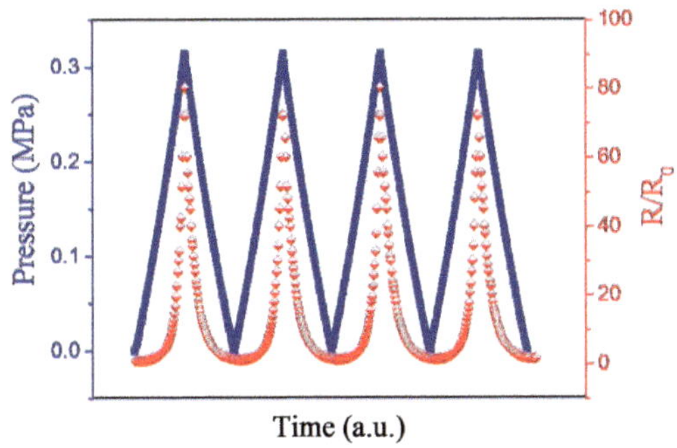

Figure 4.39 The piezo-resistive response of CNT/PDMS composites filled with 1.2 vol% CNTs.

PDMS matrix and show intensive interfacial adhesion with the matrix. Based on classical percolation theory, the σ_{AC} values of the composites were enhanced greatly with increasing f_{GCHs}. Thanks to the extraordinary advantages of the 'conductive hubs' an ultra-low f_c (~0.64 vol%) was obtained for the composite. Moreover, the highly sensitive positive PR response achieved for the composite is capable of being theoretically explained by exponential fitting to three typical stages. When $f_{GCHs} > f_c$, the PR sensitivity reduced with further increasing f_{GCHs}, due to the ever-stable conducting network. In particular, a GCH loading as small as 0.75 vol% can endow the composite with an ultrasensitive PR response, where the obtained GF and PS values were 10^3 and 0.6 kPa^{-1}, respectively. Moreover, the GCH/PDMS composites exhibited a superb consistency of mechanical–electrical responding coordination without the observation of hysteresis. Even experiencing 100 loading–unloading cycles, the relative resistance variation was still maintained at a high level. The acquired NTC behaviour from a wide testing temperature range means that GCH/PDMS composites have the potential to be applied in ambient-temperature recognition and PR sensing in extreme conditions. Furthermore, a typical film sensor made using GCHs/PDMS composite is sufficient to recognise and even distinguish different finger gestures. In addition, compared with GNP/PDMS and CNT/PDMS composites with an equivalent filler loading, the GCH/PDMS composite exhibits not only a much lower f_c value, but also higher σ_{AC} values and PR response. Therefore, this approach for constructing the conducting unit provides a feasible strategy for designing and fabricating advanced next-generation wearable electronics and flexible PR sensors.

4.6 Conclusions and Perspectives

Due to their typical advantages of high aspect ratio structures, low density and outstanding intrinsic mechanical, electrical and thermal properties, CNTs and GNPs have received continuing and widespread attention and study over the past decades. Unfortunately, current polymer-based composites cannot be fully endowed with the excellent inherent features of CNTs and GNPs. The main challenges are poor dispersion and uneven distribution of CNTs and GNPs in the matrices, which are caused by their enormous specific surface energy. In order to solve this problem, a form of all-carbon-composed micro–nano hybrid constructions of CNT arrays vertically grown onto GNP substrates was synthesised through the catalytic CVD method. The GCH construction was able to be precisely regulated by tuning critical parameters, such as temperature, hydrogen ratio and reaction time during the synthesis process. The as-prepared GCH particles are capable of being well dispersed into a thermoplastic polymer (PVDF) or thermosetting elastomer (PDMS) matrix through solvent-free, environmentally friendly and mass-producible mechanical mixing processes, to form composites with various functional properties. Due to the special construction of GCHs, existing dispersion and distribution problems of the inner matrix, such as CNT entanglements and GNP stacked aggregations, can be effectively ameliorated. Because the GCH construction endows the interface between the polymer and the fillers with an enlarged interfacial area and improved interfacial polarisation, the dielectric permittivity values of the composites increase significantly. The as-synthesised CNT arrays are capable of accelerating the formation of microcapacitor networks, and triggering variation of the crystalline phase composition in the PVDF matrix. The dielectric properties of the composites can be effectively improved for the following reasons. The typical all-carbon-composed 3D GCHs ameliorate the construction of the internal conducting network. An ultra-low percolation threshold can be obtained owing to their conductive 'hub' structures, which leads to not only significant enhancements of both dielectric and conducting properties, but also excellent flexibility of the composites. This special feature is sufficient to optimise the piezo-resistive response sensitivity of stretchable materials. Therefore, the design and synthesis of GCH constructions provides a promising strategy for dielectric and electrical conduction applications in the future.

The GCH constructions show great application potentials in dielectric and electrical conducting composites in a number of fields. Further research on their functionalities in mechanical, optical and thermal properties are needed. In order to realise practical applications on a large scale, it is necessary to strengthen the interfacial adhesion between the CNT array and the GNPs. To keep the intrinsic structure of the GCHs from being damaged and to avoid the CNT array detaching from the GNP substrates in the polymer matrix at the extremes are key issues for determining the overall performance of the composites. The interfacial interaction between GCHs and the polymer matrix are also critical, and are responsible for tuning and expanding the multifunctional properties of composites. Relevant optimisations can be carried on CNT arrays through methods such as surface functionalisation, surface modification and shell structure coating, in order to develop high performance GCH/polymer composites with multiple functionalities.

References

1. S. Iijima, *Nature*, 1991, **354**(6348), 56.
2. S. Iijima and T. Ichihashi, *Nature*, 1993, **363**, 603.
3. R. S. Ruoff and D. C. Lorents, *Carbon*, 1995, **33**(7), 925.
4. P. M. Ajayan and O. Z. Zhou, *Top. Appl. Phys.*, 2001, **80**, 391.
5. A. Hassanien, M. Tokumoto, S. Ohshima, Y. Kuriki, F. Ikazaki, K. Uchida and M. Yumura, *Appl. Phys. Lett.*, 1999, **75**(18), 2755.
6. J. W. G. Wildoer, L. C. Venema, A. G. Rinzler, R. E. Smalley and C. Dekker, *Nature*, 1998, **391**(6662), 59.
7. E. T. Thostenson, Z. F. Ren and T. W. Chou, *Compos. Sci. Technol.*, 2001, **61**(13), 1899.
8. J. A. Elliott, J. K. W. Sandler, A. H. Windle, R. J. Young and M. S. P. Shaffer, *Phys. Rev. Lett.*, 2004, **92**(9), 095501.
9. R. H. Baughman, A. A. Zakhidov and W. A. de Heer, *Science*, 2002, **297**(5582), 787.
10. C. Journet and P. Bernier, *Appl. Phys. A: Mater. Sci. Process.*, 1998, **67**(1), 1.
11. Y. Saito, K. Nishikubo, K. Kawabata and T. Matsumoto, *J. Appl. Phys.*, 1996, **80**(5), 3062.
12. Z. J. Shi, Y. F. Lian, F. H. Liao, X. H. Zhou, Z. N. Gu, Y. Zhang, S. Iijima, H. D. Li, K. T. Yue and S. L. Zhang, *J. Phys. Chem. Solids*, 2000, **61**(7), 1031.
13. A. Thess, R. Lee, P. Nikolaev, H. J. Dai, P. Petit, J. Robert, C. H. Xu, Y. H. Lee, S. G. Kim, A. G. Rinzler, D. T. Colbert, G. E. Scuseria, D. Tomanek, J. E. Fischer and R. E. Smalley, *Science*, 1996, **273**(5274), 483.
14. Y. Zhang and S. Iijima, *Appl. Phys. Lett.*, 1999, **75**(20), 3087.
15. K. Lee, S. S. Lee, J. A. Lee, K.-C. Lee and S. Ji, *Appl. Phys. Lett.*, 2010, **96**(1), 013511.
16. K. L. Choy, *Prog. Mater. Sci.*, 2003, **48**(2), 57.
17. M. Kumar and Y. Ando, *J. Nanosci. Nanotechnol.*, 2010, **10**(6), 3739.
18. A. Moisala, A. G. Nasibulin and E. I. Kauppinen, *J. Phys.: Condens. Matter*, 2003, **15**(42), S3011.
19. A. G. Nasibulin, A. Moisala, D. P. Brown and E. I. Kauppinen, *Carbon*, 2003, **41**(14), 2711.
20. H. J. Dal, A. G. Rinzler, P. Nikolaev, A. Thess, D. T. Colbert and R. E. Smalley, *Chem. Phys. Lett.*, 1996, **260**(3–4), 471.
21. D. Takagi, Y. Homma, H. Hibino, S. Suzuki and Y. Kobayashi, *Nano Lett.*, 2006, **6**(12), 2642.
22. V. Ivanov, J. B. Nagy, P. Lambin, A. Lucas, X. B. Zhang, X. F. Zhang, D. Bernaerts, G. Vantendeloo, S. Amelinckx and J. Vanlanduyt, *Chem. Phys. Lett.*, 1994, **223**(4), 329.
23. P. Nikolaev, M. J. Bronikowski, R. K. Bradley, F. Rohmund, D. T. Colbert, K. A. Smith and R. E. Smalley, *Chem. Phys. Lett.*, 1999, **313**(1–2), 91.

24. C. Bower, O. Zhou, W. Zhu, D. J. Werder and S. H. Jin, *Appl. Phys. Lett.*, 2000, **77**(17), 2767.
25. J. H. Hafner, M. J. Bronikowski, B. R. Azamian, P. Nikolaev, A. G. Rinzler, D. T. Colbert, K. A. Smith and R. E. Smalley, *Chem. Phys. Lett.*, 1998, **296**(1–2), 195.
26. E. Flahaut, A. Govindaraj, A. Peigney, C. Laurent, A. Rousset and C. N. R. Rao, *Chem. Phys. Lett.*, 1999, **300**(1–2), 236.
27. G. Che, B. B. Lakshmi, C. R. Martin, E. R. Fisher and R. S. Ruoff, *Chem. Mater.*, 1998, **10**(1), 260.
28. H. M. Cheng, F. Li, G. Su, H. Y. Pan, L. L. He, X. Sun and M. S. Dresselhaus, *Appl. Phys. Lett.*, 1998, **72**(25), 3282.
29. D. K. Smith, D. C. Lee and B. A. Korgel, *Chem. Mater.*, 2006, **18**(14), 3356.
30. R. Andrews, D. Jacques, A. M. Rao, F. Derbyshire, D. Qian, X. Fan, E. C. Dickey and J. Chen, *Chem. Phys. Lett.*, 1999, **303**(5–6), 467.
31. D. He, M. Bozlar, M. Genestoux and J. Bai, *Carbon*, 2010, **48**(4), 1159.
32. Z. F. Ren, Z. P. Huang, J. W. Xu, J. H. Wang, P. Bush, M. P. Siegal and P. N. Provencio, *Science*, 1998, **282**(5391), 1105.
33. G. F. Zhong, J. H. Warner, M. Fouquet, A. W. Robertson, B. A. Chen and J. Robertson, *ACS Nano*, 2012, **6**(4), 2893.
34. D. N. Futaba, K. Hata, T. Yamada, T. Hiraoka, Y. Hayamizu, Y. Kakudate, O. Tanaike, H. Hatori, M. Yumura and S. Iijima, *Nat. Mater.*, 2006, **5**(12), 987.
35. E. T. Thostenson, W. Z. Li, D. Z. Wang, Z. F. Ren and T. W. Chou, *J. Appl. Phys.*, 2002, **91**(9), 6034.
36. S. S. Wicks, W. N. Wang, M. R. Williams and B. L. Wardle, *Compos. Sci. Technol.*, 2014, **100**, 128.
37. V. P. Veedu, A. Y. Cao, X. S. Li, K. G. Ma, C. Soldano, S. Kar, P. M. Ajayan and M. N. Ghasemi-Nejhad, *Nat. Mater.*, 2006, **5**(6), 457.
38. X. C. Gui, Z. P. Zeng, Y. Zhu, H. B. Li, Z. Q. Lin, Q. M. Gan, R. Xiang, A. Y. Cao and Z. K. Tang, *Adv. Mater.*, 2014, **26**(8), 1248.
39. M. M. J. Treacy, T. W. Ebbesen and J. M. Gibson, *Nature*, 1996, **381**(6584), 678.
40. B. I. Yakobson and P. Avouris, *Top. Appl. Phys.*, 2001, **80**, 287.
41. E. W. Wong, P. E. Sheehan and C. M. Lieber, *Science*, 1997, **277**(5334), 1971.
42. M.-F. Yu, O. Lourie, M. J. Dyer, K. Moloni, T. F. Kelly and R. S. Ruoff, *Science*, 2000, **287**(5453), 637.
43. M.-F. Yu, B. S. Files, S. Arepalli and R. S. Ruoff, *Phys. Rev. Lett.*, 2000, **84**(24), 5552.
44. B. Peng, M. Locascio, P. Zapol, S. Y. Li, S. L. Mielke, G. C. Schatz and H. D. Espinosa, *Nat. Nanotechnol.*, 2008, **3**(10), 626.
45. B. Q. Wei, R. Vajtai and P. M. Ajayan, *Appl. Phys. Lett.*, 2001, **79**(8), 1172.
46. T. W. Ebbesen, H. J. Lezec, H. Hiura, J. W. Bennett, H. F. Ghaemi and T. Thio, *Nature*, 1996, **382**(6586), 54.
47. H. J. Dai, *Science*, 1996, **272**(5270), 1861.

48. S. Berber, Y. K. Kwon and D. Tomanek, *Phys. Rev. Lett.*, 2000, **84**(20), 4613.
49. E. Pop, D. Mann, Q. Wang, K. E. Goodson and H. J. Dai, *Nano Lett.*, 2006, **6**(1), 96.
50. M. F. De Volder, S. H. Tawfick, R. H. Baughman and A. J. Hart, *Science*, 2013, **339**(6119), 535.
51. L. M. Dai, D. W. Chang, J. B. Baek and W. Lu, *Small*, 2012, **8**(8), 1130.
52. K. Evanoff, J. Khan, A. A. Balandin, A. Magasinski, W. J. Ready, T. F. Fuller and G. Yushin, *Adv. Mater.*, 2012, **24**(4), 533.
53. C. Sotowa, G. Origi, M. Takeuchi, Y. Nishimura, K. Takeuchi, I. Y. Jang, Y. J. Kim, T. Hayashi, Y. A. Kim, M. Endo and M. S. Dresselhaus, *ChemSusChem*, 2008, **1**(11), 911.
54. T. Matsumoto, T. Komatsu, K. Arai, T. Yamazaki, M. Kijima, H. Shimizu, Y. Takasawa and J. Nakamura, *Chem. Commun.*, 2004, **7**, 840.
55. A. Le Goff, V. Artero, B. Jousselme, P. D. Tran, N. Guillet, R. Métayé, A. Fihri, S. Palacin and M. Fontecave, *Science*, 2009, **326**(5958), 1384.
56. W. A. Deheer, A. Chatelain and D. Ugarte, *Science*, 1995, **270**(5239), 1179.
57. Y. Saito and S. Uemura, *Carbon*, 2000, **38**(2), 169.
58. N. S. Lee, D. S. Chung, I. T. Han, J. H. Kang, Y. S. Choi, H. Y. Kim, S. H. Park, Y. W. Jin, W. K. Yi, M. J. Yun, J. E. Jung, C. J. Lee, J. H. You, S. H. Jo, C. G. Lee and J. M. Kim, *Diamond Relat. Mater.*, 2001, **10**(2), 265.
59. H. J. Dai, J. H. Hafner, A. G. Rinzler, D. T. Colbert and R. E. Smalley, *Nature*, 1996, **384**(6605), 147.
60. S. Akita and Y. Nakayama, *Jpn. J. Appl. Phys., Part 1*, 2002, **41**(6B), 4242.
61. P. Kim and C. M. Lieber, *Science*, 1999, **286**(5447), 2148.
62. F. D. Fim, J. M. Guterres, N. R. S. Basso and G. B. Galland, *J. Polym. Sci., Part A: Polym. Chem.*, 2010, **48**(3), 692.
63. L. M. Viculis, J. J. Mack, O. M. Mayer, H. T. Hahn and R. B. Kaner, *J. Mater. Chem.*, 2005, **15**(9), 974.
64. A. V. Yakovlev, A. I. Finaenov, S. L. Zabud'kov and E. V. Yakovleva, *Russ J Appl Chem +*, 2006, **79**(11), 1741.
65. J. P. Li, H. F. Lin, W. F. Zhao and G. H. Chen, *J. Appl. Polym. Sci.*, 2008, **109**(3), 1377.
66. H. L. Fan, L. L. Wang, K. K. Zhao, N. Li, Z. J. Shi, Z. G. Ge and Z. X. Jin, *Biomacromolecules*, 2010, **11**(9), 2345.
67. T. Ramanathan, S. Stankovich, D. A. Dikin, H. Liu, H. Shen, S. T. Nguyen and L. C. Brinson, *J. Polym. Sci., Part B: Polym. Phys.*, 2007, **45**(15), 2097.
68. S. Park and R. S. Ruoff, *Nat. Nanotechnol.*, 2009, **4**(4), 217.
69. Y. H. Wu, T. Yu and Z. X. Shen, *J. Appl. Phys.*, 2010, **108**(7), 071301.
70. M. J. Allen, V. C. Tung and R. B. Kaner, *Chem. Rev.*, 2010, **110**(1), 132.
71. H. Kim, A. A. Abdala and C. W. Macosko, *Macromolecules*, 2010, **43**(16), 6515.
72. D. Y. Cai and M. Song, *J. Mater. Chem.*, 2010, **20**(37), 7906.
73. L. W. Wang, J. B. Hong and G. H. Chen, *Polym. Eng. Sci.*, 2010, **50**(11), 2176.

74. J. J. George, A. B. Dyopadhyay and A. K. Bhowmick, *J. Appl. Polym. Sci.*, 2008, **108**(3), 1603.
75. E. Planes, J. Duchet, A. Maazouz and J. F. Gerard, *Polym. Eng. Sci.*, 2008, **48**(4), 723.
76. J. Cho, J. Y. Chen and I. M. Daniel, *Scr. Mater.*, 2007, **56**(8), 685.
77. J. Lu, L. T. Drzal, R. M. Worden and I. Lee, *Chem. Mater.*, 2007, **19**(25), 6240.
78. D. Xu, V. Sridhar, T. T. Pham and J. K. Kim, *E-Polymers*, 2008, **8**(1), 237.
79. I. H. Kim and Y. G. Jeong, *J. Polym. Sci., Part B: Polym. Phys.*, 2010, **48**(8), 850.
80. W. Kai, Y. Hirota, L. Hua and Y. Inoue, *J. Appl. Polym. Sci.*, 2008, **107**(3), 1395.
81. B. Li and W.-H. Zhong, *J. Mater. Sci.*, 2011, **46**(17), 5595.
82. C. S. Yu and B. Li, *Polym. Compos.*, 2008, **29**(9), 998.
83. L. Chen, G. H. Chen and L. Lu, *Adv. Funct. Mater.*, 2007, **17**(6), 898.
84. K. Xu, D. Erricolo, M. Dutta and M. A. Stroscio, *Superlattices Microstruct.*, 2012, **51**(5), 606.
85. Y. C. Li, S. C. Tjong and R. K. Y. Li, *Synth. Met.*, 2010, **160**(17–18), 1912.
86. J. H. Kang, C. Park, J. A. Scholl, A. H. Brazin, N. M. Holloway, J. W. High, S. E. Lowther and J. S. Harrison, *J. Polym. Sci., Part B: Polym. Phys.*, 2009, **47**(10), 994.
87. S.-E. Lee, O. Choi and H. T. Hahn, *J. Appl. Phys.*, 2008, **104**(3), 033705.
88. M. S. Sarto, A. G. D'Aloia, A. Tamburrano and G. De Bellis, *IEEE Trans. Electromagn. Compat.*, 2012, **54**(1), 17.
89. P. C. Ma, N. A. Siddiqui, G. Marom and J. K. Kim, *Composites, Part A*, 2010, **41**(10), 1345.
90. J. Yuan, S. Yao, W. Li, A. Sylvestre and J. Bai, *J. Phys. Chem. C*, 2014, **118**(40), 22975.
91. L. Ci and J. Bai, *Adv. Mater.*, 2004, **16**(22), 2021.
92. D. He and J. Bai, *Chem. Vap. Deposition*, 2011, **17**(4–6), 98.
93. D. He, H. Li and J. Bai, *Carbon*, 2011, **49**(15), 5359.
94. D. He, H. Li, W. Li, P. Haghi-Ashtiani, P. Lejay and J. Bai, *Carbon*, 2011, **49**(7), 2273.
95. W. Li, J. Yuan, Y. Lin, S. Yao, Z. Ren, H. Wang, M. Wang and J. Bai, *Carbon*, 2013, **51**, 355.
96. W. Li, D. He and J. Bai, *Composites, Part A*, 2013, **54**, 28.
97. M. Bozlar, D. He, J. Bai, Y. Chalopin, N. Mingo and S. Volz, *Adv. Mater.*, 2010, **22**(14), 1654.
98. J.-K. Yuan, W.-L. Li, S.-H. Yao, Y.-Q. Lin, A. Sylvestre and J. Bai, *Appl. Phys. Lett.*, 2011, **98**(3), 032901.
99. W. Li, D. He, Z. Dang and J. Bai, *Compos. Sci. Technol.*, 2014, **99**, 8.
100. W. Li, J. Yuan, A. Dichiara, Y. Lin and J. Bai, *Carbon*, 2012, **50**(11), 4298.
101. L. Zhu, *J. Phys. Chem. Lett.*, 2014, **5**(21), 3677.
102. P. Barber, S. Balasubramanian, Y. Anguchamy, S. Gong, A. Wibowo, H. Gao, H. J. Ploehn and H.-C. Zur Loye, *Materials*, 2009, **2**(4), 1697.
103. Z.-M. Dang, Y.-H. Zhang and S.-C. Tjong, *Synth. Met.*, 2004, **146**(1), 79.

104. P. Murugaraj, D. Mainwaring and N. Mora-Huertas, *J. Appl. Phys.*, 2005, **98**(5), 4304.
105. C. W. Nan, Y. Shen and J. Ma, *Annu. Rev. Mater. Res.*, 2010, **40**(1), 131.
106. Z.-M. Dang, J.-K. Yuan, J.-W. Zha, T. Zhou, S.-T. Li and G.-H. Hu, *Prog. Mater. Sci.*, 2012, **57**(4), 660.
107. C. W. Nan, Y. Shen and J. Ma, *Annu. Rev. Mater. Res.*, 2010, **40**(1), 131.
108. J. Y. Yi and G. M. Choi, *J. Electroceram.*, 1999, **3**(4), 361.
109. C. Li, E. T. Thostenson and T.-W. Chou, *Appl. Phys. Lett.*, 2007, **91**(22), 223114.
110. M. Fuhrer, J. Nygård, L. Shih, M. Forero, Y.-G. Yoon, H. J. Choi, J. Ihm, S. G. Louie, A. Zettl and P. L. McEuen, *Science*, 2000, **288**(5465), 494.
111. A. Buldum and J. P. Lu, *Phys. Rev. B*, 2001, **63**(16), 161403.
112. H. Stoyanov, D. Mc Carthy, M. Kollosche and G. Kofod, *Appl. Phys. Lett.*, 2009, **94**(23), 232905.
113. M. F. Gyure and P. D. Beale, *Phys. Rev. B*, 1992, **46**(7), 3736.
114. I. Balberg, D. Azulay, D. Toker and O. Millo, *Int. J. Mod. Phys. B*, 2004, **18**(15), 2091.
115. J. G. Simmons, *J. Appl. Phys.*, 1963, **34**(6), 1793.
116. N. Hu, Y. Karube, C. Yan, Z. Masuda and H. Fukunaga, *Acta Mater.*, 2008, **56**(13), 2929.
117. D. Toker, D. Azulay, N. Shimoni, I. Balberg and O. Millo, *Phys. Rev. B*, 2003, **68**(4), 041403.
118. C. W. Nan, *Prog. Mater. Sci.*, 1993, **37**(1), 1.
119. H. Zhao and J. B. Bai, *ACS Appl. Mater. Interfaces*, 2015, **7**(18), 9652.
120. J. N. Coleman, U. Khan, W. J. Blau and Y. K. Gun'ko, *Carbon*, 2006, **44**(9), 1624.
121. Y. H. Wu, T. Yu and Z. X. Shen, *J. Appl. Phys.*, 2010, **108**(7), 071301.
122. A. Dichiara, J.-K. Yuan, S.-H. Yao, A. Sylvestre and J. Bai, *J. Nanosci. Nanotechnol.*, 2012, **12**(9), 6935.
123. D. He, PhD thesis, Ecole Centrale Paris, 2010.
124. G. D. Nessim, A. J. Hart, J. S. Kim, D. Acquaviva, J. Oh, C. D. Morgan, M. Seita, J. S. Leib and C. V. Thompson, *Nano Lett.*, 2008, **8**(11), 3587.
125. W. Wasel, K. Kuwana, P. T. Reilly and K. Saito, *Carbon*, 2007, **45**(4), 833.
126. H. Li, D. He, T. Li, M. Genestoux and J. Bai, *Carbon*, 2010, **48**(15), 4330.
127. R. Prakash, A. K. Mishra, A. Roth, C. Kübel, T. Scherer, M. Ghafari, H. Hahn and M. Fichtner, *J. Mater. Chem.*, 2010, **20**(10), 1871.
128. L.-S. Fu, J.-T. Jiang, C.-Y. Xu and L. Zhen, *CrystEngComm*, 2012, **14**(20), 6827.
129. A. B. Dichiara, J. Yuan, S. Yao, A. Sylvestre, L. Zimmer and J. Bai, *J. Mater. Chem. A*, 2014, **2**(21), 7980.
130. G. Ghini, L. Luconi, A. Rossin, C. Bianchini, G. Giambastiani, S. Cicchi, L. Lascialfari, A. Brandi and A. Giannasi, *Chem. Commun.*, 2009, **46**(2), 252.
131. G. Grassi, A. Scala, A. Piperno, D. Iannazzo, M. Lanza, C. Milone, A. Pistone and S. Galvagno, *Chem. Commun.*, 2012, **48**(54), 6836.

132. Z. M. Dang, L. Wang, Y. Yin, Q. Zhang and Q. Q. Lei, *Adv. Mater.*, 2007, **19**(6), 852.
133. Z.-M. Dang, H.-Y. Wang, B. Peng and C.-W. Nan, *J. Electroceram.*, 2007, **21**(1–4), 381.
134. Z.-M. Dang, J.-P. Wu, H.-P. Xu, S.-H. Yao, M.-J. Jiang and J. Bai, *Appl. Phys. Lett.*, 2007, **91**(7), 072912.
135. Z.-M. Dang, D. Xie and C.-Y. Shi, *Appl. Phys. Lett.*, 2007, **91**(22), 222902.
136. H. Zhao, M.-H. Yang, D. He and J. B. Bai, *J. Mater. Chem. C*, 2016, **4**(38), 8911.
137. S. M. Hong, Y. W. Nam, S. S. Hwang and D. W. Chae, *Mol. Cryst. Liq. Cryst.*, 2007, **464**(1), 195–777.
138. P. Martins, C. M. Costa, M. Benelmekki, G. Botelho and S. Lanceros-Mendez, *CrystEngComm*, 2012, **14**(8), 2807.
139. P. Martins, A. C. Lopes and S. Lanceros-Mendez, *Prog. Polym. Sci.*, 2014, **39**(4), 683.
140. V. Zucolotto, J. Avlyanov, R. Gregorio and L. H. C. Mattoso, *J. Appl. Polym. Sci.*, 2004, **94**(2), 553.
141. J.-K. Yuan, S.-H. Yao, Z.-M. Dang, A. Sylvestre, M. Genestoux and J. Bai, *J. Phys. Chem. C*, 2011, **115**(13), 5515.
142. B.-H. Fan, J.-W. Zha, D. Wang, J. Zhao and Z.-M. Dang, *Appl. Phys. Lett.*, 2012, **100**(1), 012903.
143. H. Zhao, L. Zhang, M. H. Yang, Z. M. Dang and J. B. Bai, *Appl. Phys. Lett.*, 2015, **106**(9), 092904.
144. M. L. Hammock, A. Chortos, B. C. Tee, J. B. Tok and Z. Bao, *Adv. Mater.*, 2013, **25**(42), 5997.
145. R. He and P. Yang, *Nat. Nanotechnol.*, 2006, **1**(1), 42.
146. S. J. Benight, C. Wang, J. B. H. Tok and Z. Bao, *Prog. Polym. Sci.*, 2013, **38**(12), 1961.
147. S. Stassi, V. Cauda, G. Canavese and C. F. Pirri, *Sensors*, 2014, **14**(3), 5296.
148. G. Canavese, S. Stassi, C. Fallauto, S. Corbellini, V. Cauda, V. Camarchia, M. Pirola and C. F. Pirri, *Sens. Actuators, A*, 2014, **208**, 1.
149. L. M. Castano and A. B. Flatau, *Smart Mater. Struct.*, 2014, **23**(5), 053001.
150. J. H. Kong, N. S. Jang, S. H. Kim and J. M. Kim, *Carbon*, 2014, **77**, 199.
151. M. Ramuz, B. C. Tee, J. B. Tok and Z. Bao, *Adv. Mater.*, 2012, **24**(24), 3223.
152. Z. M. Dang, J. K. Yuan, S. H. Yao and R. J. Liao, *Adv. Mater.*, 2013, **25**(44), 6334.
153. J. F. Zhou, Y. H. Song, Q. Zheng, Q. Wu and M. Q. Zhang, *Carbon*, 2008, **46**(4), 679.
154. C. Yan, J. Wang, W. Kang, M. Cui, X. Wang, C. Y. Foo, K. J. Chee and P. S. Lee, *Adv. Mater.*, 2014, **26**(13), 2022.
155. S.-E. Zhu, M. Krishna Ghatkesar, C. Zhang and G. C. A. M. Janssen, *Appl. Phys. Lett.*, 2013, **102**(16), 161904.

156. Y. Hou, D. Wang, X.-M. Zhang, H. Zhao, J.-W. Zha and Z.-M. Dang, *J. Mater. Chem. C*, 2013, **1**(3), 515.
157. F. He, S. Lau, H. L. Chan and J. Fan, *Adv. Mater.*, 2009, **21**(6), 710.
158. C. Lee, L. Jug and E. Meng, *Appl. Phys. Lett.*, 2013, **102**(18), 183511.
159. S. Luo, PhD thesis, The Florida State University, 2013.
160. H. Zhao, D.-R. Wang, J.-W. Zha, J. Zhao and Z.-M. Dang, *J. Mater. Chem. A*, 2013, **1**(9), 3140.
161. L. H. Wang, T. H. Ding and P. Wang, *Sens. Actuators, A*, 2007, **135**(2), 587.
162. L. H. Wang, T. H. Ding and P. Wang, *Carbon*, 2009, **47**(14), 3151.
163. X. Song, S. Liu, Z. Gan, Q. Lv, H. Cao and H. Yan, *Microelectron. Eng.*, 2009, **86**(11), 2330.
164. P. Wang, S. Geng and T. Ding, *Compos. Sci. Technol.*, 2010, **70**(10), 1571.
165. D. J. Lipomi, M. Vosgueritchian, B. C.-K. Tee, S. L. Hellstrom, J. A. Lee, C. H. Fox and Z. Bao, *Nat. Nanotechnol.*, 2011, **6**, 788.
166. J. Hwang, J. Jang, K. Hong, K. N. Kim, J. H. Han, K. Shin and C. E. Park, *Carbon*, 2011, **49**(1), 106.
167. J.-W. Zha, W.-K. Li, J. Zhang, C.-Y. Shi and Z.-M. Dang, *Mater. Lett.*, 2014, **118**, 161.
168. T. Castillo-Castro, M. M. Castillo-Ortega, J. C. Encinas, P. J. Herrera Franco and H. J. Carrillo-Escalante, *J. Mater. Sci.*, 2011, **47**(4), 1794.

CHAPTER 5

2D Nanomaterial-based Polymer Composite Electrolytes for Lithium-based Batteries

VIDYANAND VIJAYAKUMAR,[a,b,c] MEENA GHOSH,[b,c] PARESH KUMAR SAMANTARAY,[d] SREEKUMAR KURUNGOT,[c] MARTIN WINTER*[a,e,f] AND JIJEESH RAVI NAIR*[a]

[a] Helmholtz Institute Münster (HI MS), IEK-12, Forschungszentrum Jülich GmbH, Corrensstraße 46, 48149 Münster, Germany; [b] Academy of Scientific and Innovative Research (AcSIR), Sector 19, Kamla Nehru Nagar, Ghaziabad, Uttar Pradesh-201002, India; [c] Physical and Materials Chemistry Division, CSIR-National Chemical Laboratory, Pune-411008, India; [d] International Institute for Nanocomposites Manufacturing (IINM), WMG, University of Warwick, Coventry CV4 7AL, UK; [e] Institute of Physical Chemistry, University of Münster, Corrensstraße 28/30, 48149, Münster, Germany; [f] MEET Battery Research Center, Corrensstraße 46, 48149, Münster, Germany

*Emails: m.winter@fz-juelich.de; j.nair@fz-juelich.de

5.1 Introduction

Among the primary and secondary batteries used for various applications, lithium-ion batteries (LIBs) are widely applied, from consumer electronics to hybrid (electric) vehicles and grid storage, due to their high operating voltage (>3.4 V), long life, high energy efficiency, and specific energy ($\geq$150 W h kg^{-1}).[1–4] In general, a LIB, like any other battery, is comprised of an anode, cathode, separator, and liquid electrolyte (LE).[5–11] The

Inorganic Materials Series No. 6
Two-dimensional Inorganic Nanomaterials for Conductive Polymer Nanocomposites
Edited by Chaoying Wan, Xingyi Huang and Chris Bowen

Published by the Royal Society of Chemistry, www.rsc.org

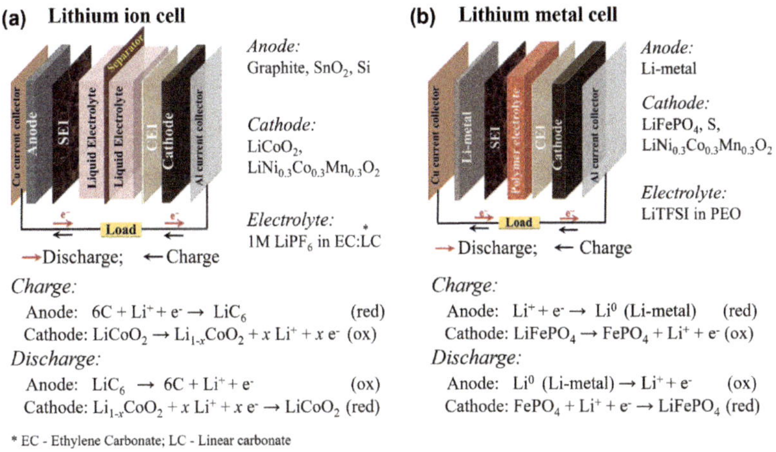

Figure 5.1 Schematic illustration of various representative components and related processes occurring at the electrodes of (a) a lithium-ion cell (liquid electrolyte) and (b) a lithium-metal cell (polymer electrolyte). Adapted from ref. 48 with permission from The Electrochemical Society, Copyright 2019.

electrochemical reactions happening at the electrodes of a LIB, along with its various components, are schematically illustrated in Figure 5.1a. Although there is an array of credible active materials investigated so far, graphite is the most commonly used anode material for LIBs.[10,12–15] Other carbon-based nanomaterials (fullerene, carbon nanotubes, graphene), silicon (Si), tin, transition metal oxides (*e.g.*, SnO_2), and transition metal chalcogenides, and even redox-active polymers[16] (*e.g.*, polypyrrole) have also been tried.[17,18] Lithiated transition metal oxides (*e.g.*, $LiNi_xMn_yCo_zO_2$, $LiCoO_2$), spinel-type $LiMn_2O_4$ and $LiNi_{0.5}Mn_{1.5}O_4$, and Li-containing olivine-type materials (*e.g.*, $LiFePO_4$), have been widely used as cathodes.[8,19–22] The anode and cathode compartments of a battery cell should be isolated by using a suitable electronically non-conducting separator. Porous single- or multilayer polyolefins and ceramic-coated polymer-based separators saturated with LEs have been used for such purposes.[23,24] Linear (*e.g.*, dimethyl carbonate) and cyclic (*e.g.*, ethylene carbonate) organic carbonates in various ratios and combinations have been used as solvent to dissolve lithium salt (*e.g.*, $LiPF_6$) to prepare liquid electrolyte.[25] 1 M $LiPF_6$ dissolved in 1 : 1 v/v ethylene carbonate (EC):dimethyl carbonate (DMC), LP50, is a classic example of a widely used organic LE.[26–29]

Depending on the kind of electrodes and the electrolyte, various types of interphases are formed at the electrode/electrolyte interfaces in a LIB cell. The interphase formed at the anode is widely called the solid electrolyte interphase (SEI),[30,31] and at the cathode it can be called the cathode electrolyte interphase (CEI).[32–34] The thickness, composition, morphology, stability, and (ionic/electronic) conduction properties of these layers determine the overall performance in such devices. Besides the electrode-active materials, a LIB electrode also contains binder(s), conduction additive(s), and current collector(s). The main feature of a binder[35,36] is its ability to hold together the electrode particles, and ensure their adhesion to the respective current collectors (Cu for the anode and Al for the cathode).[37,38] Polymers, such as carboxymethyl cellulose (CMC), styrene–butadiene rubber (SBR), poly(acrylic acid) (PAA), and poly(vinylidene difluoride) (PVDF), are widely used as binders.[39,40] As electronic conduction enhancers, carbon black with different surface areas has been employed.[41,42] At least for academic research, conductive polymers (*e.g.*, PEDOT:PSS, polypyrrole)[43] have been examined as possible alternatives.[44] Indeed, only an effective combination of electrochemically active electrode materials, binders, and electronically conductive carbons, along with a suitable current collector, can lead to excellent cycling stability, high (specific) capacity, and increased rate capability at the electrode and cell level.[9,45]

LIB technology has been the best rechargeable battery technology for decades. Still, there is a search for better performing, cheaper, and safer LIB components, and alternative battery systems. In particular, LIB manufacturers are forced to employ organic solvents as LE components, which carry safety threats related to leakage, formation of toxic components, and, in many cases, also flammability.[46,47] An upgrade in terms of cell architecture, sleek design, portability, and individual cell weight management is required for meeting the future demands on batteries.[48] Additionally, from the cathode side, the availability and high costs of the lithium precursors and transition metal compounds for use in the cathode materials are huge challenges. From the anode side, the repercussions concerning low specific capacity and energy density, and the possibility of functioning at elevated temperatures need to be addressed.

Li-metal can replace the present graphite-based electrodes, due to the energy densities and particularly the specific capacities [Li-metal (3860 $mA\,h\,g^{-1}$) *vs.* lithiated graphite (LiC_6, 339 $mA\,h\,g^{-1}$)][49,50] being enhanced. Therefore, Li-metal is thoroughly investigated as a potential candidate for rechargeable secondary batteries.[51,52] However, their safety related to usability, thermal runaway, significant reactivity due to constantly changing high surface area lithium metal (HSAL)[53–55] during battery operation, and dendrite-related short-circuits arising from inhomogeneous Li-plating and -stripping, are yet to be addressed.[56,57] Nevertheless, the use of Li-metal as anode will lead to new cell chemistries/designs, where non-lithiated cathode materials (*e.g.*, V_2O_5, S, O_2) and several other (conversion) cathodes[58] can be effectively employed in various types of rechargeable lithium-metal batteries (LMBs). This revolution in battery architecture leads to new opportunities as well as new challenges. Hence, fostering substantial research and development efforts for the resurgence of rechargeable batteries, in particular LMBs, is required.[53] Various components of an LMB are illustrated in Figure 5.1b along with the electrochemical processes happening at the respective electrodes during the charge/discharge cycling reactions.

5.2 Electrolytes

Conventional electrolytes used in LIBs are salt-in-solvent solutions in liquid-state at usable-service temperatures, and these solutions contain mobile ions generated by the dissociation of salts (solute) in solvents, such as water (aqueous) or organic molecules (non-aqueous and proton-free, also called aprotic solvents).[26] An ideal electrolyte is required to exhibit certain characteristics and those requirements are listed in Figure 5.2.[59] An adaptation of battery cell fabrication from a LE to an all-solid-state electrolyte will positively augment the overall performance of a LB. Indeed, solid-state electrolytes (SSEs) retain benefits over liquid electrolytes in terms of facile, leak-free, and low-cost fabrication, flexible and compact architecture, reliability, high energy density, reduced weight, and laminated structures at the cell level.[60] In addition, they increase the high-temperature operability of LBs, where the conventional aqueous- and organic solvent-based (non-aqueous) liquid electrolytes tend to decompose, ignite, or evaporate. SSEs are largely categorised into polymer electrolytes (PEs) and inorganic solid electrolytes (ISEs).

5.2.1 Polymer Electrolytes

A PE can be defined as a polymer in which salt is dissociated, which enables ionic conduction through the polymer matrix. Indeed, in such systems, the ion conduction generally occurs through an ion-hopping mechanism, and the process is either coupled or decoupled from the segmental motion of the polymer chains (or polymer chain dynamics).[61] Ionic conductivity in PEs depends on several factors:[62,63]

- total ion concentration,
- ion mobility (ion transport properties such as diffusion),
- ion charge (the type of salt anion/cation),

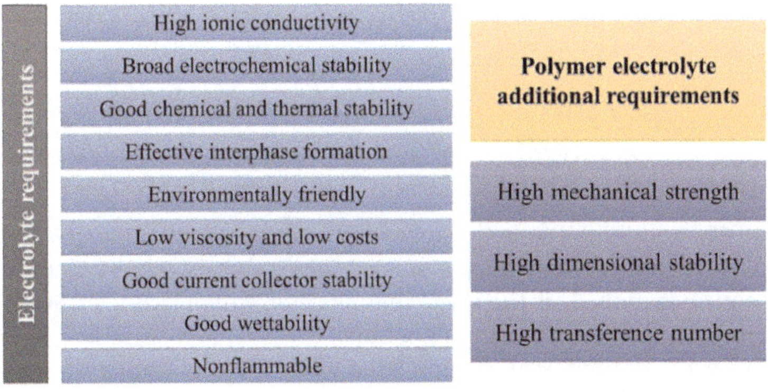

Figure 5.2 Fundamental requirements on an ideal electrolyte system. The additional needs of polymer electrolytes are indicated.
Adapted from ref. 48 with permission from The Electrochemical Society, Copyright 2019.

- polymer characteristics (*i.e.*, functional groups, viscosity, crystallinity, molecular weight, dielectric constant, *etc.*), and
- influence of additives (filler, plasticiser, *etc.*).

Indeed, the ionic conductivity is directly proportional to the number of free charge carriers and their mobility, as expressed by eqn (5.1).

$$\sigma = \Sigma n_i\, z_i\, \mu_i \tag{5.1}$$

In eqn (5.1), n_i is the number of charge carriers, z_i is the ion charge, and μ_i is the ion mobility.[64] Indeed, ionic conduction is mostly influenced by ion mobility (μ) and the diffusion coefficient (D). According to the Nernst–Einstein equation,

$$\mu = \frac{|z|eD}{k_B T} \tag{5.2}$$

$$n = \frac{\sigma k_B T}{(|z|e)^2 D} \tag{5.3}$$

In eqn (5.2) and (5.3), k_B is the Boltzmann constant and T is the absolute temperature. Ionic conductivity (σ) changes with temperature. At higher temperatures, faster segmental movements are achieved by increased polymer expansion, leading to more free voids for ion movements and reduced viscosity of the amorphous regions. In theory, ionic conductivity is explained either by the Arrhenius or by the Vogel–Tammann–Fulcher (VTF) models. The Arrhenius model of ionic conductivity implies that the ion migration predominantly occurs *via* ion hopping. In contrast, the VTF model is closely related to the free volume model, and ion transport primarily occurs through the continuous motion and relaxation of the polymer chains.[65] The equations used for these two fits are given below:

$$\sigma = A\exp\left(\frac{-E_a}{RT}\right) \quad \text{(Arrhenius equation)} \tag{5.4}$$

$$\sigma = A\exp\left(-\frac{E_a}{R(T-T_0)}\right) \quad \text{(VTF equation)} \tag{5.5}$$

In eqn (5.4) and (5.5), E_a is the activation energy, R is the universal gas constant, T is the temperature, and A is a prefactor that is related to the charge carrier concentration. T_0 is the Vogel temperature, which is at least 50 °C below the glass transition temperature (T_g). Since the 1970s, researchers have been working on developing an ideal ion-conducting PE for ambient-temperature LMBs and LIBs.[66–68] The requirements for ideal PEs are similar to those for LEs. However, they are expected to provide additional characteristics for the separator, such as the ones shown in Figure 5.2. Indeed, PEs are potential candidates to replace the presently used organic carbonate-based LEs; in principle, these solid-state systems should afford superior mechanical properties to avoid or repress the formation of HSAL deposits and related short-circuits.[69–71]

5.2.2 Polymer Electrolyte Characterisation

The properties of the PE should be characterised from a polymer point of view, as well as from an electrochemistry point of view. The various properties to be analysed and the methods used for such analyses, along with the necessary details, are given in Table 5.1.

5.2.3 Inorganic Solid Electrolytes

Inorganic materials having mobile ions are generally known as ISEs, and such materials are found in various forms such as crystalline, polycrystalline, and glassy/amorphous.[72–75] In crystalline ISEs, the ions migrate through the rigid crystal lattices *via* a hopping mechanism. However, the ion conduction mechanism in amorphous ISEs is not well understood.[72] A few ISEs, such as sulfide glasses, thio-LISICON (lithium super ionic conductor), and glass ceramics, exhibit ambient-temperature ionic conductivities of the order of 10^{-2}–10^{-3} S cm^{-1}, which are comparable to non-aqueous LEs (>10^{-3} S cm^{-1} at room temperature (RT)).[75,76] Nevertheless, high cost, tedious processing conditions, grain boundary resistances, and high charge-transfer resistances due to a chemically reactive electrode|electrolyte interface are among the critical issues to be addressed to warrant their extensive use in large-scale LBs.[77–80] A brief comparison of the three major electrolyte systems is summarised in Table 5.2.

5.2.4 Progression of Polymer Electrolytes

A macromolecular system that contains a substantial portion of repeating units as ionic or ionisable groups or both is termed as a polyelectrolyte, polymer electrolyte, or polymeric electrolyte.[81] However, a solution of lithium salt (Li^+X^-, *e.g.*, $LiPF_6$) in a polymer matrix (*e.g.*, poly(ethylene oxide) (PEO)),[82] which exhibits ionic conduction due to the mobility of lithium cations (Li^+) and anions (counter ions, X^-) are popularly called PEs among electrochemists. The Li^+-ion transport in PEO-based PEs is coupled with segmental motion and the local relaxation processes of the –EO– repeating units. Indeed, PEO is a highly crystalline material, and good polymer chain segmental mobility is achieved only in the amorphous state. Thus, the PEO-based PEs deliver an ambient-temperature ionic conductivity of $\approx 10^{-5}$–10^{-6} S cm^{-1}. Indeed, this drawback has limited the application of all-solid-state PEs in LMBs to elevated temperatures only (>60 °C).[83] Researchers have explored other possibilities to increase the ionic conductivity, especially by controlling the crystallinity of the polymer matrix by adding inorganic fillers, specific additives, or plasticisers, hence promoting low-temperature ionic conductivity.[84,85] These approaches of increasing the ionic conductivity by different means lead to the creation of

Table 5.1 The major properties and related analysis techniques used for the characterisation of PEs, along with a short description.

Measured property	Analysis technique	Details
Glass transition temperature (T_g)	DSC	To measure physical properties such as softening temperature, crystallinity, T_g, *etc.*
Crystallinity	XRD, SAXS	To measure crystallinity, morphology, and exfoliation properties, especially for polymer composites and block copolymers.
Mechanical properties	DMTA, DMA, rheology measurements.	To measure mechanical properties such as Young's modulus, elongation at break, shear modulus, *etc.* under various forces such as tensile, compression, or torsion.
Thermal stability	TGA	To measure the polymer decomposition temperature using gravimetry analysis.
Molecular weight of polymers	GPC-SEC, dynamic light scattering	To measure the molecular weight of a polymer, which determines the viscosity, length of chain, *etc.*
Ionic conductivity and electrode|electrolyte compatibility	Electrochemical impedance spectroscopy (EIS) at the desired temperature range.	With the help of an equation used to calculate ionic conductivity (σ) from the resistance (R_b) values, where A is the area, and L is the thickness of the membrane. $$\sigma = \frac{L}{R_b{*}A} \quad (5.6)$$ Resistance values such as charge transfer resistance (Rct) corresponding to the interface of a symmetric Li‖Li cell or LB cell. One can analyse the interphase layers and compatibility of PEs against various electrodes using EIS.
Chemical properties	FT-IR, Raman spectroscopy, solid-state NMR, *etc.*	To study the polymerisation reaction, interphase studies, polymer degradation, *etc.*
Transference number (ion transport properties)	Polarisation method (electrochemical method including EIS and chronoamperometry), PFG-NMR – a measurement of molecular diffusion coefficient, *etc.*	The fraction of the total current contributed by an ion in the PE is calculated using the Bruce and Vincent equation: $$T_{Li^+} = \frac{I_{ss}(V - I_0 R_0)}{I_0(V - I_{ss} R_{ss})} \quad (5.7)$$ where I_0 and I_{ss} are the initial and steady-state currents, R_0 and R_{ss} are the initial and steady-state resistance values, and V is the applied voltage. In an ideal single-ion conducting PE system, the cation transport number (T_{Li^+}) should be unity.
Electrochemical stability	CV, LSV	The onset of current and the membrane's ability to resist redox processes at the respective electrodes can be determined from the *I–V* plot.
Dendrite resistance and interphase characteristics of PEs	Galvanostatic (constant current) charge-discharge test coupled with EIS, XPS, FTIR, NMR, SEM, *etc.*	The ability of a membrane to resist against short-circuit induced by dendrites. To study morphology, chemical characteristics of the respective interphases (SEI and CEI).
Long-term cycling	Galvanostatic charge-discharge test.	The ability of a membrane to function at a certain voltage window against a cathode in a Li-metal or Li-ion cell for a long time.

Table 5.2 A comparison of the general properties, advantages, and disadvantages of various types of electrolytes used in LBs.

Electrolyte	Properties	Advantages	Present challenges
Liquid electrolyte *e.g.*, 1 M $LiPF_6$ in 1 : 1 v/v EC : DMC	Ionic conductivity: high Mechanical strength: low Flammability: high Toxicity: medium-high	High ionic conductivity, low costs, high active material utilisation, good ion transport, reaction kinetics, and established manufacturing process.	Electrolyte leakage from the battery cell stack, safety hazards due to fire and limited physical and chemical stability, electrolyte refilling, and formation are necessary for battery assembly.
Polymer electrolyte *e.g.*, LiTFSI in PEO (EO : Li ratio 20 : 1)	Ionic conductivity: low-medium Mechanical strength: medium-high Flammability: low-medium Toxicity: medium	Safety, flexibility, physical and chemical stability, leakage-free, thin-cell processing, and thermal stability.	Low ionic conductivity at RT, low ion diffusion properties and kinetics, non-uniform and thick interphase with electrodes, low transference number.
Inorganic solid electrolyte *e.g.*, sulfide ($Li_2P_2S_6$), perovskite ($Li_{3.3}La_{0.56}TiO_3$), garnet ($Li_7La_3Zr_2O_{12}$), LiPON ($Li_2P_5O_6N_5$).	Ionic conductivity: medium-high Mechanical strength: low-medium Flammability: low Toxicity: medium	Thermal stability, safety, good physical and chemical stability, leakage-free, high Li^+-ion transference number.	In part high grain boundary resistance, high activation energy for ion mobility, rigid and non-flexible structure, non-conformal interface with electrodes, in part high-temperature processing or sintering is required.

different classes of PEs. Put simply, one can state that the developed PEs fall into three main categories,[79] namely,

1. solid polymer electrolyte (SPE),
2. gel polymer electrolyte (GPE), and
3. polymer composite electrolyte (PCE).

A complete classification of polymer electrolyte systems is not provided here. However, the details can be found elsewhere.[68,86,87] A short overview of the three major classes of PEs is given in the following sections.

5.2.4.1 *Solid Polymer Electrolytes*

An ion-conducting solid/dry medium that serves as an ionic conductor and an electronic separator to isolate the electrodes to avoid an electronic short-circuit is called an SPE. In an SPE, the polymer host can solubilise (act as a solvent) single or multiple types of salt,[88] without the help of any types of liquid, such as plasticisers, solvents, *etc.* In general, SPEs can be divided into two classes according to the preparation process, namely, the 'salt-in-polymer' (the concentration of the polymer is higher than the selected salt) and the 'polymer-in-salt' (the concentration of the salt is higher than the selected polymer matrix) systems. The majority of research since the discovery of PEs has been devoted to the development of salt-in-polymer systems.[48,66] PEO-based SPEs, which contain various kinds of lithium salts, have been in focus since the first report of ionic conduction in a polymer matrix by Wright *et al.* in 1973.[89] This discovery led to the foundation of intensive research activities regarding the promising aspects of PE-based LMBs, also called lithium-metal polymer batteries (LMPBs), for the next generation of electrochemical power sources.

5.2.4.1.1 Lithium Salts. For any kind of electrolyte, ion conduction, for instance through salts, is an integral component, and the same is true also for all types of PEs. Various salt structures and chemistries are used for the preparation of PEs, and their characteristics are studied in detail. The basic parameters that decide the characteristics of the PE from the salt point of view are given in Figure 5.3, where the fundamental requirements of lithium salts are summarised.

Various kinds of lithium salts[90] with different anion chemistries are shown in Figure 5.4. Indeed, during the last 35 years, several new salt chemistries based on borates, phosphates, imides, *etc.* have been studied in SPEs and LEs. Every salt anion has its own merits and demerits.[28,91,92] In most cases, especially for LEs, more than one type of lithium salt has been used. Nevertheless, the second or third salt is used in small additive amounts, unlike the primary salt, which is present in the tens of moles to several moles range. In particular, $LiPF_6$, due to its high ionic conductivity in carbonate-based LE systems (10 mS cm^{-1} at 20 °C), has been the dominant Li source for LE compositions for almost three decades.

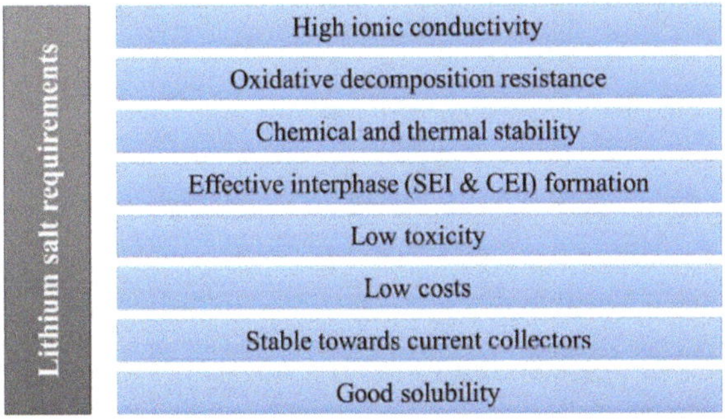

Figure 5.3 Requirements for lithium salts for application in LBs.

$LiPF_6$ in electrolyte solution decomposes above 50 °C to form LiF and PF_5, and further to HF and POF_3 in the presence of even a small amount of moisture.[47,93,94] These reactions at elevated temperature lead to the development of thermally stable salt systems for application in SPEs and, in this respect, the most studied salts[95,96] are LiTFSI, $LiClO_4$, and $LiSO_3CF_3$, with $LiClO_4$ being known to be rather unsafe. However, very recent trends indicate that other types of salts are producing interesting results, especially those having organic anions (Figure 5.4). Additionally, LiFSI, LiDFOB, LiTDI, $LiBF_4$, LiBOB, *etc.* have been used as additives in lower amounts.

5.2.4.1.2 Polymer Matrices. The polymer matrix is an integral component of an SPE membrane, which determines the physicochemical and electrochemical characteristics and the performance in an LMPB cell. Various kinds of polymer backbone/chain architectures have been used for the development of SPEs, namely, homopolymers, copolymers, random polymers, block copolymers, crosslinked networks, *etc.* Some of the most common polymer architectures are given in Figure 5.5.[48]

Several polymer matrices are employed as hosts for the lithium salt, and most of them are based on a homopolymer architecture. The main polymers that have been studied so far include, PEO,[97] poly(acrylonitrile) (PAN),[98] poly(methyl methacrylate) (PMMA),[99] poly(vinyl chloride) (PVC), poly(vinylidene difluoride) (PVDF),[100] and poly(vinylidene difluoride-*co*-hexafluoropropylene) (PVDF–HFP).[101] In addition, many other polymer matrices, such as poly(ethyleneimine), poly(carbonates), poly(vinyl acetate) (PVAc), poly(vinyl carbonate) (PVCA), poly(oligoethyleneoxy phosphazene), polysiloxanes, carboxymethyl cellulose (CMC), *etc.* have been studied.[102–104] Indeed, these polymers have been used in blends as well; for example, in several research works, PEO is blended with PAN, PMMA, or PVDF to take advantage of the features of both. The physical properties, advantages, and disadvantages of the most commonly used polymer matrices for the production of SPEs are summarised in Table 5.3. Additionally, the chemical structures of the major polymer matrices that have been widely employed in SPEs are given in Figure 5.6.

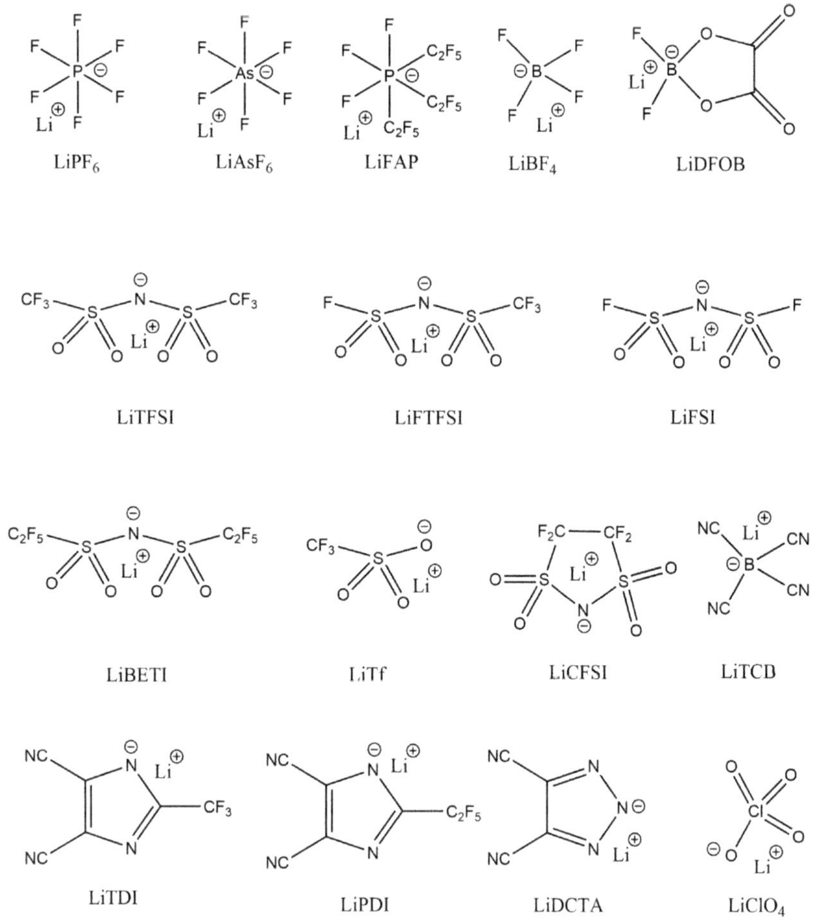

Figure 5.4 Chemical structures of various lithium salt anions: lithium hexafluoro phosphate ($LiPF_6$), lithium hexafluoroarsenate ($LiAsF_6$), $LiPF_3(CF_2CF_3)_3$ (LiFAP), lithium tetrafluoroborate ($LiBF_4$), lithium difluoro(oxalato)borate (LiDFOB), lithium bis(trifluoromethanesulfonyl)imide (LiTFSI), lithium (fluorosulfonyl) (trifluoromethanesulfonyl) imide (LiFTFSI), lithium bis(fluorosulfonyl)imide (LiFSI), lithium bis(pentafluoroethanesulfonyl)-imide (LiBETI), lithium triflate (CF_3SO_3Li, LiT_f), lithium carbonylbis-(fluorosulfonyl)imide (LiCFSI), lithium tetracyano borate ($LiB(CN)_4$, LiTCB), lithium 2-trifluoromethyl-4,5-dicyanoimidazole (LiTDI), lithium 4,5-dicyano-2-(pentafluoroethyl)imidazole (LiPDI), lithium 4,5-dicyano-1,2,3-triazolate (LiDCTA), and lithium perchlorate ($LiClO_4$).

Other emerging polymer architectures in terms of chemical structure, such as single-ion-conducting polyelectrolytes (SIC-PEs),[114–117] have been extensively studied globally for improving the ionic transport of SPEs, especially, the transference number.[118] Anionic polyelectrolytes or polymeric single-ion conductors are composed of a polymer backbone, where an anionic moiety is a repeating unit, which carries a lithium counter ion that can move freely.[119]

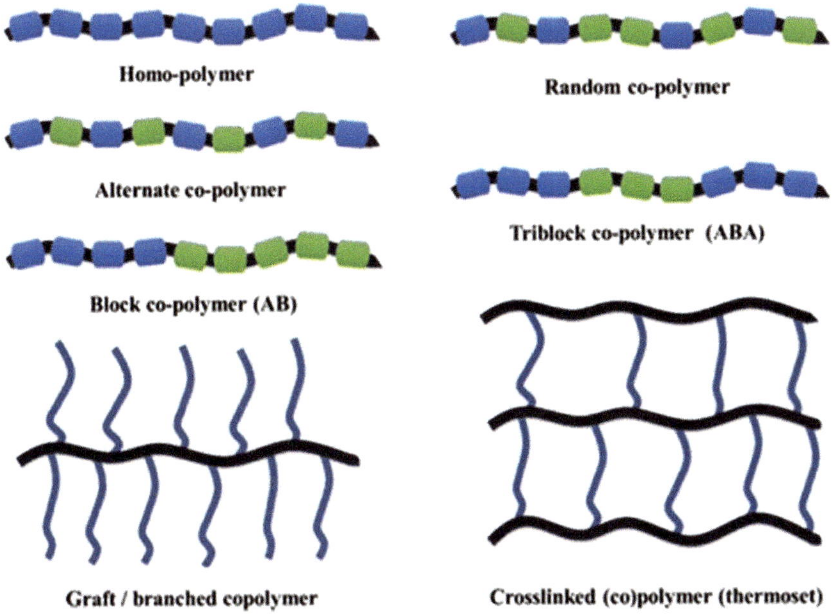

Figure 5.5 Polymer chain architectures commonly employed as SPE constituents in lithium-metal polymer batteries.
Adapted from ref. 48 with permission from The Electrochemical Society, Copyright 2019. Adapted from ref. 330 with permission from Elsevier, Copyright 2017.

Such systems deliver good ionic conductivity and high a Li^+-ion transference number (t_{Li^+}) close to unity.[114] In this respect, various block/random/cross-linked (co)polymer electrolyte systems[120] have been investigated recently, and they have the potential to replace the conventional polymer backbones that have been used so far in all forms of SPEs. Many research groups around the world are focusing on exploring new polymer chemistries that meet the requirements of an ideal polymer matrix for developing the next generation of SPEs, and such works are expected to deliver promising results soon.[117,121–124]

5.2.4.2 Gel Polymer Electrolytes

Since the first report, there has been an enormous amount of research work to improve the overall properties of SPEs, such as ionic conductivity, electrochemical stability, ion transport, and mechanical characteristics. However, due to the high crystallinity and slow polymer chain dynamics, these polymer matrices exhibit a very low ionic conductivity at RT ($\sim 10^{-6}$ to 10^{-5} S cm^{-1}). To an extent, the challenges related to the low ionic conductivity in SPEs are overcome by the introduction of GPEs.[125]

In a GPE, a solvent (liquid plasticiser) and salt are added into the polymer matrix, which accelerates ion mobility due to an increase in the amorphous regions of the macromolecular system, polymer chain mobility, and salt

Table 5.3 Summary of the properties of the major polymer matrices that have been employed in SPEs for lithium-metal polymer batteries.[105–113]

Polymer acronym	Crystallinity/%	T_g/°C	T_m/°C	Advantage	Disadvantage
PEO	83	−67	65	Low cost, easy availability, good compatibility with lithium metal.	High crystallinity, low oxidation stability, incompatibility with high-voltage cathodes.
PAN	28–34	125	319	Good mechanical strength, good heat resistance, good flame retardancy, good chemical stability.	Bad interphase formation with lithium-metal, poor processability, brittle, high T_g.
PMMA	—	105	130	Biocompatible, 100% recyclable, transparent, good impact resistance, easy jellification, good scalability.	Limited heat (80 °C) and chemical resistances, unstable interphase formation with lithium metal, rigid, high T_g.
PVC	10	−40	171	Low cost, easy to upscale, fire resistant, commercially available.	Bad processability, plasticisers are needed for processing, Li-metal incompatible, narrow electrochemical stability window.
PVDF	35–70	−38 to −42	171	High dielectric constant ($\varepsilon = 8.4$), high salt solubility, high anodic stability, high mechanical strength, exceptional chemical resistance, good processability, commercially available.	Incompatible against lithium metal, expensive, strong solvents are required for processing.
PVDF–HFP	10–30	−130 to −50	100 to 143	Highly amorphous compared with PVDF, improved solubility over PVDF, solubility in a wide range of solvents, lower T_g than PVDF, good chemical resistance.	Incompatible against lithium-metal, expensive.

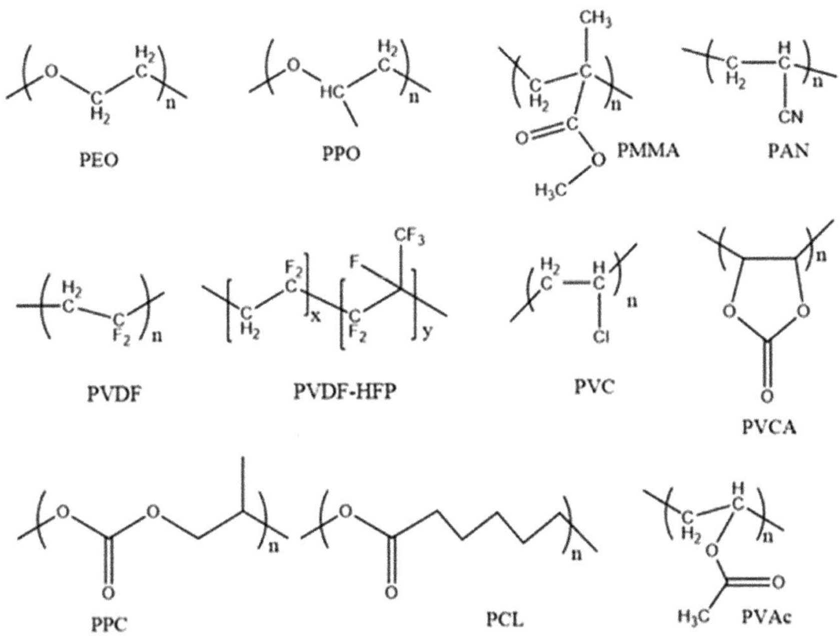

Figure 5.6 Chemical structures of polymer matrices that have been largely used for the preparation of SPEs. Poly(ethylene oxide), PEO; poly(propylene oxide), PPO; poly(methyl methacrylate), PMMA; poly(acrylonitrile), PAN; poly(vinylidene difluoride), PVDF; poly(vinylidene difluoride-*co*-hexafluoropropylene), PVDF-HFP; poly(vinyl chloride), PVC; poly(vinyl carbonate), PVCA; poly(propylene carbonate), PPC; poly(caprolactone), PCL; and poly(vinyl acetate), PVAc.

dissociation.[126] Organic solvents and room-temperature ionic liquids (RTILs) have been used for plasticising the polymer matrix.[122,127–130] Indeed, RTILs are organic salts with large cations, which are molten salts at RT, and are directly used as liquid electrolytes or plasticisers in electrochemical devices.[122,131,132] In the case of organic solvents, linear and cyclic carbonates,[133] lactones, esters, ethers,[134,135] sultones, nitriles, *etc.* have been widely used,[136–141] yet, most commonly used organic plasticisers are mixtures of organic carbonates. In a GPE, the ionic conduction predominantly occurs through the plasticiser, solvent, or ionic liquid that is incorporated into the polymer matrix. Even though the mechanical properties are inferior to those of SPEs, GPEs exhibit higher ionic conductivity and superior electrochemical performances at RT.[142] Due to the advantages of ionic conductivity, most of the available reports on PEs or LBs are dominated by GPEs. The structures of the most common organic plasticisers are given in Figure 5.7 and the structures of the most important types of organic cations of RTILs investigated as solvents/plasticisers in GPEs for LMPBs are given in Figure 5.8. These cations, together with the anions of lithium salts presented in Figure 5.4, form the most common RTILs that have been investigated in LBs.

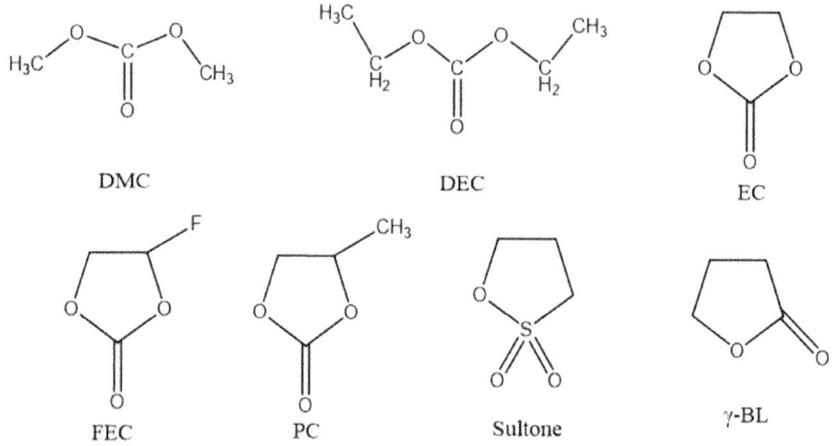

Figure 5.7 Chemical structures of the most commonly used plasticisers for GPEs.

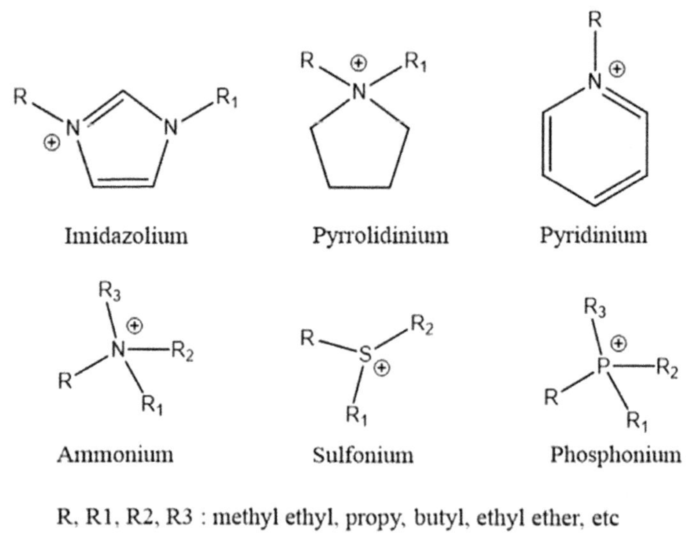

Figure 5.8 The structures of the most important types of organic cations of the RTILs investigated as plasticisers in GPEs for LMBs.[143]

Most of the lithium salts used in SPEs have also been evaluated in GPEs, however, the plasticiser content varies over a wide spectrum. In some cases, the role of the polymer was just to jellify the electrolyte system to avoid leakage, whereas in other cases, the researchers attempted to impart mechanical properties (*e.g.*, Young's modulus) to these GPEs; hence, the plasticiser content was limited to lower values. However, when the plasticiser content is low, the ionic conductivity is significantly hampered. Most of the polymer matrices, such as PEO, PMMA, PVDF, polysiloxane, polycarbonate, and blends of the same, that have been employed in SPEs have also been employed in GPEs. Recently, new varieties of polymer matrices have been employed as part of finding the most suitable polymer matrix for GPEs, and research is still ongoing.

5.2.4.3 Polymer Composite Electrolytes

A PCE is also an ionically conducting and electronically insulating system, which should be a multiphase material that forms domains of different sizes. Therefore, a PCE is an electrolyte as well as a separator that could exist in the gel- or truly solid-like form.[79,85] The addition of (nano) particles into the PE system can modify several physicochemical and electrochemical characteristics, leading to numerous possibilities for the development of high energy and high capacity LMPBs. A polymer composite is a composite in which at least one of the constituting components is a polymer (*IUPAC Recommendations 2004*).[144] It is developed to advance the inherent and applied characteristics of the polymer system, namely, thermal, mechanical, and electrical properties, and processing conditions.[145] Indeed, in a polymer nanocomposite (PNC) at least one of the phase domains is composed of particles in the order of nanometres (*IUPAC Recommendations 2008*).[146] Additionally, several types of fabrications and designs, such as fibre, foam, sheet-like architectures, *etc.*, can be developed to exploit the functions and performance of PNCs. The basic components of any PCE include polymer matrix (*e.g.*, PEO, PVDF, PMMA, PVDF–HFP, *etc.*), lithium salt (*e.g.*, LiTFSI), fillers, and several target-specific additives. The commonly used polymer matrices and lithium salts are described in Section 5.2.1.

5.2.4.3.1 Fillers. The particle dispersion, their distribution in the polymer matrix, particle aspect ratio, the particle/polymer matrix interface, concentration, networking of particles, and the percolation threshold are fundamental features in the designing of PNCs and related PCEs. Indeed, uniform dispersion of particles along with optimum interfacial interactions with the polymer matrix are required for achieving enhanced mechanical properties and effective intercalation. The fillers are either added into the polymer matrix or generated *in situ* using several processes. Generally, PCEs are synthesised mainly by four techniques as given in Table 5.4, and, depending on the type of the method and materials used, three types of microstructure can be achieved, namely, unintercalated (or microcomposite), intercalated (and/or flocculated), or exfoliated (or delaminated).[147] Intercalation is a process in which the polymer chains are inserted between the interlayers of the filler; subsequently, the *d*-spacing of layered nanomaterials of the polymer composite is increased (usually $\approx$20–80 Å), whereas exfoliation is a process in which the layered nanomaterials are delaminated, hence the nanolayers are suspended in the polymer matrix.[148] In the case of exfoliation, the *d*-spacing (>10 nm) between the layers is significantly expanded, and the intercalation forces between the nanolayers of the fillers are diminished.

In general, the fillers added into a polymer composite can influence the characteristics of the PCE, depending on whether they take part in ionic

Table 5.4 Summary of various polymer composite processing techniques along with their advantages and disadvantages.[149,150]

Process	Advantage	Disadvantage
Melt intercalation	Most widely used for thermoplastic PNCs, no solvents are used, ecofriendly, easy to upscale, and economical.	High-temperature process, the polymer matrix may degrade, and functionalised filler damage.
In situ polymerisation	Polymerisation can be carried out using various sources, such as heat, radiation, initiator diffusion, or cationic exchange, better exfoliation than melt, and exfoliation adsorption.	Processing conditions influence the uniformity of the PNC.
Template synthesis (sol–gel technology)	Form the filler in an aqueous solution or gel, and used for the synthesis of the double-layer hydroxide-based nanocomposite.	Less compatible with layered silicates, high-temperature process, polymer degradation, filler aggregation or agglomeration, and very high filler loading is challenging.
Exfoliation adsorption (polymer or prepolymer intercalation)	Suitable for water-soluble polymers, easy processing, and up-scalable.	Solution-based process, incomplete solvent evaporation after the process, and leads to solvent-related impurities and not ecofriendly.

conduction (active) or they induce physicochemical changes (*e.g.*, amorphous nature) to the polymer matrix (passive).[150] Some examples of active fillers include lithium garnets, perovskites, *etc.*, which take part in the ion conduction process; hence, ionic conductivity and ion transport properties are enhanced. When active fillers are used in PCEs, in some cases, the sensitivity of these materials leads to the formation of insulating components, such as LiOH, Li_2CO_3, *etc.* In some cases, even though there may be a change in the physical properties of the polymer matrix by the addition of active fillers, the main reason for their incorporation into the electrolyte system is to increase lithium-ion concentration and enhance the t_{Li^+} values. Some examples of the most widely used passive fillers are Al_2O_3, ZrO_2, SiO_2, *etc.*,[151,152] which may enhance the ionic conductivity and transport properties by acting as a solid plasticiser. The filler softens the polymer backbone by reducing the T_g, or crystallinity. The passive fillers improve the ionic conductivity of PCEs mainly by (1) increasing the amorphous nature of the polymer matrix and (2) creating new ion-conducting pathways through the interaction (Lewis acid–base interactions) of ionic species from the Li-salt and surface functional groups of the fillers.[150] Table 5.5 summarises the advantages and disadvantages of the active and passive fillers used for the synthesis of PCEs, along with some of the most common examples.

Dissanayake *et al.*[154,155] used the Lewis acid–base theory to explain the increase in ionic conductivity arising from the interactions among the polymer, salt, and

Table 5.5 The advantages and disadvantages of fillers (both active and passive) used for the synthesis of PCEs.[150,153]

Filler	Examples	Advantages	Disadvantages
Active	$Li_{1.5}Al_{0.5}Ge_{1.5}(PO_4)_3$ (LAGP), $Li_5La_3Ta_2O_{12}$ (LLTO), $Li_{3.25}Ge_{0.25}$-$P_{0.75}S_4$ (LGPS).	Ionic conduction occurs through the particles, possess Li^+-ions for ionic conduction, good ionic conductivity, and high cation transference number.	Moisture sensitive, in part CO_2 sensitive, incompatible with Li-metal, stringent synthesis protocols, and high amount of lithium concentration required.
Passive	ZrO_2, SiO_2, BaO_2, $Ca_3(PO_4)_2$	Li^+-ion conduction occurs through the particle surface. No direct contribution to Li^+-ion concentration. Easy to prepare, inexpensive, and forms a good interface with Li-metal. Enhanced transport properties and ionic conductivity, with *in situ* syntheses.	Moderate contribution to ionic conduction and dispersions are possible with solvent-based processes only.

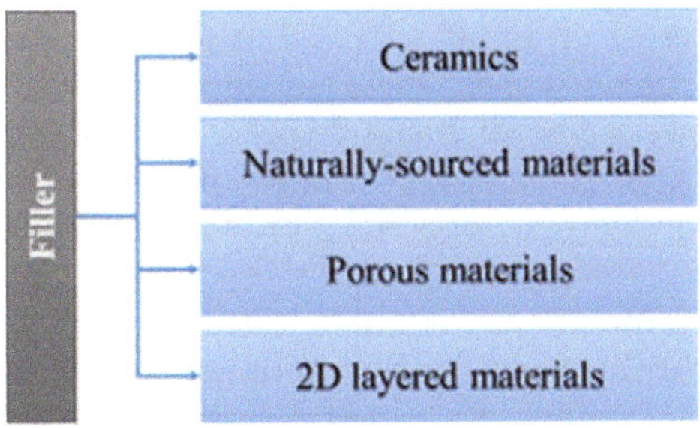

Figure 5.9 A simplified classification of the types of fillers used in PCEs for LBs.

the surface groups of the fillers. Among the metal oxide fillers, especially for Al_2O_3 nanoparticles, the acidic form tends to deliver the highest ionic conductivity. Besides, the physicochemical and electrochemical characteristics of the PCEs are improved by the addition of inorganic fillers; in some cases, these fillers can act as an impurity scavenger (*e.g.*, H_2O).[156] However, an optimum amount of fillers is required to achieve the required properties, whereas, an excess amount may completely freeze the polymer chain segmental motion resulting in lower ionic conductivity, as ionic conduction predominantly occurs through the polymer matrix. Hence, fine-tuning is necessary while using (nano)fillers for synthesising PCEs. The commonly used fillers can be classified into four subclasses, as shown in Figure 5.9 and the detailed explanation is given below.

5.2.4.3.1.1 Ceramics. Ceramics are rigid materials consisting of metals bonded to C, N, or O, involving an infinite three-dimensional (3D) network of sintered crystalline grains. The term ceramic is generally applied to the family of inorganic and non-metallic materials that are exposed to high temperature during manufacture or use (*IUPAC Recommendation 2008*).[146] As mentioned above, ceramics can have a passive or active role in Li^+-ion conduction. These materials can be further divided into several classes; depending on the number of metal types present, they can be called simple or mixed metal oxides. Simple metal oxides, such as Al_2O_3, MgO, SiO_2, ZnO, and ZrO_2, have often been used in PCEs.[157,158] These fillers can be synthesised externally (*ex situ*) and added into a PE mix to obtain the PCE. In some cases, the nanomaterials can be synthesised within (*in situ*) the PCE matrix by adding nanomaterial precursors. For example, TiO_2 nanoparticles can be *in situ* generated from its precursors using a sol–gel process.[159] A typical reaction mixture contains precursors, lithium salt, and a polymer matrix, which after mixing proceeds with several reactions (hydrolysis and condensation) to form a PCE, as shown below:

$$\text{Hydrolysis:}\quad Ti(OC_2H_5)_4 + 4H_2O \rightarrow Ti(OH)_4 + 4C_2H_5OH \tag{5.8}$$

$$\text{Condensation:}\quad Ti(OH)_4 \rightarrow TiO_2 + 2H_2O \tag{5.9}$$

There are similar approaches to produce PCEs using inorganic–organic copolymers (organically modified ceramics, ORMOCERs),[160] which exhibit ionic conductivities of up to 10^{-4} S cm^{-1} at RT. In addition to the simple metal oxides mentioned above, there are mixtures of metal oxides where more than one metal type of metal is present. These mixed oxides (*e.g.*, $BaTiO_3$, $MgAl_2O_4$, $MgAl_2SiO_6$, *etc.*) have been applied as (nano)fillers in PCEs, where the ionic conductivity, interfacial properties, and ion transport characteristics are considerably improved.[161,162]

The active inorganic fillers (ceramic electrolytes or ISEs) contain mobile Li^+-ions in the structure. These materials can be packed into three major types, namely, oxides, phosphates, and sulfides. Perovskites, such as $Li_{0.33}La_{0.55}7TiO_3$ (LLTO),[163] and garnets, such as $Li_7La_3Zr_2O_{12}$ (LLZO),[164] are good examples of oxide-based solid-state Li^+-ion conductors. The La:Li ratio, the grain size, and the grain boundary composition play a crucial role in determining the ionic conductivity and electrochemical characteristics. Representative examples of phosphates include LAGP, LATP, and $LiTi_{0.5}Zr_{1.5}(PO_4)_3$[165] and among the phosphates, zirconium-based phosphates have the lowest ionic conductivity.[76] The sulfides, such as $Li_2S–P_2S_5$, $Li_2S–Ga_2S_3–GeS_2$, *etc.*[80] are glass ceramics, which are partially crystalline and show higher ionic conductivity at lower temperatures. Nevertheless, the crystalline form of sulfides, such as $Li_{3.25}Ge_{0.25}P_{0.75}S_4$ (LGPS),[166] also exhibits high ionic conductivity close to that of $Li_2S–P_2S_5$.

5.2.4.3.1.2 Naturally Sourced Materials. Cellulose-derived materials,[167–169] such as CMC, and their nanosized forms,[170] such as cellulose microfibrils (MFCs)[171] and whiskers, have been used in PCEs as reinforcing agents. Such materials can be termed as organic fillers.[172] Also, plant-derived materials,[173] such as carrageenan, xanthan gum, and animal-derived chitin, chitosan, *etc.*[174] have been used in PCEs to obtain improved mechanical and electrochemical properties.[175]

5.2.4.3.1.3 Porous Materials. Ordered mesoporous materials[176] have a high surface area, tunable and well-ordered microstructure and pore size; hence, they have found applications in catalysis, separation, and electronics. The most studied porous materials are from the zeolite family (MCM-41 and SBA-15). SBA-15 possesses uniform, long, and connected tubular channels having pore size ranges from 5 to 30 nm. Hence, these materials have been employed to enhance interfacial stability, ionic conductivity, lithium transport properties, and electrochemical stability in PCEs.[177] Metal organic frameworks (MOFs) are microporous solids with very high surface area, which are made of an infinite network of metal centres connected by organic linkers through metal–ligand coordination.[178] MOFs are applied in sensors, catalysis, gas storage, purification, and separation, in addition to their use in PCEs as fillers[179] (*e.g.*, a mixture of PEO + LiTFSI + Al-BTC-MOF (aluminium benzenetricarboxylate-MOF), which demonstrated two orders of magnitude increase in ionic conductivity at RT). Besides, modified MOFs have also been employed as solid-state Li^+-ion conductors for LMBs in which ionic conductivity values exceeding 0.3 mS cm^{-1} are achieved along with improved lithium transference numbers.[180,181]

5.2.4.3.1.4 2D Layered Materials. Two-dimensional layered nanomaterials have also been used as fillers in PCEs, among them layered silicates have been studied the most, especially for LMB applications. A detailed discussion of 2D layered nanomaterials and their application in PCEs, or as a separate surface protection thin layer in LMBs, is given in the following sections. Basic details about the most commonly used 2D nanomaterials, including graphene, boron nitride, MXenes, 2D layered clay, transition metal dichalcogenides, and phosphorene, are given in the respective sections.

5.3 Polymer Composite Electrolytes Based on 2D Layered Nanomaterials

A detailed discussion of various 2D layered nanomaterials has been included in other chapters; hence, such details or introductions are avoided in the following sections. Additionally, in this chapter, in particular in Section 5.2, a variety of polymer composites and their preparation, types of fillers, and their roles in the polymer matrix, the measurement techniques, processing methods, *etc.* are also briefly covered. Hence, in the context of this chapter, a detailed discussion on 2D layered nanomaterial-based PCEs is provided; thus, the following sections will summarise the important studies on PCEs based on graphene, boron nitride, MXenes, 2D layered clay, transition metal dichalcogenides, and phosphorene for LBs. In this respect, various electrolyte systems that contain gel, solid, mixed composites, *etc.* will be covered in detail. Additionally, 2D layered nanomaterials are also used as a surface protection layer (interlayer) between the anode and the electrolyte to avoid HSAL-based dendritic short-circuits; therefore a special focus on such applications of polymer composites is also summarised in the following sections.

5.3.1 Graphene in Lithium Batteries

Graphene in its pristine state is composed of a single layer of sp^2-arranged carbon atoms in a honeycomb crystal lattice and is a monolayer isolate of 3D graphite, wherein π-stacks of such sheets hold the lamellae of graphite in place with an interlayer *d*-spacing of 3.34 Å between each sheet. Owing to its narrow interlayer spacing, high thermal conductivity (5000 $W\,m^{-1}\,K^{-1}$), excellent intrinsic carrier mobility (200 000 $cm^2\,V^{-1}\,s^{-1}$), exemplary optical transmittance (97.7%), high Young's modulus (~1 TPa), and high surface area (2675 $m^2\,g^{-1}$), graphene and its derivatives, such as graphene oxide (GO), are some of the most researched 2D materials, with diverse applications.[182–191]

Graphene can be isolated from highly oriented pyrolytic graphite using micromechanical exfoliation, solvent-mediated exfoliation, or it can be obtained *via* chemical vapour deposition (CVD) and epitaxial growth methods.[192–198] GO is the oxidised form of a few layers of graphene with carboxyl (>COO−) and some hydroxyl (−OH) groups present at the edges, with epoxide and remaining hydroxyl moieties on the basal plane of the sheet.[199] It can be synthesised from graphite using Brodie's method, Hummers' method, or Staudenmaier's method.[200–202] Brodie's method was the first reported method to form GO from graphite, and uses a combination of fuming HNO_3 and $KClO_3$ as oxidising agents. Staudenmaier's method, on the other hand, uses a mixture of two-thirds of H_2SO_4 and one-third of HNO_3 along with $KClO_3$ for single-step complete oxidation of graphite. However, both methods led to the formation of explosive ClO_2 gas and a reaction time of four days, which is not an industrially feasible process. In 1958, Hummers and Offermann came up with a process for industrial-scale preparation of GO from graphite, popularly known as the Hummers' method. In this method, H_2SO_4, along with $NaNO_3$, is used as an intercalating agent for graphite, while $KMnO_4$ is used for oxidising the intercalated graphite. By using $KMnO_4$ as a strong oxidising agent, the reaction time reduces from four days to a few hours. Further, no explosive ClO_2 gas is formed in this process, and, since fuming HNO_3 is replaced with $NaNO_3$, acid fog is also eliminated.[203]

Graphene nanosheets (GNSs) have been shown to exhibit exemplary electrochemical performance as an active electrode material in LBs.[204–207] In several approaches, authors have reported very high specific capacities, above 1000 $mA\,h\,g^{-1}$ even at very high current densities (*e.g.*, 100 $mA\,g^{-1}$). Additionally, graphene has been used as a wrapping material to tightly hold nanoparticles, which is later used as an anode in LBs.[208] The atomically flat layers of mesoporous graphene provided a high surface area for Li^+-ion adsorption and intercalation while orthogonally directed open mesopores of the nanosheets offered efficient transport of ion diffusion, which delivered excellent Li^+-ion storage capacity. It is critical to note that high surface area contacts between GNSs and the electrolytes may result in a thick SEI layer in the first charge/discharge cycle resulting in a low coulombic efficiency[209] in many cases. It is well known that the electronic conductivity in GNSs is mainly due to long-range π-conjugated networks. Moreover, due to the functionalisation of graphitic carbons to GO, some of these networks get disrupted, which leads to a decrease in carrier mobility and concentration.[210] However, these defects do not potentially limit the use of GO as an electrode material. The hydrophilic oxygen functionalities, such as hydroxyl, carboxyl, and the epoxide linkages in GO, act as anchors for nanoparticles such as metals/metal oxides to attach onto the GO surface and its edges, which yields better interaction and better manifestation of properties compared to graphene.[211]

Graphene-based polymer composites have also been investigated as a host for superior physical and chemical sulfur confinement in S‖Li cells. Recently, Tan *et al.*[212] reported a hybrid composite of graphene aerogel decorated with MnO_2 in a poly(vinylpyrrolidone)–poly(vinyl alcohol) (PVP – PVA) interpenetrating network (IPN) (see Figure 5.10). It was observed that the initial discharge capacity of the composites was 1492 mA h g^{-1} at 0.1 C. Further, it retained a capacity of 755 mA h g^{-1} during 250 cycles at 0.2 C and only had 0.06% capacity decay per cycle during the next 500 cycles

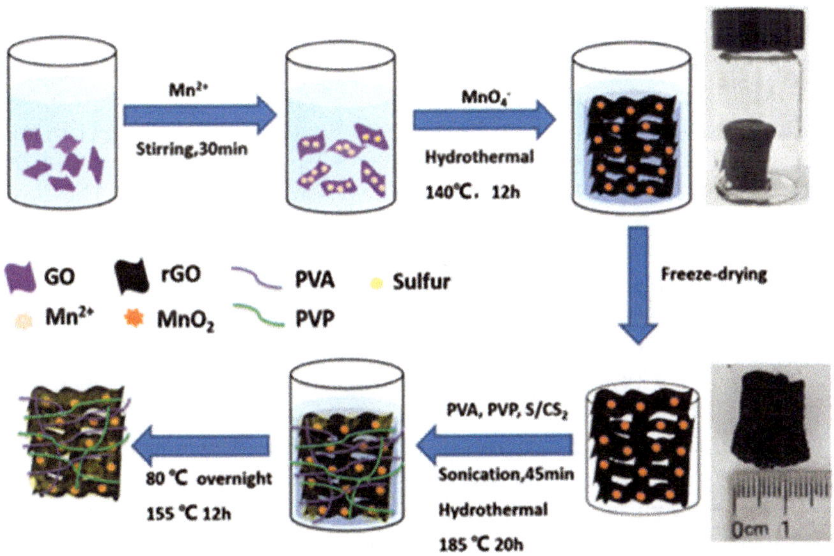

Figure 5.10 Schematic representation of the hybrid aerogel polymer composite with graphene-decorated MnO_2 particles.
Reproduced from ref. 212 with permission from Elsevier, Copyright 2019.

at 1 C. The next two sub-sections explore in detail the use of graphene and GO as a filler in PCEs and as an interlayer material in LMBs.

5.3.1.1 Graphene as a Filler in PCEs

Functionalised GO has been used as a filler in PCEs and some of the recent results highlighting their influence in LBs are given here. Shim *et al.*[213] fabricated a PCE by using a hybrid of an inorganic and organic branched/graft co-polymer matrix based on poly(ethylene glycol) methyl ether methacrylate (PEGMA) and 3-(3,5,7,9,11,13,15-heptaisobutylpentacyclo-[9.5.1.1.^{3,9}1.^{5,15}1^{7,13}]-octasiloxane-1-yl) propyl methacrylate (MA-POSS), where poly(ethylene glycol) (PEG)-grafted GO is used as a nanofiller in the LIB. The PCE with 0.2 wt% of modified GO enhanced the ionic conductivity from 0.011 mS cm^{-1} to 0.21 mS cm^{-1}. Indeed, the modified GO increases the amount of dissociated lithium salt in the electrolyte, which mainly arises from the Lewis acid–base interaction between the modified GO and the ions of the lithium salt.

Ye *et al.*[214] synthesised three types of polymer-functionalised graphene (polymer-FG), namely, PEG-FG_{brush}, polymeric ionic liquid (PIL)-functionalised GO brushes (PIL(TFSI)-FG_{brush}), and PIL(TFSI)-FG with different structural designs and morphologies (see Figure 5.11). Later, these functionalised nanofillers were incorporated into the PEO/$LiClO_4$-based SPE to fabricate a PCE to optimise the ion transfer conditions. It was observed that polymer-FG with PIL brushes significantly increased the Li^+-ion conductivity (O : Li = 8 : 1) of the PEO-based PCE by more than two orders of magnitude at 30 °C. Additionally, a 300% increment in tensile strength was achieved for the PCE with 0.6 wt% of PIL(TFSI)-FG_{brush}. Finally, the $LiFePO_4$|PIL(TFSI)-FG_{brush}-PCE|Li cell delivered a specific capacity of 156 mA h g^{-1} in the first cycle at 60 °C.

In a study by Liu *et al.*,[215] a PVDF/graphene-based gel-PCE (g-PCE) was obtained *via* a non-solvent-induced phase separation technique, and the effect of graphene on the ionic conductivity was investigated. The ionic conductivity of the PVDF-based GPE was 1.85 mS cm^{-1}, which doubled to 3.61 mS cm^{-1}, even with the incorporation of 0.002 wt% of graphene. Furthermore, the porosity of the PE improved from 57% in the PVDF-only GPE to 88% in the PVDF/graphene-based g-PCE. Additionally, the electrolyte uptake of the PVDF-only PE was increased from 186% to 470% in the PVDF/graphene-based g-PCE. The electrochemical stability window was also high (4.7 V *vs.* Li|Li^+), and the $LiCoO_2$|PCE|Li cell was cycled at RT for 100 cycles.

Jia *et al.*[216] used GO as a nanofiller in a PAN/$LiClO_4$ system to obtain a flexible PCE. Addition of 1 wt% GO led to an increase in ionic conductivity by one order of magnitude to obtain 0.4 mS cm^{-1} at 30 °C, a reduction in activation energy from 2.31 to 1.03 eV, and an increase in tensile modulus from 37 to 80 MPa was also observed. Furthermore, the 1.0 wt% GO-based PCE showed a satisfactory average discharge capacity of 166 mA h g^{-1} at 0.2 C in a $LiFePO_4$‖Li cell, which was better than the performance in the GO-free PE-type cell (136 mA h g^{-1}).

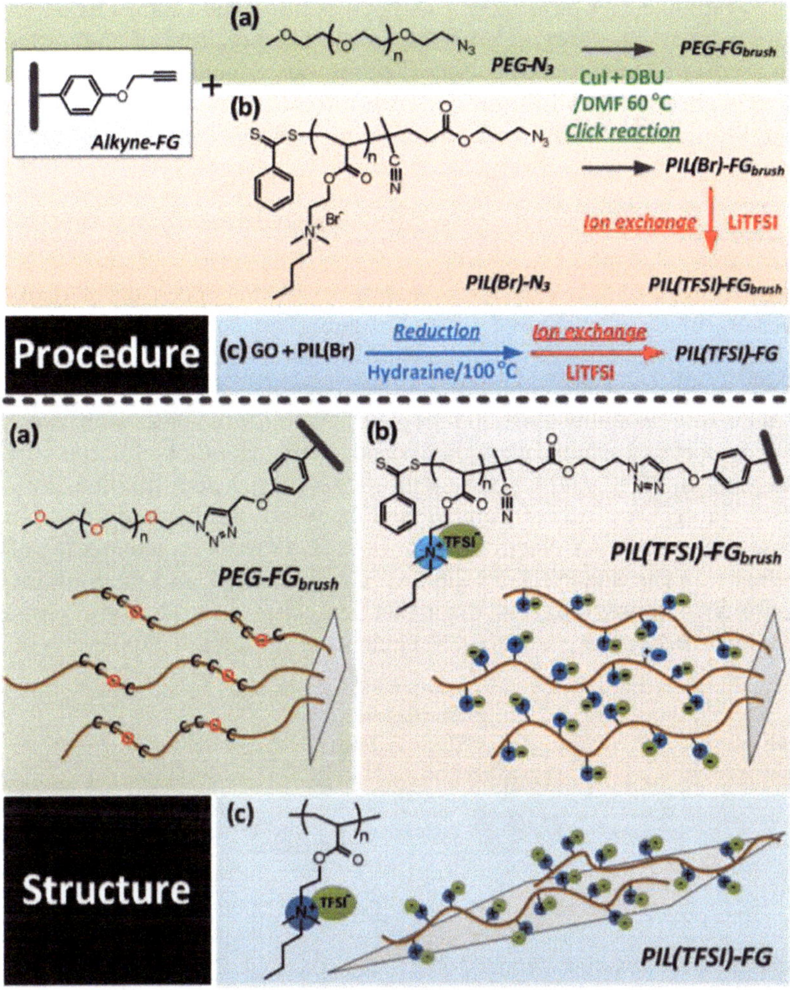

Figure 5.11 Scheme illustrating the synthesis and structure of polymer-functionalised GO brush (polymer-FG_{brush}) and PIL(TFSI)-FG.
Reproduced from ref. 214 with permission from the Royal Society of Chemistry.

To improve the ionic conductivity of highly loaded GO in a PIL matrix, Yang *et al.*[217] synthesised poly(1-propyl-3-vinyl-imidazolium) chloride *via* free radical polymerisation. The PE loaded with a high concentration of GO. Using the ice-templating technique, they obtained 3D polymer nanocomposite electrolytes. It was observed that the PIL-based PCE exhibits a high ionic conductivity of 0.31 mS cm^{-1} with 50 wt% GO loading, which is more than seven times higher than that of the pristine PIL electrolyte without any GO loading. However, they did not demonstrate any cycling behaviour in laboratory-scale test cells. In general, there have been attempts to use graphene, GO, or reduced GO (rGO) as fillers in PCE for LBs. However, future work is necessary to understand the full potential of such attempts with respect to their electrochemical properties, including oxidation stability, resistance towards lithium dendrite formation, *etc.* The results demonstrated in the above works are encouraging and must be pursued further.

5.3.1.2 *Graphene as an Interlayer in LMBs*

In addition to graphene as a filler in PCEs, it has also been used as a protection film in LMBs. Very few research works have been devoted to this, however, the results achieved are encouraging and some of these results are highlighted here. For example, Wu *et al.*[218] fabricated a multilayered porous carbon nanocomposite containing silica. For this, they embedded Si nanoparticles in a poly(vinylpyrrolidone) (PVP) matrix, which was later pyrolysed to form porous carbon layers composed of nitrogen-doped carbon framework, carbon black, and carbon nanotubes (CNTs). The combined Si-porous carbon layer was further sandwiched between flexible graphene sheets before using it as an electrode. Figure 5.12 shows the schematics of the synthesis. In the first cycle, a solid SEI layer is formed, and the multilayered Si-porous carbon/graphene electrode exhibits a maximum reversible capacity of 1020 mA h g^{-1} with 75% capacity retention after 100 cycles of use.

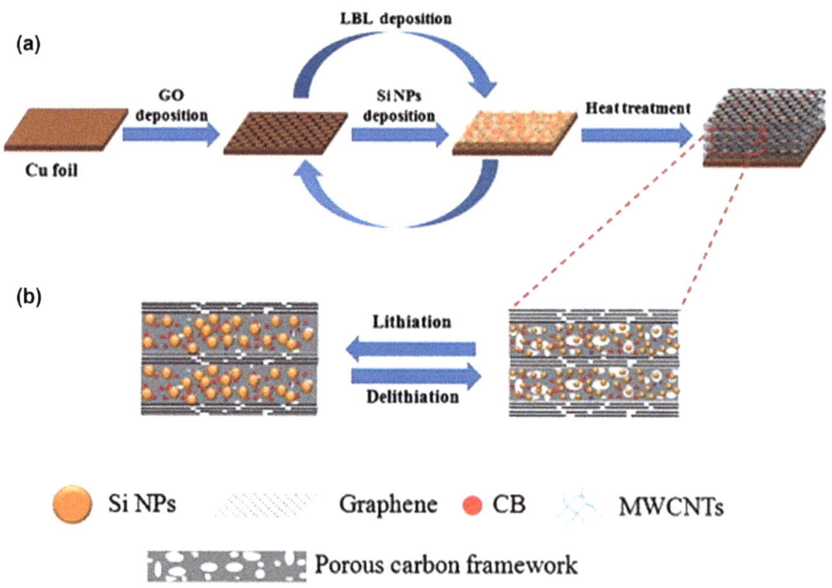

Figure 5.12 Schematic illustration of (a) alternate deposition of GO and Si-containing solutions, as well as subsequent heat treatment (600 °C) under Ar gas flux. During the heat treatment (GO sheets undergo reduction to form rGO and the PVP undergoes conversion to form porous carbon). (b) The structural changes occurring in the electrode during the lithiation/delithiation processes.
Reprinted from ref. 218 with permission from Elsevier, Copyright 2015.

To inhibit the fast capacity degradation and safety concerns of S||Li batteries, Cheng *et al.*[219] reported a polysulfide-induced SEI layer having a graphene framework structure with Li depositing in the pores as the anode. It was observed that the graphene-based framework is superior in inhibiting dendrites after 70 h of lithiation, while a cell with a copper foil-based material short-circuited in just 4 h of lithiation at a current density of 0.5 $mA\,cm^{-2}$. Additionally, the graphene-modified Li-anode with the SEI layer induced by the polysulfide-containing electrolyte is able to increase the coulombic efficiency up to ~97% for more than 100 cycles.

Gao *et al.*[220] reported a molecular-level SEI design for Li-metal anode by using a reactive polymer composite of poly(vinyl sulfonyl fluoride-ran-2-vinyl-1,3-dioxolane)–GO as the SEI precursor. This polymer–inorganic SEI layer is stable at elevated temperature and exhibits excellent surface passivation properties, uniformity, and mechanical strength while exhibiting efficient Li deposition. Additionally, the artificial polymer–inorganic SEI layer helped in achieving stable cycling of $LiNi_{0.5}Mn_{0.3}Co_{0.2}O_2$ (NMC532)||Li cells under lean electrolyte and high capacity conditions, which demonstrates that graphene-based artificial SEI layer formation would be an interesting approach to achieve safer batteries, especially when Li-metal is used as an anode. In summary, graphene is an interesting choice, both as a filler in PCEs and as a surface protection layer to design artificial interphasial layers. Research in this area is worth pursuing, especially for the development of safer LMPBs.

5.3.2 Boron Nitride-based PCEs

Boron nitride (BN) is isoelectronic with carbon. It has an equal number of boron and nitrogen atoms in its structure and is found in various crystalline forms. BN can be produced in both amorphous and crystalline forms.[221] The well-defined crystallographic structures of BN include hexagonal (h-BN, graphitic), rhombohedral BN (r-BN), cubic polymorph (c-BN, diamond-like), and wurtzite BN (w-BN, lonsdaleite-like). BN is also found in other forms, such as fullerene-type (BN fullerenes), one-dimensional single- or multi-walled BN nanotubes (BNNTs), and atomic layers.[222,223] Among the various crystalline forms, h-BN is the most widely investigated, the most stable, and the softest among all other polymorphic forms.[224] BN nanosheets, or BNNS, are the single layers of h-BN,[225] and BNs with a higher aspect ratio with a width of <50 nm are known as BN nanoribbons, or BNNRs.

The number of layers in h-BN can be calculated from transmission electron microscopy (TEM) and atomic force microscopy (AFM) images. With the recent upsurge in the interest in two-dimensional (2D) layered materials, h-BN has demonstrated promise due to its characteristics, which are similar to graphene, and which have prompted researchers to call it 'white' graphite. In particular, h-BN has fascinated researchers due to its thermal and chemical stabilities, low dielectric constant, and intrinsic electrical insulation characteristics; hence, it has been proposed for applications in insulating, magnetic, and thermally conducting composite materials. The characteristics and functionalities of h-BN can be tailor-made by hybridisation, functionalisation, substitution, and doping. Such tunability leads to the development of and intense research in various forms of h-BNs, including single- and few-layer h-BN nanosheets, for applications such as photonics, electronic packaging, biomedicines, fuel cells, *etc.*, which are all the subject of current research.[226] B and N atoms of a 2D h-BN are arranged alternately to form a honeycomb structure through sp^2 hybridisation, where the B–N bond length is 1.45 Å and the interlayer spacing is 0.333 nm (in graphite it is 0.335 nm).[227] Three sp^2 hybrids are formed by each B and N atom with the arrangement resembling a graphene-like layered structure where the adjacent layers are held together by weak van der Waals forces.[228]

From a practical viewpoint, h-BNs have been prepared by several methods, including mechanical separation, CVD, aqueous solvent thermal synthesis, solvent stripping, *etc.* However, c-BN can be produced in powder form using high-pressure high-temperature synthesis and low-pressure deposition, such as physical vapour deposition (PVD) and plasma CVD,[222] which finds application in wear-resistant ceramics, lubricants, cutting tool applications, *etc.* Interestingly, c-BN exhibits the second highest thermal conductivity, after that of diamond.[221]

The applications of every form of BN are different; h-BN can possibly be used as a filler material to produce polymer composites and applied in hazardous ultraviolet environments and as field emitters.[222] Other homologs of BN (atomically thin BN nanoribbons) are expected to have tunable electronic conductivity with potential applications in optoelectronics and spintronics.[229] Some forms of BN, such as h-BN, demonstrate superior oxidation (stable up to 900 °C in the air) and intercalation resistance than graphite, and, hence, the production of nanosheets of BN by conventional routes such as exfoliation is difficult. Recently, fabrication of BNNSs has been attempted by several non-conventional methods, including mechanical exfoliation, liquid-phase sonication, chemical exfoliation using boric acid and urea, and CVD; however, the large-scale production of BNNSs (single layer) with precise characteristics is still a challenge, as the obtained products are in a multilayer form. However, following the approaches used for graphene production from carbon nanotubes, unwrapping multi-walled BNNTs (Figure 5.13) has been attempted to obtain few- and single-layered BNNSs/BNNRs. These demonstrated improved conductivity; hence, they are useful for functional semiconductor-based applications in optoelectronic and spintronic devices.

Similar to graphene, BN is used in LBs as electrode additives, interfacial layers to protect against short-circuits induced by lithium dendrites, and as a filler in PCEs both in gel and solid-state systems. BN in the composite form has also been used for thermal conductivity applications, either BN alone or with graphene, and excellent results were achieved.[230] In some cases, researchers were able to fine-tune the orientation of BN nanosheets using magnetic fields, and hence improved thermal conductivity values are achieved.[231] In addition, BN has also been used as a filler in electrodes to prepare binder-free anodes for application in LIBs.[232] Indeed, in one such

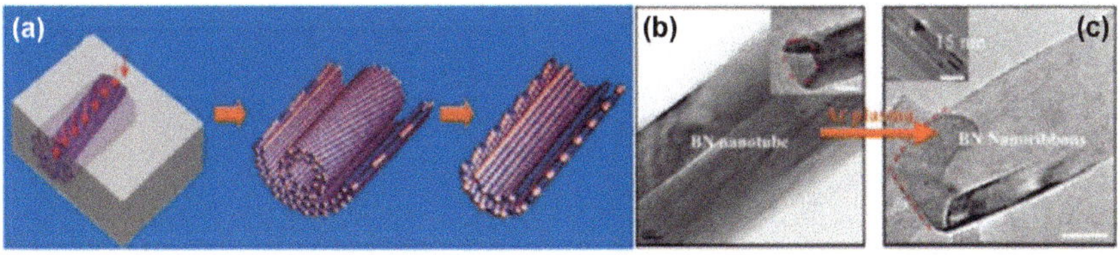

Figure 5.13 Schematic representation of unwrapped multi-wall BNNTs. (a) The stepwise unwrapping and exfoliation of BNNT walls by plasma etching to form BNNRs. (b,c) TEM images showing the morphology change of BNs before and after the etching process. Insets: the morphology of BNNT before (b) and after (c) the etching process (scale bar = 20 nm). Adapted from ref. 229 with permission from American Chemical Society, Copyright 2010.

attempt, a composite anode film is produced using reduced graphene oxide (rGO) as electroactive material and BN as a binder. The binder-free composite anode film (rGO/2 wt% BN) exhibited a specific capacity of 278 $mA h g^{-1}$ at a high current density of 100 $mA g^{-1}$ with good capacity retention for more than 200 cycles. Thus, BN has various applications in LBs; it can be used as an additive in the anode, as electrolyte, or even as an interfacial layer. However, in the context of this chapter, we have included the application of BNs as part of a polymer electrolyte segment.

5.3.2.1 *Boron Nitride as a Filler in PCEs*

Aydin *et al.* prepared gel-based PCEs (g-PCEs) *via* an electrospinning approach. Porous polymer composites (PPCs) in the form of nanofibres have been prepared using a PAN-based polymer matrix in which 10 wt% of h-BN was added as a filler. Afterwards, these PPCs were activated for ionic conduction by a swelling step using 1 M $LiPF_6$ in EC : EMC (1 : 1) v/v solution and their application in $LiCoO_2$-based batteries was studied. They observed that h-BN influences the ionic conductivity, whereas physicochemical characteristics of the PPCs are influenced by the PAN ratio, the resulting fibre diameter and porosity, and the amount of liquid. The activated PPCs show ionic conductivities > 10^{-3} $S cm^{-1}$ at RT, high oxidation stability up to 4.7 V *vs.* $Li|Li^+$, and high specific capacity (144 $mA h g^{-1}$) in $LiCoO_2$|g-PCE|Li cells. In this case, the nanofibre-based PPC just acts as a reservoir for the liquid electrolyte and does not actively participate in any ionic conduction process. Hence this membrane is a separator rather than an electrolyte.[233]

Interestingly, Rodrigues *et al.* investigated a polymer-free composite electrolyte gel (paste) using a simple single-step process where a stoichiometric mixture of h-BN, 1-methyl-1-propylpiperidinium bis(trifluoromethylsulfonyl)-imide (PP13-TFSI), and LiTFSI are mechanically homogenised. For comparison, they measured the ionic conductivity using bentonite clay. The final gel composite electrolyte (Figure 5.14a) demonstrated excellent ionic conductivity (3 $mS cm^{-1}$, Figure 5.14b), wide electrochemical stability up to

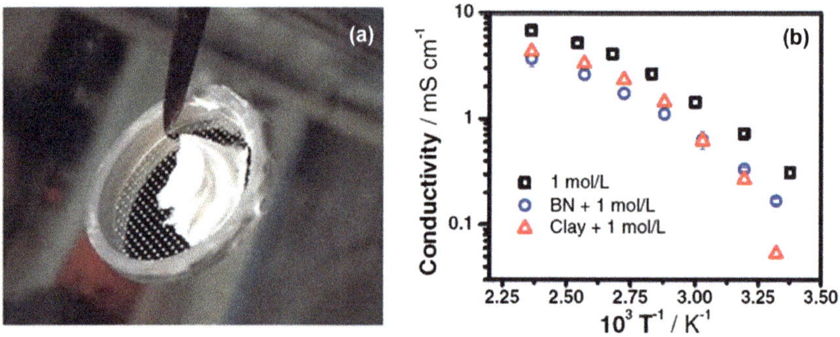

Figure 5.14 (a) Digital image confirming the unchanged h-BN-based composite electrolyte gel after exposing to 120 °C for 20 days. (b) Ionic conductivity *vs.* temperature plot of the h-BN- and clay-based composite electrolytes against 1 mol L^{-1} solution of LiTFSI in PP13-TFSI.
Reproduced from ref. 234 with permission from John Wiley & Sons, Copyright 2016 WILEY-VCH Verlag GmbH & Co. KGaA, Weinheim.

5 V *vs.* $Li|Li^+$, and extended thermal stability up to 150 °C. Later, the electrolyte gel was tested in lithium-rich manganese oxide ($Li_{1+x}Mn_{2-x}O_4$, LMO)|electrolyte gel|lithium titanate ($Li_4Ti_5O_{12}$, LTO) full-cells, which achieved >150 mA h g^{-1} at 120 °C.[234]

Shim *et al.*[235] demonstrated the ability of a PVDF–HFP-based g-PCE for lithium dendrite suppression using perfluoropolyether (PFPE)-functionalised few-layer h-BN nanoflakes (BNNFs) for application in S‖Li and Air‖Li batteries. They attempted to explore also the possibility of utilising the Lewis acid character of B in BN, which can interact with Lewis bases. Hence, the Li^+-ion transport values can be enhanced by trapping the anions of the lithium salt.[236] They also demonstrated that optimum physicochemical and electrochemical characteristics using small amounts of (0.5 wt%) PFPE-functionalised BNNFs could be achieved. The functionalisation is carried out to increase compatibility between the polymer matrix and the BNNFs. Indeed, a sonication-assisted exfoliation process is used to prepare BNNFs and, afterwards, a non-covalent functionalisation is carried out using PFPE-functionalised pyrene molecules. The polymer composite (Figure 5.15) is prepared *via* blade casting of a mixture of PFPE-functionalised BNNFs and PVDF-HFP. The porous polymer composite membrane obtained after drying is activated by a swelling process in 1 M LiTFSI in EC:DEC (1 : 1 vol%) to produce gel-PCEs. The optimised g-PCEs demonstrated good ionic conductivity (0.8 mS cm^{-1}), superior Li^+-ion transference number (0.62), high Young's modulus (27.4 GPa), and effective suppression of Li dendrites for nearly 2000 hours at 1.0 mA cm^{-2} (25 °C) in LMBs.

Zhang *et al.* proposed a facile and cost-effective route for the production of a polymer/salt hybrid electrolyte (or PCE),[237] where 2D layered BN is applied as a filler in the PCE to increase ionic conductivity, mechanical characteristics, and electrode|electrolyte interfacial contact. They have produced

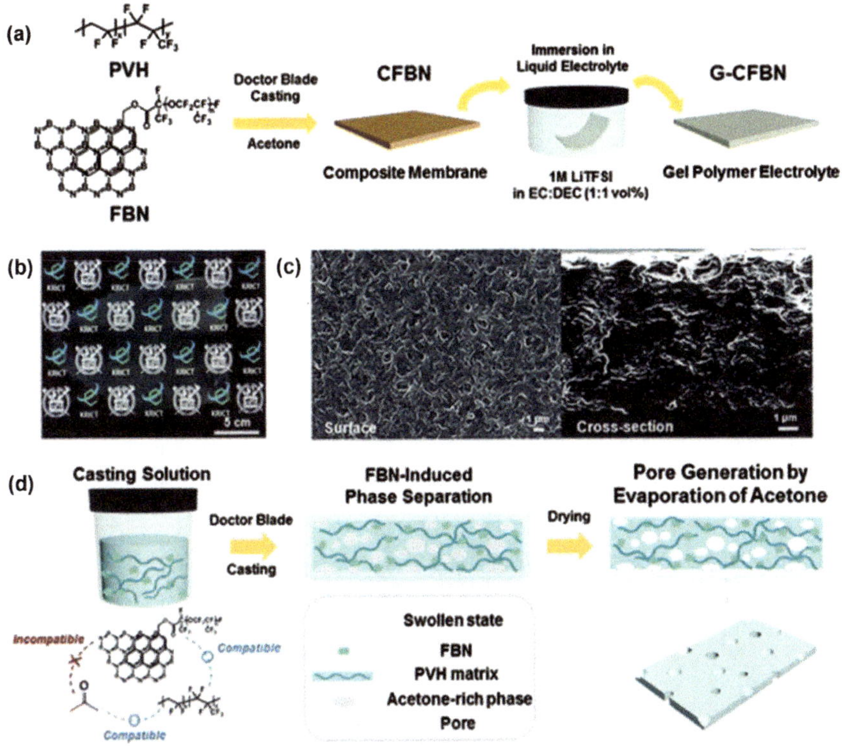

Figure 5.15 (a) Schematic representation of the preparation of PVDF-HFP (mentioned as PVH in the figure)-based gel-PCE (G-CFBNs) using PFPE-functionalised BNNFs (FBNs). (b) Digital image of the porous polymer composite (CFBN, 0.5 wt% FBN) before activation. (c) Topographical (left) and cross-sectional (right) SEM images of CFBN (0.5 wt% FBN). (d) Proposed mechanism behind the autonomous pore formation in CFBN by phase separation.
Reproduced from ref. 235 with permission from the Royal Society of Chemistry.

PCEs with a small amount of BN (1 wt%) filler (see Figure 5.16) and the resulting PCE demonstrated an ionic conductivity of >1 mS cm^{-1} at RT, along with moderate hardness (~5 MPa), Young's modulus (~130 MPa), high oxidation stability, lower impedance, and lithium dendrite growth resistance for several hundreds of hours without any visible short-circuits. Additionally, the PCE in the $LiFePO_4|PCE|Li$ cell configuration demonstrated improved long-term galvanostatic cycling stability with a coulombic efficiency of 98%.

A commercially available separator was coated with BNNSs using a spray-coating technique to produce a thermally conducting polymer composite, which, after swelling in the organic carbonate-based LE, was employed in LMBs as a PCE and as a separator (see Figure 5.17). This approach increased the cycling stability of a Cu‖Li cell when the lithium plating/stripping process was carried out at a current rate of 0.5 mA cm^{-2} and 1.0 mA cm^{-2} to achieve a

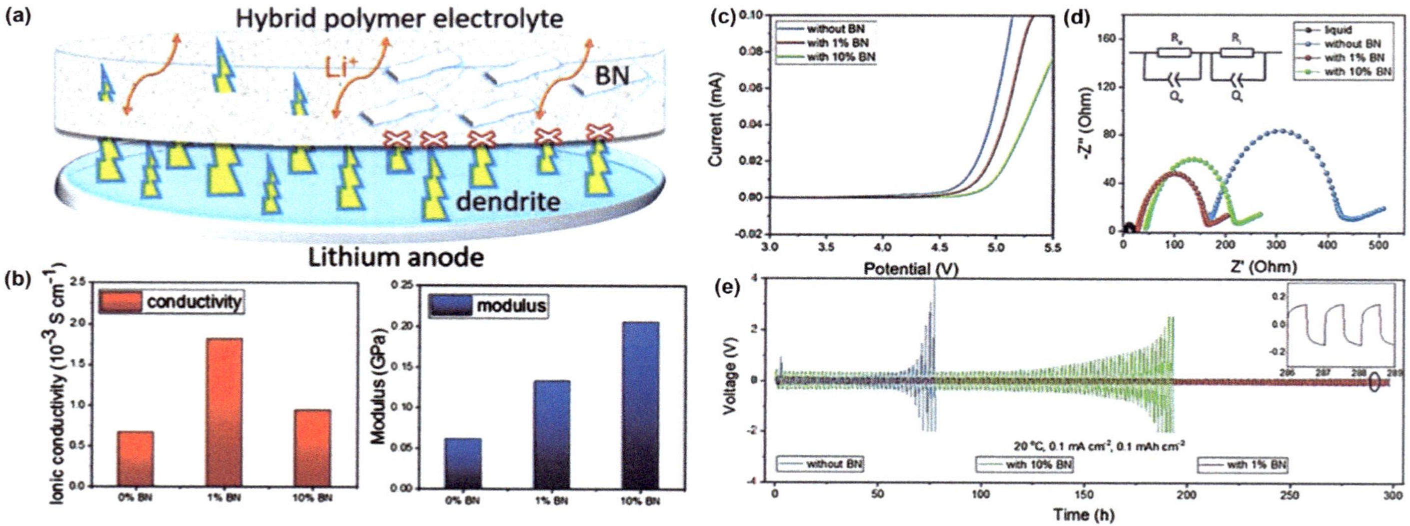

Figure 5.16 (a) A scheme showing the role of BN in the PCE (hybrid polymer electrolyte) to suppress the dendrite growth during cycling. (b) Showing the impact of BN on ionic conductivity and modulus. (c) Showing the oxidation stability of the PCE using LSV measurements. (d) Impedance spectra (inset: fitted equivalent circuit) of the lithium symmetric cells against LE and the PCE with various BN content. (e) Long-term striping/plating cycling (current density = 0.1 mA cm^{-2}) of the Li||Li cells using the PCEs. Inset: plating/stripping curve of the PCE with 1 wt% BN between 286 and 289 hours. Reproduced from ref. 237 with permission from Elsevier, Copyright 2019.

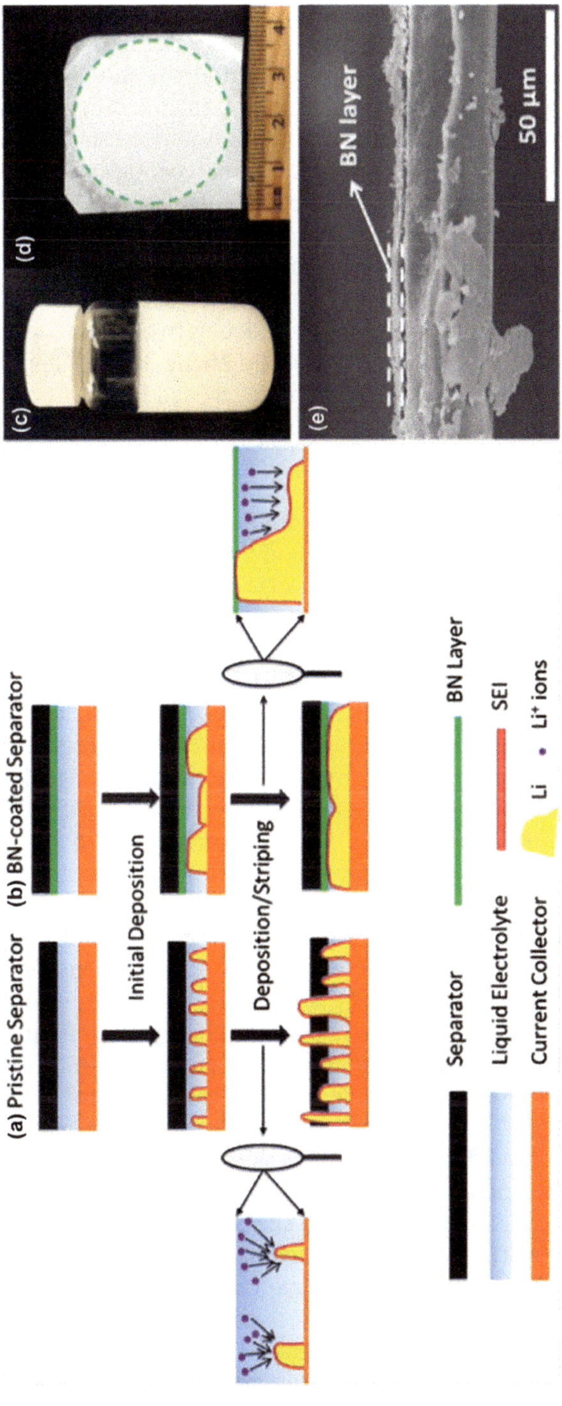

Figure 5.17 Schematic representation of lithium plating/stripping process occurring in pristine separator and BN-coated separator. (a) Repeated lithium plating/stripping processes leading to HSAL ('dendrite') growth, and (b) BN coating leads to a uniform lithium plating/stripping process. Photographs of (c) BN suspended PVDF/NMP solution and (d) BN-coated separator (coated area is marked with the green circle). (e) Cross-section of BN-coated separator by SEM. Adapted from ref. 238 with permission from American Chemical Society, Copyright 2015.

deposition capacity of 1.0 mA h cm^{-2}. A homogeneous Li-deposition was confirmed by improved coulombic efficiency of 92% after 100 cycles, at 0.5 mA cm^{-2} using a BN-coated separator. The cycling efficiency decayed rapidly when a pristine separator was used under the same conditions.[238] The authors observed that it is important to initially deposit lithium to a smaller surface area. Additionally, a homogeneous thermal distribution is necessary for achieving long-term cell cycling. The proposed method of spray coating and related roll-to-roll technologies, and the thermally conductive separator to avoid the thermal runaway reaction, are encouraging for further research in BN-based PCEs.

Similar to the use of h-BN in LEs for producing gels, it has also been used with ISEs as a mixture to enhance several electrolyte characteristics, such as ionic conductivity, higher mechanical strength, and better chemical and electrochemical stability. In one of the first examples, Zhang *et al.*[239] demonstrated that h-BN can be used (1 wt% h-BN) as an effective additive with garnet-type $Li_{6.25}Al_{0.25}La_3Zr_2O_{12}$ (LALZO) ISEs during sintering to produce solid-state electrolytes (Figure 5.18) that exhibit good ionic conductivity (>0.06 mS cm^{-1}), surface hardness (~0.5 GPa), and Young's modulus (5.6 GPa), along with enhanced chemical stability against air (reduced formation of surface Li_2CO_3). The composite electrolyte also demonstrated good cycling behaviour in a $LiCoO_2$|h-BN-LALZO|Li cell, which delivered a discharge capacity of 120 mA h g^{-1}.

5.3.2.2 Boron Nitride as an Interlayer in LMBs

Similar to the use of 2D h-BN in PCEs, they have also been employed as a thin interlayer between Li-metal and electrolyte so that Li-dendrite growth can be suppressed. Yan *et al.* proposed an innovative electrode design by directly growing 2D h-BN atomic crystal layers and graphene on Cu current collectors (see Figure 5.19). In this electrode, Li^+ ions are capable of diffusing through the point and line defects of these atomic crystal layers during the plating process, hence a Li-metal layer is formed between the ultra-thin 2D h-BN layers and the Cu foils. These 2D interlayers give excellent interfacial protection for newly deposited Li-metal against electrolyte components, and hence remarkable chemical resistance, mechanical strength, and flexibility are obtained. They have observed dendrite-free yet smooth Li-metal deposition for over 50 cycles, with a coulombic efficiency ~97% in LE at a current density of 2.0 mA cm^{-2} and an areal capacity of 5.0 mA h cm^{-2}.[240]

In another approach by Shen *et al.*, 2D BNNSs are chemically exfoliated from bulk BN and coated using a dry-pressing transfer process onto the surface of a PEO-based polymer electrolyte (see Figure 5.20). The formation of a robust interfacial layer improved the physicochemical characteristics of the PEO-based electrolyte. The physical characterisation by various microscopic methods (TEM, SEM) showed that the dry-pressing transfer process of the BNNS layer to the PEO-based PE was successful, which resulted in a continuous dense surface layer having a thickness of 150 nm. This dense BNNSs surface layer coating imparted good mechanical stability and homogeneous Li^+-ion flux distribution to the PEO-based PE.[241] The BNNS-coated PEO

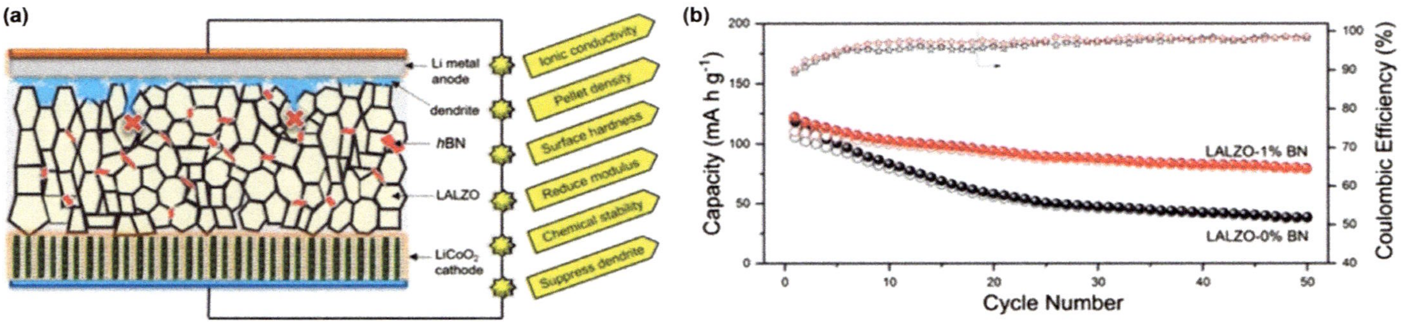

Figure 5.18 (a) Schematic representation of an ISE-based solid-state $LiCoO_2$|h-BN-LALZO|Li cell, where h-BN is used as a filler, which highlights the properties achieved due to the addition of h-BN in the electrolyte mixture. (b) Galvanostatic charge–discharge cycling in $LiCoO_2$|h-BN-LALZO|Li cells with 0 wt%-BN-LALZO and 1 wt%-BN-LALZO (C rate = 0.1; V = 3.0–4.2 V). Adapted from ref. 239 with permission from American Chemical Society, Copyright 2019.

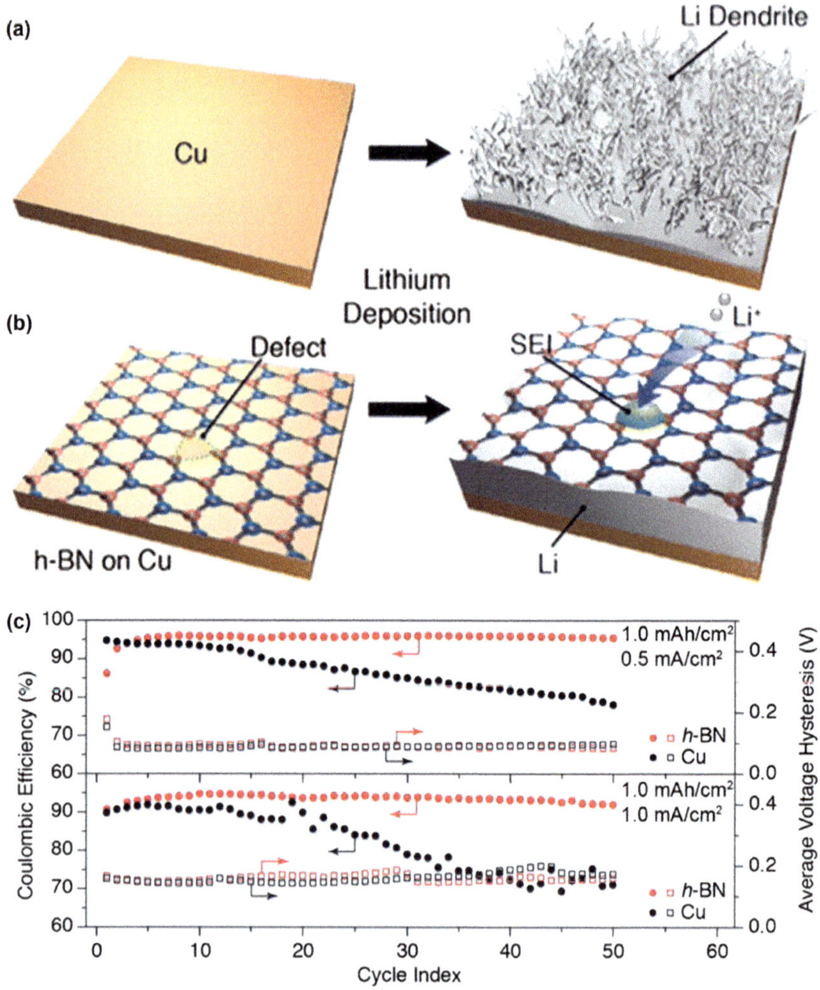

Figure 5.19 Schematic representation of Li-deposition and h-BN film characterisation. (a) Li-metal deposition on bare Cu substrate, where a large amount of Li-dendrite growth is shown due to the lack of confinement. (b) The defects on h-BN film (grown on Cu foil) serve as a channel for Li^+-ions during Li deposition, and the deposition occurs between the h-BN and Cu surface. (c) Coulombic efficiency and voltage hysteresis *vs.* cycle index for bare and h-BN-coated copper foil-based cells (cell configuration = Cu‖Li, current rate = 0.5 and 1.0 mA cm^{-2}, liquid electrolyte = 1 M $LiPF_6$ in EC:DEC).
Adapted from ref. 240 with permission from American Chemical society, Copyright 2014.

electrolytes showed an ionic conductivity of 0.2 mS cm^{-1} at 60 °C. The electrolyte demonstrated stable Li-plating/stripping and delivered a specific discharge capacity of 110 mA h g^{-1} at 2 C for 200 cycles in a $LiFePO_4$| BNNS-coated PEO-electrolyte |Li cell.

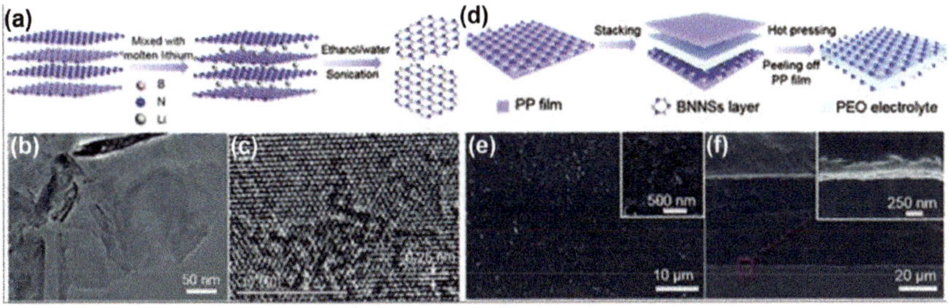

Figure 5.20 (a) Scheme showing the BNNS exfoliation process. (b) TEM and (c) HRTEM images of BNNS. (d) Scheme of the BNNS transfer process onto a PEO film. SEM images of the surface (e) and cross-section (f) of the BNNS-coated PEO electrolyte; insets (e,f) respective magnified SEM images. Reproduced from ref. 241 with permission from the Royal Society of Chemistry.

BN has been used as an interlayer for protecting the electrodes of S‖Li cells from violent reactions with polysulfide electrolytes, which is one of the issues of ISE-based polysulfides.[242–244] Indeed, the all-solid-state S‖Li battery is considered to be one of the most promising storage systems for future applications due to their high theoretical energy density, easy availability, low cost, and enhanced safety due to the use of low-voltage cathodes. On the other hand, inorganic sulfidic electrolytes are interesting candidates due to their very high RT ionic conductivity. However, the S‖Li cell based on $Li_7P_3S_{11}$ solid electrolyte has serious issues at the electrode|solid electrolyte interface.[245] Indeed, the reaction between Li-metal and $Li_7P_3S_{11}$ produces a non-lithium-ion-conducting layer, which could increase the internal resistance and induce capacity fading due to the Li^+-ion transfer and transport inhibition during the charge/discharge process. In this respect, Xu *et al.* proposed h-BN (2 wt%) doping in the $Li_7P_3S_{11}$ solid electrolyte. Indeed, the BN nanoflakes could partially separate the $Li_7P_3S_{11}$ solid electrolyte from Li-metal, hence improving the interfacial compatibility.[246]

Finally, Fan *et al.* attempted a similar approach for S‖Li batteries to address inherent problems, such as the rapid capacity fading due to the shuttling of soluble polysulfides between the electrodes. In this case, an interlayer was made on the cathode surface by a thin film coating of functionalised BNNSs/graphene (see Figure 5.21). The interlayer, in this respect, reduced the charge-transfer resistance and the shuttling problem. The as-assembled S‖Li cell, where a sulfur-containing porous carbon nanotube (60 wt% of sulfur) is used as the cathode, demonstrated good specific capacity (initial specific capacity of 1100 $mA\,h\,g^{-1}$ at 3 C) and cycling stability for over 1000 cycles (cycle decay ~0.0037%/cycle).[247] Hence, in general, various types of BNs have been employed in LMBs, either as a filler or as a surface protection layer. Both approaches require interesting and judicious selection of materials and mehods, which will deliver increased Li^+-ion transference number, dendrite resistance, and long-term cycling possibilities. Therefore, BN-based PCEs are worth pursuing further for the next generation of energy storage devices, including LMPBs.

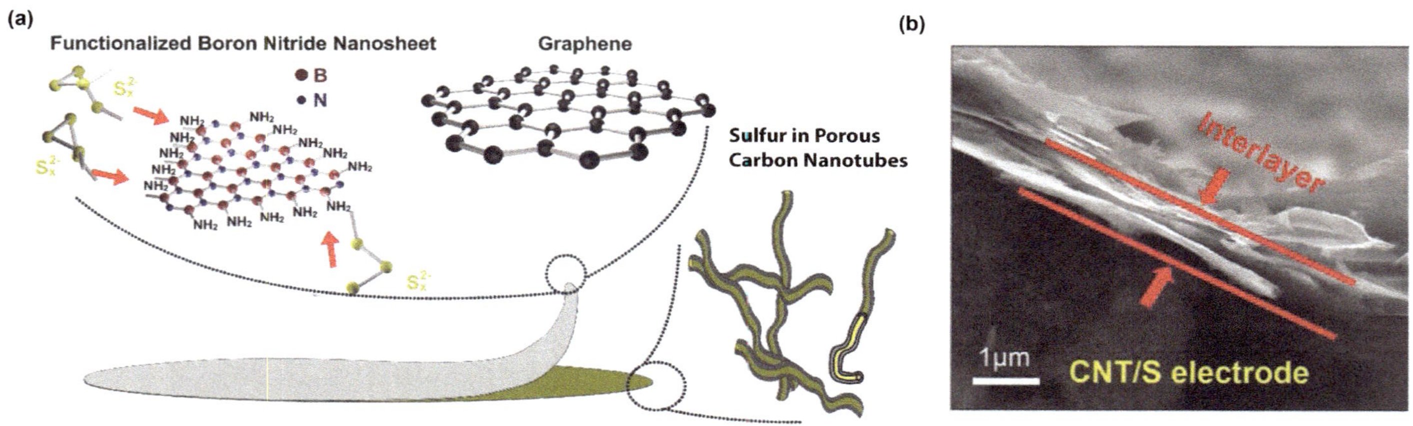

Figure 5.21 (a) Schematic representation of an S‖Li cell with a FBN/G interlayer. (b) SEM cross-section image of FBN/G interlayer and cathode.
Reproduced from ref. 247 with permission from John Wiley & Sons, Copyright 2017 WILEY-VCH Verlag GmbH & Co. KGaA, Weinheim.

5.3.3 Layered Clay-based PCEs

Generally, the term clay refers to materials that develop plasticity when in combination with solvents such as water.[248] Clay minerals are 2D layered materials that are mainly classified into two types, namely, (1) cationic clay minerals (CCMs) and (2) anionic clay minerals (ACMs). This section focuses on a detailed discussion on 2D clay minerals as nanofillers in PCEs for LB applications.

5.3.3.1 Cationic Clay Minerals

CCMs are layered silicates (also called phyllosilicates or sheet silicates), which are compounds based on silicon, oxygen, and other elements.[249] They are 2D materials mainly consisting of two basic structural units: tetrahedral and octahedral sheets. The tetrahedral stacks (single, double, or multiple layers) of silicate (SiO_4) tetrahedrons where each SiO_4 units is linked by sharing three corners, in turn form a hexagonal array in two dimensions.[250] The fourth corner of the tetrahedron is covalently bonded to an adjacent octahedron of aluminium, magnesium, or iron consisting of six-fold coordination with oxygen and hydroxyl groups (see Figure 5.22).

Depending on the arrangement of the tetrahedral and octahedral sheets, CCMs can be classified into three main types of 1 : 1, 2 : 1, and 2 : 1 : 1. In the case of the 1 : 1 clay mineral, each tetrahedral sheet is connected to one octahedral sheet.[250,255] The 2 : 1 structure involves each octahedral sheet connected to two tetrahedral sheets. Finally, in the 2 : 1 : 1 structure, each octahedral sheet is adjacent to another and connected to two tetrahedral sheets. Further, the valency of the octahedrally coordinated central metal atom dictates whether the CCM is a trioctahedral clay (contains divalent metal ions such as Fe^{2+} and Mg^{2+}) or a dioctahedral clay (contains trivalent metal ions such as Al^{3+}).[256,257] For general understanding, the classification and examples of various types of silicate-based clay minerals are summarised in Figure 5.23.

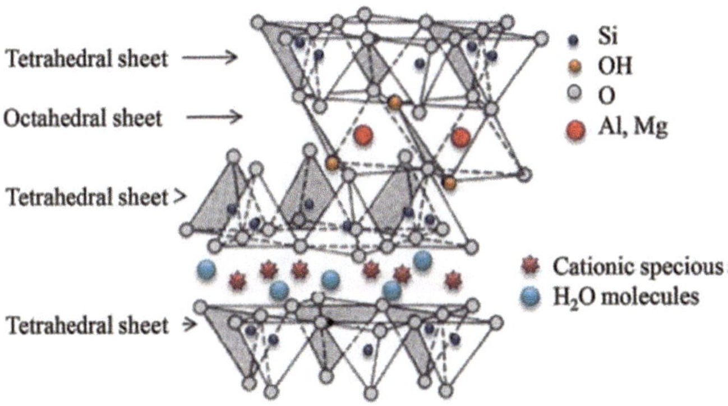

Figure 5.22 Smectite structure of the 2 : 1 phyllosilicate clay mineral. The two tetrahedral sheets are sandwiched between an octahedral sheet with exchangeable cations and water molecules in the interlayer galleries. Reproduced from ref. 251 with permission from the Royal Society of Chemistry.

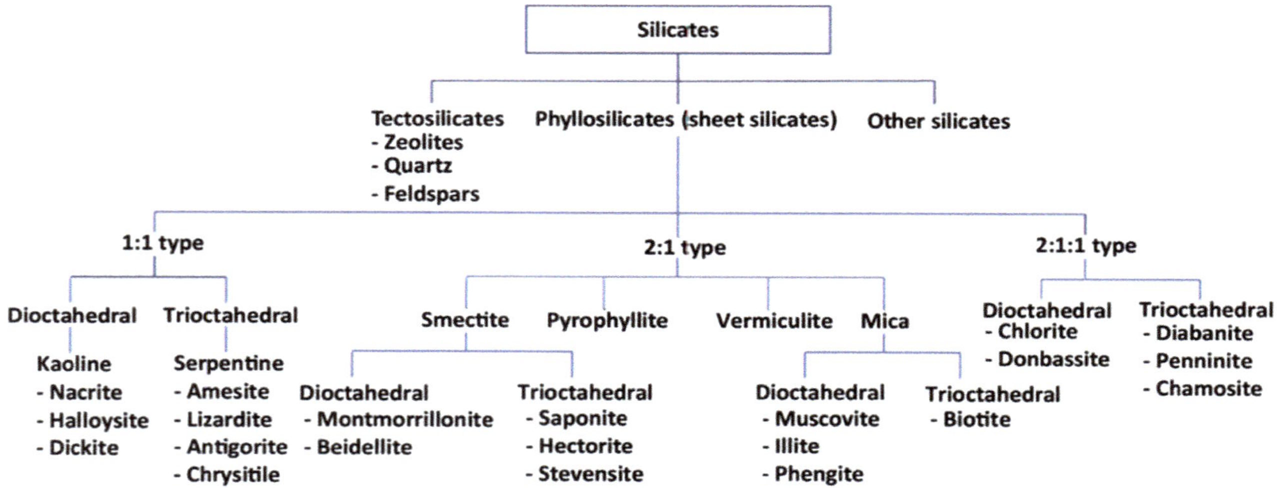

Figure 5.23 Classification of the silicate-based clay minerals.[251–254]
Adapted from ref. 252, https://doi.org/10.1080/24749508.2017.1361128, under the terms of the CC BY 4.0 license, http://creativecommons.org/licenses/by/4.0/.

Table 5.6 CCMs and their cation exchange capacity values. Adapted from ref. 259 with permission from Elsevier, Copyright 2013.

CCM type	Cation exchange capacity (cmol(+) kg^{-1})
Kaolinite	3–15
Halloysite · 2 H_2O	5–10
Halloysite · 4 H_2O	40–50
Montmorillonite	70–120
Vermiculite	130–210
Illite	10–40
Micas (biotite, muscovite)	Up to 5
Chlorite	10–40
Sepiolite, palygorskite	20–30

The term 'cationic clay mineral' is used to describe silicate-based clay minerals with a net negative charge, which is, in fact, a metaphor for their high cation exchange capacity. This negative charge is a result of two contributing factors: (1) charge arising due to the substitution of Al^{3+}, Si^{4+}, Mg^{2+}, and Fe^{3+} ions, which is regarded as the permanent or structural charge of the clay mineral, and (2) charge arising from the broken edges of the clay (this charge is pH dependent).[251,258] To balance the net negative charge, inorganic cations (Li^+, Na^+, Ca^{2+}, *etc.*) occupy the interlayer spaces of natural clay minerals along with water molecules.

The cation exchange capacity of several CCMs is summarised in Table 5.6.[259,260] It should be noted that in certain clay minerals in the kaoline group, such as illite and chlorite, the permanent charge from the ion substitution is zero. Therefore, the major contribution to the net charge is from the pH-dependent broken edges. At acidic pH, such clay minerals may exhibit anionic exchange capacity.[261] For instance, kaoline clay minerals exhibiting phosphate ion exchange are known.[262]

Modification of the clay minerals with organic solvents,[263] organic cations (quaternary amines),[264] and other substances (such as polymers)[265,266] is also adopted to tune their physical and chemical properties.[267] The accommodation of solvents, ions, *etc.*, leads to swelling and shrinkage, making CCMs different from rigid silicate minerals (tectosilicate) and zeolites. In the past several years, such modified CCMs have been employed in a variety of applications. A few examples include biomedical (drug delivery, regenerative medicines, *etc.*), rheological control agent (cleaning products, toothpaste, personal care products, *etc.*), film formation (adhesives, paints, protective coatings for automotive and wood, *etc.*), aerospace, construction, petroleum, wastewater treatment, *etc.*[268–270] Recently, with advances in material technology, the possibility of using CCMs as fillers in the organic polymer matrix has been extensively explored,[250,271] This has resulted in the realisation of several polymer/clay nanocomposites leveraging the utility of CCMs in the aforementioned applications, as well as denoting their entry into battery research in the form of PCEs.[272]

PEs with inorganic fillers are well known as PCEs for battery applications.[80,273] CCMs have recently received tremendous attention within the battery community, owing to their natural occurrence and abundance. In other words, CCMs can be called as naturally occurring nanomaterials (nanoclays) that are handy for PCE application. The unique features allowing CCMs to be used in PCEs for LBs are listed below.

1. CCMs can accommodate inorganic ions such as Li^+ and Na^+ along with their high cation exchange capacity, which is very important when employed as PCEs in alkali metal-ion batteries. The intercalated structure of CCMs can emulate nanometric channels for the transport of cations.
2. CCMs exhibit features such as intercalation and exfoliation. Polymer chains can intercalate between the interlayer spacing of CCMs by disrupting the covalent interactions between the layers, leading to exfoliation of individual clay layers.[274] Depending on the organisation of the polymer and clay minerals, three main types of composites can be expected: (1) phase-separated composite (2) intercalated nanocomposite, and (3) exfoliated nanocomposite (Figure 5.24).[250,275–277] It is reported that exfoliated CCMs can help in decreasing the ion-pairing effect in PCEs.[278]

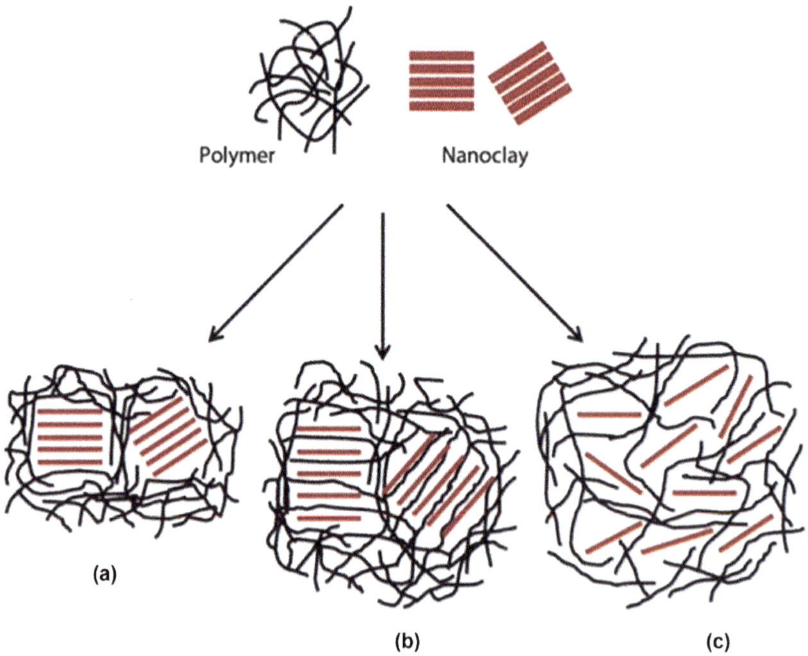

Figure 5.24 Different types of the polymer/clay mineral composites: (a) phase-separated polymer composite, (b) intercalated polymer nanocomposite, and (c) exfoliated polymer nanocomposite.
Reproduced from ref. 276 with permission from Elsevier, Copyright 2012.

3. The intercalation of the polymer chains between the layers helps in preventing polymer recrystallisation. In other words, the amorphous character of the PEs can be improved with the incorporation of CCMs, which facilitate the ionic conductivity.[274]
4. Due to the net negative charge of CCMs, the migration of anions is restricted between the clay layers. This can help in realising PCEs with high cation transference number.[80,279]
5. When polymers are intercalated between the CCM sheets, the external stimuli, such as temperature, potential, and mechanical, do not directly influence the polymer chains. This helps in improving the physicochemical properties (thermal, mechanical, and electrochemical stability) of PCEs compared to conventional PEs.[274]

The polymer/clay composites are generally prepared by three methods, namely (1) solution blending, (2) melt-blending, and (3) *in situ* polymerisation.[250,275] All three of these methods (see Table 5.4 for details) can be extended towards the preparation of PCEs as well (both CCM- and ACM-based PCEs). In the case of the solution-blending method (Figure 5.25a), the clay mineral is first dispersed in a solvent (H_2O, $CHCl_3$, toluene, *etc.*) that can intercalate between its layers. The polymer and electrolyte salt should be necessarily soluble in the solvent used. While mixing of the polymer and clay solution, the solvent molecules in the interlayers are displaced by the polymer chains. Cations can also be intercalated in the interlayer galleries.

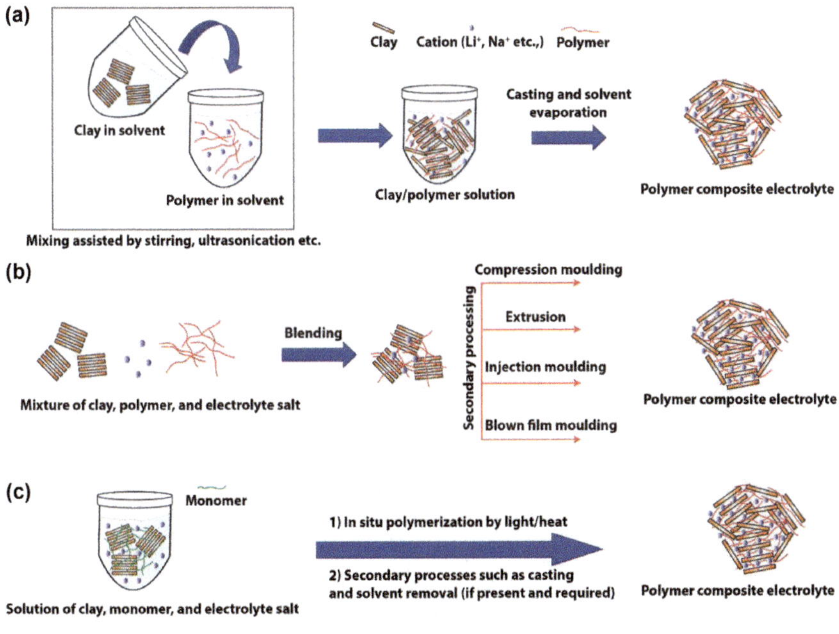

Figure 5.25 Most commonly used clay-based PCE preparation techniques: (a) solution blending, (b) melt blending, and (c) *in situ* polymerisation.[250,276]

This is followed by the removal of the solvent by evaporation or precipitation. The second method of melt-blending (Figure 5.25b) is a solvent-free method. In this case, the polymer/clay mineral composites are prepared at a temperature above the softening point of the polymer. Compared to the former method of solution blending, this method ensures better dispersion of the polymer and clay. When PCEs are prepared using the melt-blending process, electrolyte salt should be able to withstand the blending temperature without decomposition. The *in situ* polymerisation method (Figure 5.25c) is easy compared to the melt and solution-blending processes. Here, the clay mineral is first dispersed in the monomer solution. Later, the polymerisation is triggered by external sources, such as light or heat in the presence of a suitable initiator, so that the polymer chains can be formed between the layers as well as over the individual clay particles. By using this method, PCEs with high nanofiller loading can be achieved. In many cases, this technique is also applied in the solvent-free form if the monomer can disperse/solubilise clay, polymer, and salt.

Among the several CCMs summarised in Figure 5.23, smectite clays are extensively employed for PCE preparation. They include montmorillonite (MMT), saponite, laponite, hectorite, *etc.* Among them, MMT ($(Na, Ca)_{0.33}(Al, Mg)_2(Si_4O_{10})(OH)_2 \cdot nH_2O$)[279] is the most widely used owing to its natural abundance, high surface area, and efficient intercalation/exfoliation chemistry. MMT particles are plate-shaped particles having a thickness of about 1 nm and an average diameter close to 1 μm, which results in a high aspect ratio.

Wang *et al.* recently reported the capability of the MMT minerals to act as the Lewis base centre to interact with the Li^+-ion in the PEO-based PCE.[280] They employed the solution-casting method for preparation of the solid-state PCEs, where LiTFSI and $Li_{6.4}La_3Zr_{1.4}$–$Ta_{0.6}O_{12}$ (LLZTO) act as the Li^+-ion sources. The Li^+–MMT interaction helps in weakening the Li^+–PEO coordination, thus improving the Li^+-ion transport. Ionic conductivity values of 0.02 and 4.7 mS cm^{-1} are achieved at 30 and 70 °C, respectively. It should be noted that the ionic conductivity achieved here is higher than that of the pristine MMT (0.00006 mS cm^{-1}) or Li^+-ion-intercalated MMT (0.02 mS cm^{-1}) as reported elsewhere.[281] From X-ray diffraction (XRD) analysis, the reduction in the crystallinity of PEO with the incorporation of MMT is also proven. Interestingly, the oxidation stability of PCE is also improved from 3.9 to 4.6 V *vs.* Li|Li^+ with the incorporation of MMT. The improvement in the oxidation stability is attributed to the removal of impurities such as water from the electrode|electrolyte interface owing to the intercalation properties of MMT. The $LiFePO_4$|MMT-PCE|Li cell displayed a specific capacity of 150 mA h g^{-1} at 0.08 C (at 70 °C), which is higher than the MMT-free counterpart (120 mA h g^{-1}).

Apart from PEO, other polymers, such as PMMA and PVDF, are also used for the preparation of CCM-based PCEs. Srinivasan *et al.* used solution blending for realising Li^+-ion-conducting PVDF–MMT PCEs.[276] Since, CCMs are generally hydrophilic, in many cases, surface modification of the clay is required to make it organophilic so that its dispersion in an organic medium like PVDF will be facile. In this work, the CCMs are treated with quaternary amines to make them organophilic, utilising an ion-exchange reaction. Later, the CCM solution is made in dimethylacetamide (DMAc) solvent with the help of ultrasonication-induced delamination of the individual clay layers. Blending of the clay solution with the PVDF solution followed by phase-inversion method-assisted casting leads to the porous polymer/CCM composite membranes. However, to induce ionic conductivity, activation of membranes with a liquid electrolyte (1 M $LiPF_6$ in EC : DEC) is necessary to produce the g-PCE. An increase in the clay content (1–4 wt%) in g-PCE leads to characteristic XRD peaks associated with MMT that diminish/widen, indicating exfoliation of the clay platelets and intercalation of the polymer chains. High ionic conductivity of the order of 10^{-3} S cm^{-1} is achieved for the g-PCEs at 30 °C. The better interfacial stability of the g-PCEs and Li-metal surface, compared to the MMT-free GPE, is also proved by the EIS analysis. The $LiMn_2O_4|PVDF\text{–}MMT_{(2wt\%)}|Li$ cell during cycling (3.5–4.3 V) displayed a specific capacity of 120 mA h g^{-1} (0.1 C) at ambient temperature, surpassing the MMT-free GPE-based cell, with good cycling stability.

Recently, an interesting phenomenon of anisotropic ionic conduction in PEO–MMT PCE was reported by Tominaga *et al.*[282] Here, an *in situ* polymerisation method was employed for preparing the PCEs. In this work, a strong magnetic field was used to orient the MMT layers in parallel and in perpendicular orientation inside the PEO matrix. The results indicate that PCEs with parallel-oriented MMT display higher ionic conductivity compared to the perpendicular-oriented one at 30 °C. The obtained ionic conductivity value for the parallel-oriented MMT (0.012 mS cm^{-1}) is six times higher than for the neat PEO–Li. In a report by Park *et al.*, Na^+-ions in MMT were replaced by Li^+-ions and used for PEO-based PCE preparation.[283] At 25 °C, an ionic conductivity of ~10^{-6} S cm^{-1} was attained, compared to 10^{-8} S cm^{-1} in the case of the filler-free counterpart. Additionally, a high Li^+-ion transference number of 0.55 was also achieved for the PCE.

Saponite $[Ca_{0.25}(Mg,Fe)_3((Si,Al)_4O_{10})(OH)_2 \cdot n(H_2O)]$, a trioctahedral clay, has also been employed for the preparation of a PEO-based PCE.[284] Before use, the Na^+-ions intercalated in saponite are replaced with Li^+-ions by base treatment (LiOH). Compared to the low ionic conductivity of PEO–Li (of the order of 10^{-8} S cm^{-1}), the PEO-intercalated saponite-based PCE prepared through the solution-blend process (2 hours) exhibits high ionic conductivity values of ~4.1 mS cm^{-1} at 25 °C. Later, PEO-saponite PCE was prepared using the melt-blend process as well. A melt-blend PCE with 95 wt% saponite in the PEO displays a low ionic conductivity of 1.4 mS cm^{-1} compared to the solution-blend process. The high ionic conductivity of the solution-blended PCE arises from the optimum intercalation of PEO-chains between the saponite layers. The report also claims a Li^+-ion transference number of 0.99 for the solution-blended PCE.

Another recent report suggests the utility of employing vermiculite for PCE preparation.[285] The vermiculite CCMs are vertically aligned using temperature gradient freeze-drying. Later, the vertically aligned vermiculite (VAVS) is infiltrated into the PEO-based SPEs (VAVS-PCE) to prepare PCEs. The VAVS-PCE exhibited high ionic conductivity of ~0.2 mS cm^{-1} at 25 °C, surpassing that of non-vertically aligned vermiculite-based PCE (VS-PCE) and conventional PEO–Li membrane. Interestingly, the Li^+-ion transference number of VAVS-PCE is as high as 0.5, whereas VS-PCE and PEO-Li fall behind with values of 0.25 and 0.12, respectively. The interfacial stability of Li-metal and VS-PCE was proved using plating/stripping analysis in a Li‖Li cell (stable over 1300 h). Also, the LFP|VAVS-PCE|Li cell displays a specific capacity of 167 mA h g^{-1} at 0.1 C (35 °C). In another report from the same group, the potential of few-layered vermiculite for improving properties such as ionic conductivity, mechanical properties, and electrochemical and thermal stability of PEO-based SPE has also been proved.[286]

Muscovite, a dioctahedral mica, is converted to another kind of mica known as lepidolite utilising doping with Li^+-ions.[287] The potential of lepidolite to be used as a solid-state Li^+-ion conductor is demonstrated by preparing a composite with PEO. The intrinsic ionic conductivity of lepidolite helps in realising PCEs even in the absence of lithium salt. For instance, at 100 °C, the Li-salt-free PCE exhibited an ionic conductivity of 0.015 mS cm^{-1}. With the inclusion of $LiClO_4$ in the PCE, the ionic conductivity can be improved further to 0.16 mS cm^{-1} at 60 °C. With a high content of lepidolite (48 wt%), a high Li^+-ion transference number of 0.72 is also achieved. Alternatively, the PCE was used for solid-state LFP|PCE|Li cell fabrication and cycled at 60 °C. The cell displayed a maximum capacity of 120 mA h g^{-1} (0.15 C) at the 25th cycle.

This section summarises several aspects of 2D CCMs as a filler in PCEs. The layered structure of the clay with a net negative charge helps in the transport of cations, which is very important in terms of alkali metal-ion batteries. The high cation exchange capacity of CCMs helps in achieving a high cation transference number. All the know-how related to the preparation of polymer/nanocomposites can be applied to PCEs as well. The intercalation of the polymer chains in the interlayer galleries is very important in facilitating ion transport in the CCM-based PCEs. The alignment of CCMs in PCE also influences ionic conductivity. The incorporation of CCM can help in improving the mechanical, thermal, and electrochemical properties of polymer electrolytes. So far, only low-voltage cathodes, such as LFP, are employed for the fabrication of LMBs with CCM-based PCEs. Therefore, it is important to tune and further improve their properties so that they can be used along with high-voltage cathodes based on Ni, Co, *etc.* Additionally, the application of CCM PCEs can be extended to sodium- and other post-lithium-ion battery chemistries.

5.3.3.2 Anionic Clay Minerals

Similar to CCMs, which have a cation exchange capacity, ACMs can exchange anions.[288] Layered double hydroxides (LDHs) are ACMs possessing a layered structure with a net positive charge. LDHs are studied for applications as catalysts, electrocatalysts, photocatalysts, and luminescence materials. In the last 10 years, these materials have been widely used for applications in batteries and supercapacitors, in addition to their use in the catalytic production of hydrogen and oxygen. Various transition metal-based LDHs with a hydrotalcite-like structure have been studied as electrode materials for LBs, and earth-abundant metal hydroxides are promising due to their low cost and high specific capacity.[289] LDH materials have several advantages, such as low cost, tunability, chemical stability, unique anion exchange, and intercalation properties, and facile synthesis methods.

LDHs are 2D materials having the ability to exchange anions. The LDH structure is described by the ideal formula $[M^{II}{}_xM^{III}{}_{1-x}(OH)_2]_{intra}[A^{m-}{}_{x/m} \cdot nH_2O]_{inter}$. Here, M^{II} (Mg^{2+}, Mn^{2+}, Fe^{2+}, Co^{2+}, *etc.*) and M^{III} (Al^{3+}, Fe^{3+}, *etc.*) are metal cations and A is the anion (halide ions, NO_3^-, CO_3^{2-}, SO_4^{2-}, *etc.*). The terms, 'intra' and 'inter' represent the intralayer domain and interlayer space, respectively. This structure is referred to as the natural mineral, hydrotalcite. The structure of LDH contains edge-sharing, brucite ($Mg(OH)_2$)-like octahedron ($M(OH)_6$) layers. The positive charge in the layers of LDH arises from the partial substitution of M(II) to M(III), and this positive charge is balanced by negative ions. A few special cases of LDH where M(II) is substituted with Li^+ and M(III) with M^{4+} cations arc also known. The schematic representation of the LDH structure, as well as the anion exchange capacity of several LDHs, are presented in Figure 5.26a[290] and b,[291] respectively. Due to the extremely low cation exchange capacity of LDH, its application in cation-conducting PCEs is rarely reported.[150] The methods explained in Figure 5.25 for the preparation of clay–polymer composites can also be applied to polymer–LDH composite preparation. Similar to CCMs, LDHs can also lead to the formation of different types of composites, as shown in Figure 5.24.

To tune the surface properties of LDH, its modification using anionic surfactants has often been attempted.[292,293] For instance, anionic surfactant-modified LDHs ($Mg_2Al(OH)_6$(R–SO_4) and $Mg_2Al(OH)_6$(R–SO_4)(R′-R″-PO(OH)), are used for the preparation of PEO-based Li^+-ion-conducting PCEs.[294] It has been shown that compared to sulfonate-based LDH (SLDH), the phosphonate-counterpart (PLDH) is effective in replacing the CO_3^{2-} impurities intercalated in LDHs. Therefore, PLDH is much prone to exfoliation, whereas SLDH displays aggregated LDH sheets. This exfoliated structure in PLDH is the reason for the enhanced compatibility towards PEO. The exfoliated layers also help in inhibiting the crystallisation of PEO, favouring the conduction (0.015 mS cm^{-1} at 30 °C) of Li^+-ions (Figure 5.27). In another work from the same group, oligoethylene oxide oligomer was converted to its phosphonate and used for modifying LDH (OMLDH).[295] A t_{Li^+} value of 0.42 was achieved for the PCEs (with PEO/$LiClO_4$) along with an ionic conductivity of 0.016 mS cm^{-1} (at 30 °C).

Hydrocalcite ($[Mg_{0.67}Al_{0.33}(OH)_2][(CO_3)_{0.17} \cdot mH_2O]$)-based PCE exhibiting Li^+-ion conduction was reported by Bhat *et al.*, with poly(ethylene glycol) (PEG) as the polymer matrix and $LiClO_4$ as the conducting salt.[296] The PCE with 3.6 wt% loading of LDH exhibited an RT ionic conductivity of

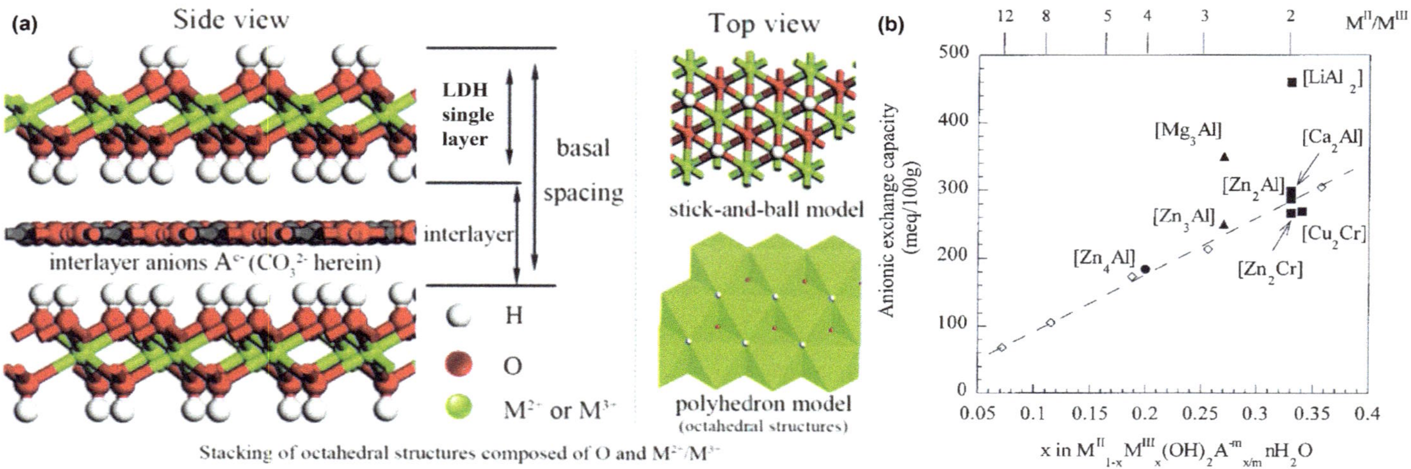

Figure 5.26 (a) General representation of LDH structure. (Reproduced from ref. 290 with permission from the Royal Society of Chemistry). (b) Anion-exchange capacity *vs.* layer charge plot for common LDHs. Adapted from ref. 291 with permission from American Chemical Society, Copyright 2001.

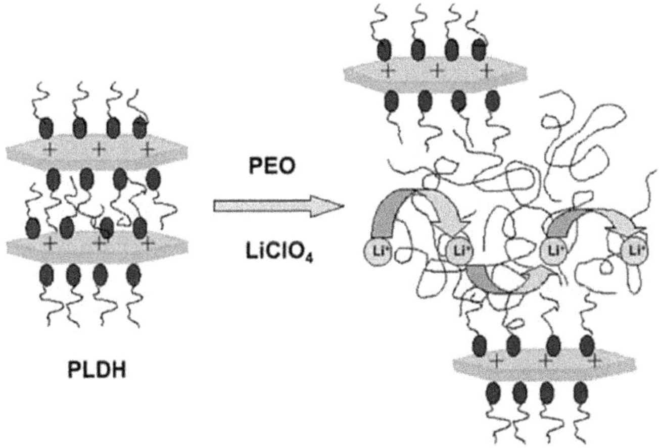

Figure 5.27 Schematic representation of exfoliated PLDHs dispersed in PEO matrix and related Li^+-ion interactions in the amorphous region.
Reproduced from ref. 294 with permission from Springer Nature, Copyright 2003.

0.01 mS cm^{-1}. Very recent work demonstrates the potential of LDH in PCEs to trap polysulfide anions in the S‖Li batteries for improving the lifecyle.[297] Here, a polymer separator (polypropylene) was modified with a composite of LDH (NiFe) and nitrogen-doped graphene (NG). Although this cannot be directly considered as an example for a PCE based on LDH and polypropylene, this work opens up the possibility for realising PCEs with anion trapping, and, hence, controlling polysulfide shuttling in S‖Li batteries. Despite the very few numbers of reports on LDH-based PCEs for battery applications, the high anion exchange capacity of LDHs has been heavily adapted for designing PCE membranes for anion-exchange membrane fuel cell applications.[298–300]

5.3.4 2D Transition Metal Dichalcogenide (2D TMC)-based PCEs

2D TMCs are a family of nanomaterials with an atomically thin sheet-like structure having a large lateral dimension, which are similar to graphite. In 2010, the Nobel Prize recognition for graphene opened up the opportunity for developing new classes of 2D layered materials, with 2D TMCs are of specific interest. The thickness of the monolayers of 2D TMCs varies in the range of 0.6–0.7 nm.[301] Owing to their unique structural properties, this class of nanomaterials possesses high surface area, good mechanical strength, and abundant active sites. The common representation for 2D-TMCs is MX_2, where M and X are the transition metal from groups 4–10 and the group 16 element of the periodic table, respectively. In bulk layered MX_2, the X–M–X layers are bonded together through weak van der Waals interactions along the crystallographic *c*-axis.[302] Depending on the surface energy, the stacked layers can be delaminated using mechanical exfoliation techniques such as ultrasonic treatment. For instance, the surface energy of MoS_2 and WS_2 (65–75 mJ m^{-2}) is close to that of graphene (65–120 mJ m^{-2}); therefore, they are cleavable under similar conditions to that of graphene.[303,304] Besides, depending on the elemental combinations, the number of layers, the presence of doping atoms, and the vacancies in the lattice structure, the band gap of these materials varies from 0 to 2 eV. Therefore, semimetal and semiconducting characteristics are commonly observed in 2D-TMCs.[305]

In the past few decades, the 2D-TMC-based nanomaterials have gathered significant interest in energy-related applications due to their tunable physicochemical properties.[307] Most of this research is focused on utilising 2D-TMCs as electrode materials for water electrolysers, supercapacitors, and Li-ion batteries (LIBs).[308–310] To improve the electrochemical properties, composites of 2D-TMCs with different carbon morphologies, such as graphene, carbon nanotube, and amorphous carbon, are also developed. Apart from that, MoS_2 is utilised as the source of sulfur for lithium–sulfur batteries, instead of the conventionally used elemental sulfur cathode.[311] During the discharge of the cell at low voltage (0.01 V *vs.* $Li|Li^+$), the intercalation of Li^+-ion into the MoS_2 cathode could form metallic Mo and Li_2S, which eventually acts as the host for Li^+-ions. The layered structure of 2D-TMCs favours the co-intercalation of alkali metals (such as Li^+ and Na^+) and organic donors through the charge-transfer interaction. In 1999, Yang *et al.* reported a 2D-TMC–polymer composite for the first time,[312] which was prepared by combining water-soluble polymer (such as PEO, PEG, and PVP) with MoS_2, WS_2, and $MoSe_2$. Later, a similar concept was extended for the preparation of organic–inorganic hybrid polymer electrolytes by incorporating 2D-TMCs as the inorganic filler in the polymer matrix. In line with this, PAN-intercalated MoS_2 polymer composite (Figure 5.28) was prepared by treating $LiMoS_2$ with a colloidal suspension of PAN in benzene.[306] The enhanced interlayer spacing of MoS_2 indicates the successful intercalation of the PAN strands. The conductivity of the $Li_{0.6}MoS_2(PAN)_{1.2}$ composite electrolyte was found to be 0.33 mS cm^{-1} at 25 °C, which is higher than that of the pristine MoS_2 (2.1 μS cm^{-1}). Furthermore, the MoS_2-based polymer electrolyte with the PEO matrix ($Li_{0.1}$ $MoS_2(PEO)_1$) showed improved conductivity of 6.6 mS cm^{-1}. In another report, composites of $NbSe_2$ intercalated with different polymers, such as PVP, PEO, and PEG, were also reported.[313] However, the application of 2D-TMC-polymer composites as the electrolyte in LIBs is rare.

Figure 5.28 Schematic showing the intercalation of the PAN chains into the MoS_2 layers.[306]
Adapted from ref. 306 with permission from the Royal Society of Chemistry.

Another interesting property of 2D-TMCs is the phase-change characteristics that are associated with the electronic properties of the material. It is observed that during lithium intercalation, the semiconducting 2H phase of MoS_2 could spontaneously be transformed into the metallic 1 T phase resulting in enhanced conductivity. Therefore, 2H MoS_2 is investigated as a surface protective layer in LMBs,[314] which not only serves as a barrier between the metallic lithium anode and electrolyte, but also offers uniform flux of Li^+-ions near to the anode surface, thus suppressing the dendritic growth of lithium.

Similarly, the intercalation of PEO in the interlayer space of 2D transition metal oxides (such as MoO_3) improves the ion transport network of Li^+-ions.[316] In a recent report, Li and co-workers[315] have demonstrated the excellent Li^+-ion transport property of PEO-based PCEs by introducing MnO_2 nanosheets as the 2D fillers (Figure 5.29). Due to its strong oxidising nature, birnessite MnO_2 exhibits good lithiophilic character and is proved to facilitate the uniform plating of lithium. Therefore, when MnO_2 is present in the polymer composite, it eases the diffusion of Li^+ and enhances the segmental motion of PEO, thereby contributing to a better ion transport mechanism. In fact, upon addition of LiTFSI salt, the PEO/LiTFSI/MnO_2 electrolyte with 5 wt% MnO_2 loading (PEO/LiTFSI/5% MnO_2) shows improved ionic conductivity (0.21 $mS\,cm^{-1}$ at 60 °C) compared to PEO/LiTFSI (0.13 $mS\,cm^{-1}$ at 60 °C). PEO/LiTFSI/5% MnO_2 electrolyte was found to be stable up to 4.5 V *vs.* $Li|Li^+$, which is 1.8 V higher than that of PEO/LiTFSI. The plating and stripping of lithium in the Li‖Li cell with PEO/LiTFSI/5% MnO_2 electrolyte occur at low overpotential, and the cell could sustain over 800 h without failure. More importantly, the $LiFePO_4$‖Li full-cell with the MnO_2-based PCE displays better electrochemical performance in terms of specific capacity, charge-transfer resistance, coulombic efficiency, and cycling stability compared to the PEO/LiTFSI-based counterpart. The overall characteristics of the 2D-TMC-based electrolyte systems and their application in LBs are reported in Table 5.7.

5.3.5 2D MXene-based PCEs

MXenes, another class of 2D materials, are transition metal carbides, nitrides, or carbonitrides; they are represented as $M_{n+1}X_nT_z$, where M is the transition metal, X is a carbon (in the case of carbides) or nitrogen (in the case of nitrides), and T stands for the terminating groups (O, F, or OH).[317] Gogotsi *et al.* reported the first MXene, $Ti_3C_2T_x$, in 2011. Since then, several new MXene compositions have been developed. Along with the unique morphology, the high electronic conductivity and terminal groups present on the surface of MXene induce hydrophilicity that enables their versatile application, similar to graphene.

Like the 2D-TMCs, MXenes are mostly studied as the electrode material in the capacitor and LIB anode. However, owing to the 2D sheet-like structure, they are used as inorganic fillers for designing PCEs. It is observed that the

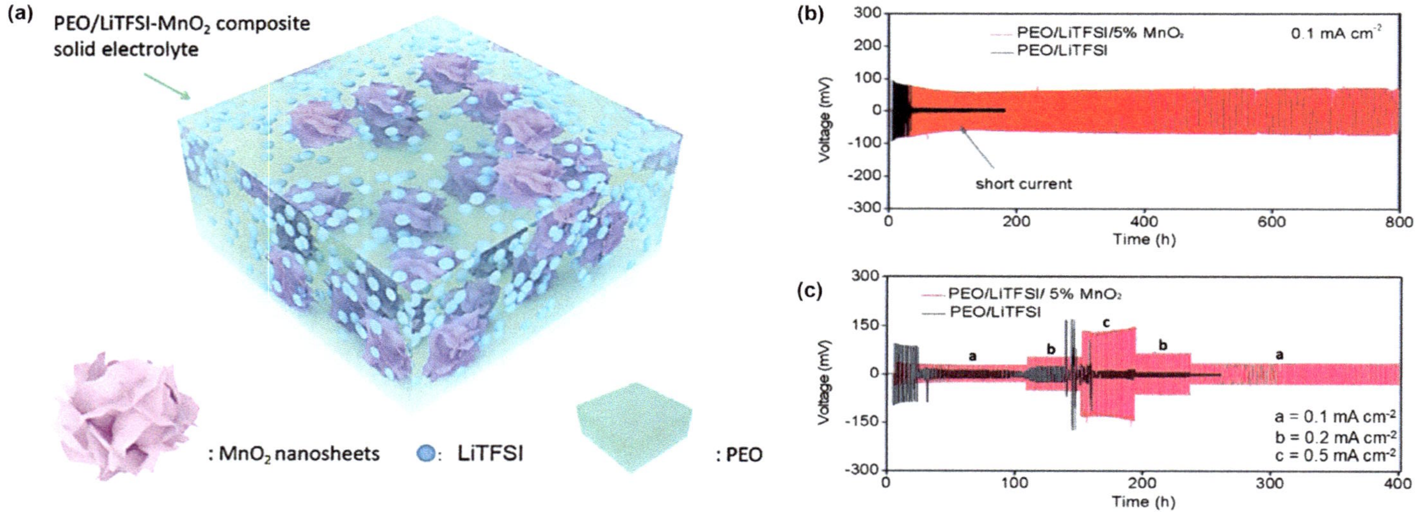

Figure 5.29 (a) Schematic representation of the solid-state PCE (PEO/LiTFSI/MnO_2). (b) Voltage *vs.* time profile of the Li‖Li cell using the solid-state PCE and SPE (PEO/LiTFSI) at 60 °C (current density = 0.1 mA cm^{-2}). (c) Voltage *vs.* time profile of a Li‖Li cell constructed using the solid-state PCE and SPE at different current rates as marked in the figure. Reproduced from ref. 315 with permission from the Royal Society of Chemistry.

Table 5.7 Different types of 2D layered nanomaterials and their application in LBs.

Material	Conductivity ($mS\,cm^{-1}$)	Cell	Cell voltage (V *vs.* Li/Li^+)	Application in LB	Ref.
$Li_{0.6}MoS_2$–$(PAN)_{1.2}$	0.33, 25 °C	$Li_{0.6}MoS_2(PAN)_{1.2}$ | $LiClO_4$-PC-EC|Li	~2.85	Electrolyte and electrode material	306
2H-MoS_2	—	CNT-S|LiTFSI-$LiNO_3$-DME-DOL|Li-MoS_2	1.5–3.0	Surface protection layer on the Li-metal	314
MoS_2	—	MoS_2|$LiPF_6$-EC-DMC|Li	0.01–3.0	Cathode in S‖Li	311
$(PEO)_x$–$[Na(H_2O)]_{0.25}$–MoO_3	—	Li|$LiClO_4$-PC|$(PEO)_x$ $[Na(H_2O)]_{0.25}$ MoO_3	2.3–3.7	Cathode material	316
PEO–LiTFSI–MnO_2	0.21, 60 °C	$LiFePO_4$|PCE|Li	2.5–4.0	Electrolyte–PCE	315
PEO_{20}–LiTFSI–$MXene^{0.05}$	0.022, 28 °C	$LiFePO_4$|PCE|Li	2.5–4.0	Electrolyte–PCE	318
MXene–$mSiO_2$–ePPO	0.46	$LiFePO_4$|PCE|Li	2.5–4.0	Electrolyte–PCE	319
Phosphorene	—	O_2|1 M-LiTFSI-DMAc|Li $LiCoO_2$|1.3 M $LiPF_6$-EC:DEC|Li	2.0–4.5 3.0–4.4	Surface protection layer on Li-metal	326

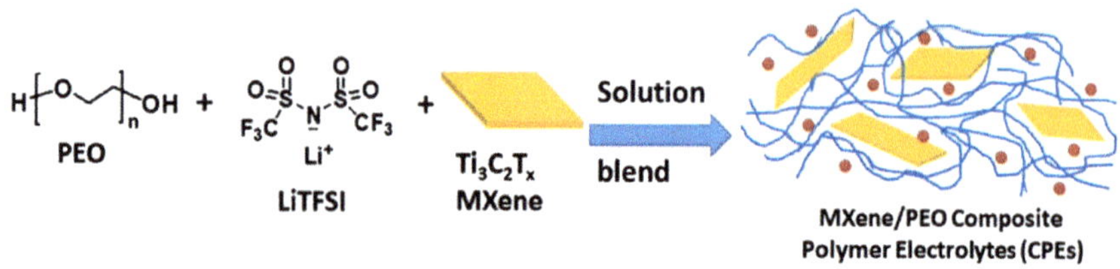

Figure 5.30 Schematic illustration of the fabrication steps of the MXene/PEO-based PCE.
Reproduced from ref. 318 with permission from the Royal Society of Chemistry.

hydrophilic surface of MXenes could favour the interaction between the MXene and the polymer matrix, which in turn reduces the crystallisation of the polymer and improves the mobility of the ions assisted by the segmental motion of the polymer chain. For example, Pan *et al.*[318] have reported $Ti_3C_2T_x$/PEO PCE (see Figure 5.30) containing LiTFSI salt (PEO_{20}–LiTFSI–$MXene^{0.05}$) showing an ionic conductivity of 0.022 mS cm^{-1} (at RT). At 60 °C, the ionic conductivity of the same electrolyte is increased to 0.69 mS cm^{-1}, a value much higher than that of the filler-free PEO_{20}–LiTFSI (0.36 mS cm^{-1}). However, with higher MXene content (PEO_{20}–LiTFSI–$MXene^{0.1}$), the ionic conductivity of the electrolyte is found to be reduced to 9.47×10^{-6} S cm^{-1} (at 28 °C) due to the tortuous pathway for ion transport and higher crystallinity. The Li^+-ion transference number (t_{Li^+}) of the PEO_{20}–LiTFSI–$MXene^{0.05}$ composite is 0.17, and a slight change in t_{Li^+} value with different MXene concentration is observed, which is attributed to the interaction of fillers with the cations and anions present in the composite electrolyte.

Recently, Shi *et al.*[319] have used MXene-based mesoporous silica nanosheets to facilitate Li^+-ion conduction in poly(propylene oxide) (PPO)-based CPEs. To tune the electrical conductivity of MXene, a sandwich-like structure of MXene and silica nanosheets was designed by hydrolysing tetraethyl orthosilicate (TEOS) on both the surfaces of MXene–Ti_3C_2. The functional groups (F and OH) present on the MXene/SiO_2 nanosheets undergo Lewis acid–base interaction with the $TFSI^-$ ions as well as the PPO chains enhancing the mobility of Li^+-ion in the electrolyte. As a result, with 30 wt% LiTFSI content and 2 wt% MXene/SiO_2, the composite electrolyte achieves high ionic conductivity of 0.46 mS cm^{-1}, which is two times higher than that of the previously reported SiO_2/PPO composite. Additionally, the electrolyte shows good anodic stability up to 4.3 V *vs.* Li|Li^+ and reversible plating/stripping of lithium for 2000 h in Li||Li symmetric cell that outperforms the SiO_2-containing MXene-free PPO electrolytes. Even though MXenes are an interesting system and a lot of new types being discovered, there are very few studies on MXene-based PCEs available in the literature. Therefore, focused research is necessary to exploit the full potential of MXene-based PCEs in LMBs.

5.3.6 Phosphorene-based PCEs

Phosphorene is an exfoliated 2D nanosheet of layered black phosphorous, which was first identified in 2014 by two different research groups.[320–322] Since then, this 2D material has attracted attention in various applications, such as electrocatalysis, photocatalysis, field-effect transistors, and energy storage. Depending on the number of layers, the thickness of the phosphorene nanosheet can vary from 2 to 10 nm. In earlier works, a scotch-tape-based microcleavage technique has been adopted to obtain phosphorene from the layered black phosphorous. Later, other techniques, such as liquid-phase exfoliation, ultrasonication combined with mechanical milling, and electrochemical exfoliation, were developed for the large-scale production of phosphorene.[323,324]

At room temperature, phosphorene can slowly react with Li-metal to form lithium phosphide (Li_3P), a chemically stable and mechanically strong compound with good ionic conductivity.[325] Although, Li_3P is proposed to be a potential electrolyte for solid-state devices, the utility of phosphorene in the lithium-battery electrolyte is yet to be explored. However, phosphorene is employed as a surface protection layer over the Li-metal for suppressing electrolyte decomposition and dendrite growth in LMBs. For instance, Kim and co-workers[326] have reported phosphorene-coated Li-metal for an $O_2\|Li$ battery. The voltage *vs.* time profile of the Li‖Li symmetric cell with phosphorene coating shows low-voltage polarisation and better cycling stability compared to that of a bare Li-metal-based cell. The X-ray photoelectron spectroscopy study evidences the presence of Li_3P on the surface of Li-metal. The same strategy could also improve the cycling performance of the $LiCoO_2\|Li$ cell in 1.3 M $LiPF_6$/EC:DEC electrolyte within the voltage range of 3.0–4.4 V *vs.* $Li|Li^+$. The transformation of phosphorene into Li_3P involves multiple lithiation steps. Therefore, it is being used as the anode material in LIBs, and transforms into Li_3P after complete discharge of the cell, delivering a high theoretical capacity of 2596 $mA\,h\,g^{-1}$. However, the anisotropic diffusion of Li^+ ions in phosphorene results in large volume expansion and poor cycling performance of the cell.[327]

5.3.7 Other 2D Nanomaterial-based PCEs

In addition to the materials mentioned above, the elements of group 14 in the periodic table can be present in various allotropic forms, such as silicene, germanene, and stanene. Owing to their structural similarity with few-layered graphene, they are also used to study the Li^+-insertion mechanism.[328,329] However, none of them are directly employed as the inorganic filler for preparing polymer composite electrolytes for LIBs, so far.

5.4 Conclusions

In this chapter, the latest developments of PCEs based on 2D layered nanomaterials as fillers have been explored, especially as separators in LBs.

Research work on the 2D layered nanomaterial-based PCEs is emerging as an important topic for LBs, particularly as a filler component that imparts interesting physicochemical and electrochemical properties to PEs. As demonstrated in this chapter, the 2D layered nanomaterials are employed in LBs in multiple dimensions, such as additives or active materials in electrodes, and fillers in polymer composites, to produce PCEs. In addition, surface protection layers have been built over the cathode or anode of the battery system to tune the interphasial properties. In the case of LMBs, these 2D layered nanomaterials have been used as an ultra-thin interlayer over the Li-metal anode and polymer/liquid electrolyte to avoid high surface area lithium ('dendrite')-induced short-circuits. The research in this direction needs a well-directed approach to achieve the set targets of solid-state LMBs, such as low cost, high safety, design flexibility, processability, eco-friendliness, and durability. Thus, from an overall point of view, it can be expected that a deep-rooted understanding of the 2D layered nanomaterial-based PCEs will deliver encouraging results, which could change the way energy storage devices are conceived in the future.

Glossary

ACM	Anionic clay minerals
AFM	Atomic force microscopy
BNNF	Boron nitride nanoflakes
BNNS	Boron nitride nanosheets
BP	Boiling point
CCMs	Cationic clay minerals
CEI	Cathode electrolyte interphase
CMC	Carboxymethyl cellulose
CV	Cyclic voltammetry
DEC	Diethyl carbonate
D_{X^-}	Diffusion coefficient of anion
D_{Li^+}	Diffusion coefficient of Li^+ ion
DMA	Dynamic mechanical analysis
DMAc	Dimethylacetamide
DMC	Dimethyl carbonate
DMTA	Dynamic mechanical thermal analysis
DSC	Differential scanning calorimetry
DTA	Differential thermal analysis
E_a; A	Activation energy; prefactor
EC	Ethylene carbonate
EIS	Electrochemical impedance spectroscopy
FBN	Functionalised boron nitride
FT IR	Fourier-transform infrared spectroscopy
GPC–SEC	Gel permeation chromatography–size exclusion chromatography
GPE	Gel polymer electrolyte

g-PCE	Gel polymer composite electrolyte
HRTEM	High-resolution transmission electron microscopy
HSAL	High surface area lithium
ISE	Inorganic solid electrolyte
LAGP	Lithium aluminium germanium phosphate, $Li_{1.5}Al_{0.5}Ge_{1.5}(PO_4)_3$
LATP	Lithium aluminium titanium phosphate, $Li_{1.4}Al_{0.4}Ti_{1.6}(PO_4)_3$
LC	linear carbonate
LCO	$LiCoO_2$
LDHs	Layered double hydroxides
LFP	$LiFePO_4$
LGPS	$Li_{10}GeP_2S_{12}$
LIB	Lithium-ion battery
LiBOB	Lithium bis(oxalato borate)
LiDFOB	Lithium difluoro(oxalato)borate
LiFSI	Lithium bis(fluorosulfonyl)imide
LiFTFSI	Lithium (fluorosulfonyl) (trifluoromethanesulfonyl)imide
LISICON	Lithium superionic conductor
LiTCB	Lithium tetracyano borate
LiTDI	4,5-Dicyano-2-(trifluoromethyl)imidazole-1-ide
LiTf	Lithium trifluoromethanesulfonate (lithium triflate)
LiTFSI	Lithium bis(trifluoromethanesulfonyl)imide
LLTO	Lithium lanthanum titanate, $Li_{0.5}La_{0.5}TiO_3$
LMB	Lithium-metal battery
LMPB	Lithium-metal polymer battery
LMO	Lithium manganese oxide, $LiMn_2O_4$
LSV	Linear sweep voltammetry
MCM-41	Mobil composition of matter no. 41
MMT	Montmorillonite
MP	Melting point
NMC	$LiNi_xMnyCo_zO_2$, $x+y+z=1$, ($x:y:z=1:1:1$, $6:2:2$ or $8:1:1$, *etc.*)
NG	Nitrogen-doped graphene
NMP	*N*-methyl-2-pyrrolidone
NMR	Nuclear magnetic resonance
ORMOCERs	ORganically MOdified CERamics
PAA	Poly(acrylic acid)
PAN	Poly(acrylonitrile)
PC	Polymer composite
PCE	Polymer composite electrolyte
PE	Polymer electrolyte
PEDOT	Poly(3,4-ethylenedioxythiophene)
PEG	Poly(ethylene glycol)
PEO	Poly(ethylene oxide)
PFPE	Perfluorpolyether

PFG NMR	Pulsed-field gradient nuclear magnetic resonance
PIS	Polymer-in-salt
PLDH	Phosphonate-LDH
PMMA	Poly(methyl methacrylate)
PNC	Polymer nanocomposite
PPC	Polypropylene carbonate
PPO	Polypropylene oxide
PSS	Polystyrene sulfonate
PVAc	Polyvinyl acetate
PVC	Polyvinyl chloride
PVCA	Polyvinyl carbonate
PVDF	Polyvinylidene difluoride
PVDF-HFP	Polyvinylidene difluoride-*co*-hexafluoropropylene
PVP	Poly(vinyl pyrrolidone)
R; σ	Universal gas constant; ionic conductivity
R_{ct}	Charge-transfer resistance
RT	Room temperature
RTIL	Room-temperature ionic liquid
SBA-15	Santa Barbara amorphous type
SBR	Styrene–butadiene rubber
SEI	Solid electrolyte interphase
SIP	Salt-in-polymer
SLDH	Sulfonate-LDH
SPE	Solid polymer electrolyte
SSE	Solid-state electrolyte
TBT	Tetra butyl titanate
TEOS	Tetraethyl orthosilicate
T_g	Glass transition temperature
TGA	Thermogravimetric analysis
ΔV	Voltage change
VAVS	Vertically aligned vermiculite
VTF	Vogel–Tammann–Fulcher
XPS	X-ray photoelectron spectroscopy
XRD	X-ray diffraction
SAXS	Small-angle X-ray scattering

References

1. M. V. Reddy, G. V. Subba Rao and B. V. R. Chowdari, *Chem. Rev.*, 2013, **113**, 5364–5457.
2. M. Li, J. Lu, Z. Chen and K. Amine, *Adv. Mater.*, 2018, **30**, 1800561.
3. P. Meister, H. Jia, J. Li, R. Kloepsch, M. Winter and T. Placke, *Chem. Mater.*, 2016, **28**, 7203–7217.
4. J. Betz, G. Bieker, P. Meister, T. Placke, M. Winter and R. Schmuch, *Adv. Energy Mater.*, 2019, **9**, 1803170.
5. M. Winter and R. J. Brodd, *Chem. Rev.*, 2004, **104**, 4245–4269.

6. A. Manthiram, *ACS Cent. Sci.*, 2017, **3**, 1063–1069.
7. R. Schmuch, R. Wagner, G. Hörpel, T. Placke and M. Winter, *Nat. Energy*, 2018, **3**, 267–278.
8. M. Winter, B. Barnett and K. Xu, *Chem. Rev.*, 2018, **118**, 11433–11456.
9. P. Roy and S. K. Srivastava, *J. Mater. Chem. A*, 2015, **3**, 2454–2484.
10. M. N. Obrovac and V. L. Chevrier, *Chem. Rev.*, 2014, **114**, 11444–11502.
11. M. V. Reddy, G. V. Subba Rao and B. V. R. Chowdari, *Chem. Rev.*, 2013, **113**, 5364–5457.
12. M. Winter, J. Besenhard, J. Albering, J. Yang and M. Wachtler, in *Progress in Batteries and Battery Materials*, 1998, vol. 17, ed. L. Breveglieri, pp. 208–213.
13. J. O. Besenhard and M. Winter, *ChemPhysChem*, 2002, **3**, 155–159.
14. H. Jia, R. Kloepsch, X. He, M. Evertz, S. Nowak, J. Li, M. Winter and T. Placke, *Acta Chim. Slov.*, 2016, 470–483.
15. J. Yang, *Electrochem. Solid-State Lett.*, 1999, **2**, 161–163.
16. J. Kim, J. H. Kim and K. Ariga, *Joule*, 2017, **1**, 739–768.
17. M. Winter, K.-C. Möller and J. Besenhard, *Lithium Batteries*, Kluwer Academic Publishers, 2004, pp. 144–194.
18. L. Zolin, J. R. Nair, D. Beneventi, F. Bella, M. Destro, P. Jagdale, I. Cannavaro, A. Tagliaferro, D. Chaussy, F. Geobaldo and C. Gerbaldi, *Carbon*, 2016, **107**, 811–822.
19. J. Kasnatscheew, M. Evertz, R. Kloepsch, B. Streipert, R. Wagner, I. Cekic Laskovic and M. Winter, *Energy Technol.*, 2017, **5**, 1670–1679.
20. F. Wu and G. Yushin, *Energy Environ. Sci.*, 2017, **10**, 435–459.
21. M. S. Whittingham, *Chem. Rev.*, 2004, **104**, 4271–4301.
22. Y. Arinicheva, M. Wolff, S. Lobe, C. Dellen, D. Fattakhova-Rohlfing, O. Guillon, D. Böhm, F. Zoller, R. Schmuch, J. Li, M. Winter, E. Adamczyk and V. Pralong, *Advanced Ceramics for Energy Conversion and Storage*, Elsevier, 2020, pp. 549–709.
23. H. Lee, M. Yanilmaz, O. Toprakci, K. Fu and X. Zhang, *Energy Environ. Sci.*, 2014, **7**, 3857–3886.
24. J. Holtmann, M. Schäfer, A. Niemöller, M. Winter, A. Lex-Balducci and S. Obeidi, *J. Appl. Electrochem.*, 2016, **46**, 69–76.
25. I. Cekic-Laskovic, N. von Aspern, L. Imholt, S. Kaymaksiz, K. Oldiges, B. R. Rad and M. Winter, *Top. Curr. Chem.*, 2017, **375**, 37.
26. K. Xu, *Chem. Rev.*, 2014, **114**, 11503–11618.
27. X. Fan, L. Chen, X. Ji, T. Deng, S. Hou, J. Chen, J. Zheng, F. Wang, J. Jiang, K. Xu and C. Wang, *Chem*, 2018, **4**, 174–185.
28. M. Amereller, T. Schedlbauer, D. Moosbauer, C. Schreiner, C. Stock, F. Wudy, S. Zugmann, H. Hammer, A. Maurer, R. M. Gschwind, H. D. Wiemhöfer, M. Winter and H. J. Gores, *Prog. Solid State Chem.*, 2014, **42**, 39–56.
29. T. Böttcher, B. Duda, N. Kalinovich, O. Kazakova, M. Ponomarenko, K. Vlasov, M. Winter and G. V. Röschenthaler, *Prog. Solid State Chem.*, 2014, **42**, 202–217.

30. E. Peled, *J. Electrochem. Soc.*, 1979, **126**, 2047.
31. M. Winter, *Z. Phys. Chem.*, 2009, **223**, 1395–1406.
32. D. R. Gallus, R. Wagner, S. Wiemers-Meyer, M. Winter and I. Cekic-Laskovic, *Electrochim. Acta*, 2015, **184**, 410–416.
33. Y. Qian, P. Niehoff, M. Börner, M. Grützke, X. Mönnighoff, P. Behrends, S. Nowak, M. Winter and F. M. Schappacher, *J. Power Sources*, 2016, **329**, 31–40.
34. W. Li, A. Dolocan, P. Oh, H. Celio, S. Park, J. Cho and A. Manthiram, *Nat. Commun.*, 2017, **8**, 14589.
35. R. Amin-Sanayei and W. He, *Advanced Fluoride-Based Materials for Energy Conversion*, Elsevier, 2015, pp. 225–235.
36. M. Wachtler, M. R. Wagner, M. Schmied, M. Winter and J. O. Besenhard, *J. Electroanal. Chem.*, 2001, **510**, 12–19.
37. P. Meister, X. Qi, R. Kloepsch, E. Krämer, B. Streipert, M. Winter and T. Placke, *ChemSusChem*, 2017, **10**, 804–814.
38. E. Kramer, S. Passerini and M. Winter, *ECS Electrochem. Lett.*, 2012, **1**, C9–C11.
39. D. Bresser, D. Buchholz, A. Moretti, A. Varzi and S. Passerini, *Energy Environ. Sci.*, 2018, **11**, 3096–3127.
40. J. Landesfeind, A. Eldiven and H. A. Gasteiger, *J. Electrochem. Soc.*, 2018, **165**, A1122–A1128.
41. X. Qi, B. Blizanac, A. DuPasquier, P. Meister, T. Placke, M. Oljaca, J. Li and M. Winter, *Phys. Chem. Chem. Phys.*, 2014, **16**, 25306–25313.
42. X. Qi, B. Blizanac, A. DuPasquier, M. Oljaca, J. Li and M. Winter, *Carbon*, 2013, **64**, 334–340.
43. P. R. Das, L. Komsiyska, O. Osters and G. Wittstock, *J. Electrochem. Soc.*, 2015, **162**, A674–A678.
44. I. Cho, J. Choi, K. Kim, M.-H. Ryou and Y. M. Lee, *RSC Adv.*, 2015, **5**, 95073–95078.
45. Y. H. Kwon, M. M. Huie, D. Choi, M. Chang, A. C. Marschilok, K. J. Takeuchi, E. S. Takeuchi and E. Reichmanis, *ACS Appl. Mater. Interfaces*, 2016, **8**, 3452–3463.
46. M. Armand and J.-M. Tarascon, *Nature*, 2008, **451**, 652–657.
47. S. Nowak and M. Winter, *J. Electrochem. Soc.*, 2015, **162**, A2500–A2508.
48. J. R. Nair, L. Imholt, G. Brunklaus and M. Winter, *Electrochem. Soc. Interface*, 2019, **28**, 55–61.
49. C. Daniel and J. O. Besenhard, *Handbook of Battery Materials*, Wiley-VCH Verlag GmbH & Co. KGaA, Weinheim, Germany, 2011.
50. M. Winter and J. O. Besenhard, *Handbook of Battery Materials*, Wiley, 1998, pp. 383–418.
51. R. Korthauer, *Lithium-Ion Batteries: Basics and Applications*, Springer Berlin Heidelberg, Berlin, Heidelberg, 2018.
52. J. B. Goodenough, *Nat. Electron.*, 2018, **1**, 204.
53. M.-H. Ryou, Y. M. Lee, Y. Lee, M. Winter and P. Bieker, *Adv. Funct. Mater.*, 2015, **25**, 834–841.

54. J. Heine, P. Hilbig, X. Qi, P. Niehoff, M. Winter and P. Bieker, *J. Electrochem. Soc.*, 2015, **162**, A1094–A1101.
55. J. Becking, A. Gröbmeyer, M. Kolek, U. Rodehorst, S. Schulze, M. Winter, P. Bieker and M. C. Stan, *Adv. Mater. Interfaces*, 2017, **4**, 1700166.
56. R. Wagner, N. Preschitschek, S. Passerini, J. Leker and M. Winter, *J. Appl. Electrochem.*, 2013, **43**, 481–496.
57. G. Bieker, M. Winter and P. Bieker, *Phys. Chem. Chem. Phys.*, 2015, **17**, 8670–8679.
58. F. Wu and G. Yushin, *Energy Environ. Sci.*, 2017, **10**, 435–459.
59. H. Lee, M. Yanilmaz, O. Toprakci, K. Fu and X. Zhang, *Energy Environ. Sci.*, 2014, **7**, 3857–3886.
60. Q. Li, J. Chen, L. Fan, X. Kong and Y. Lu, *Green Energy Environ.*, 2016, **1**, 18–42.
61. G. Pistoia, *Lithium-Ion Batteries*, Elsevier, 2014.
62. C. Masquelier, *Nat. Mater.*, 2011, **10**, 649–650.
63. M. Park, X. Zhang, M. Chung, G. B. Less and A. M. Sastry, *J. Power Sources*, 2010, **195**, 7904–7929.
64. A. Arya and A. L. Sharma, *J. Mater. Sci.*, 2020, **55**, 6242–6304.
65. K. M. Diederichsen, H. G. Buss and B. D. McCloskey, *Macromolecules*, 2017, **50**, 3831–3840.
66. M. Armand, *Solid State Ionics*, 1994, **69**, 309–319.
67. W. H. Meyer, *Adv. Mater.*, 1998, **10**, 439–448.
68. V. Di Noto, S. Lavina, G. A. Giffin, E. Negro and B. Scrosati, *Electrochim. Acta*, 2011, **57**, 4–13.
69. X.-B. Cheng, R. Zhang, C.-Z. Zhao and Q. Zhang, *Chem. Rev.*, 2017, **117**, 10403–10473.
70. W. Xu, J. Wang, F. Ding, X. Chen, E. Nasybulin, Y. Zhang and J.-G. Zhang, *Energy Environ. Sci.*, 2014, **7**, 513–537.
71. Z. Li, J. Huang, B. Yann Liaw, V. Metzler and J. Zhang, *J. Power Sources*, 2014, **254**, 168–182.
72. T. Famprikis, P. Canepa, J. A. Dawson, M. S. Islam and C. Masquelier, *Nat. Mater.*, 2019, **18**, 1278–1291.
73. J. W. Fergus, *J. Power Sources*, 2010, **195**, 4554–4569.
74. Z. Zhang, Y. Shao, B. Lotsch, Y.-S. Hu, H. Li, J. Janek, L. F. Nazar, C.-W. Nan, J. Maier, M. Armand and L. Chen, *Energy Environ. Sci.*, 2018, **11**, 1945–1976.
75. J. C. Bachman, S. Muy, A. Grimaud, H.-H. Chang, N. Pour, S. F. Lux, O. Paschos, F. Maglia, S. Lupart, P. Lamp, L. Giordano and Y. Shao-Horn, *Chem. Rev.*, 2016, **116**, 140–162.
76. C. Cao, Z.-B. Li, X.-L. Wang, X.-B. Zhao and W.-Q. Han, *Front. Energy Res.*, 2014, **2**, 25.
77. X. Yu and A. Manthiram, *Energy Environ. Sci.*, 2018, **11**, 527–543.
78. P. Knauth, *Solid State Ionics*, 2009, **180**, 911–916.
79. A. Manuel Stephan and K. S. Nahm, *Polymer*, 2006, **47**, 5952–5964.
80. P. Yao, H. Yu, Z. Ding, Y. Liu, J. Lu, M. Lavorgna, J. Wu and X. Liu, *Front. Chem.*, 2019, 7, 522.

81. M. Hess, R. G. Jones, J. Kahovec, T. Kitayama, P. Kratochvíl, P. Kubisa, W. Mormann, R. F. T. Stepto, D. Tabak, J. Vohlídal and E. S. Wilks, *Pure Appl. Chem.*, 2006, **78**, 2067–2074.
82. Z. Xue, D. He and X. Xie, *J. Mater. Chem. A*, 2015, **3**, 19218–19253.
83. P. Lightfoot, M. A. Mehta and P. G. Bruce, *Science*, 1993, **262**, 883–885.
84. J. Weston and B. Steele, *Solid State Ionics*, 1982, **7**, 75–79.
85. F. Croce, G. B. Appetecchi, L. Persi and B. Scrosati, *Nature*, 1998, **394**, 456–458.
86. F. Gray and M. Armand, *Handbook of Battery Materials*, Wiley-VCH Verlag GmbH & Co. KGaA, Weinheim, Germany, 2011, pp. 627–656.
87. P. V. Wright, *Electrochim. Acta*, 1998, **43**, 1137–1143.
88. J. R. Nair, I. Shaji, N. Ehteshami, A. Thum, D. Diddens, A. Heuer and M. Winter, *Chem. Mater.*, 2019, **31**, 3118–3133.
89. D. E. Fenton, J. M. Parker and P. V. Wright, *Polymer*, 1973, **14**, 589.
90. R. Younesi, G. M. Veith, P. Johansson, K. Edström and T. Vegge, *Energy Environ. Sci.*, 2015, **8**, 1905–1922.
91. R. W. Schmitz, P. Murmann, R. Schmitz, R. Müller, L. Krämer, J. Kasnatscheew, P. Isken, P. Niehoff, S. Nowak, G.-V. Röschenthaler, N. Ignatiev, P. Sartori, S. Passerini, M. Kunze, A. Lex-Balducci, C. Schreiner, I. Cekic-Laskovic and M. Winter, *Prog. Solid State Chem.*, 2014, **42**, 65–84.
92. T. Böttcher, B. Duda, N. Kalinovich, O. Kazakova, M. Ponomarenko, K. Vlasov, M. Winter and G. V. Röschenthaler, *Prog. Solid State Chem.*, 2014, **42**, 202–217.
93. D. Aurbach, K. Gamolsky, B. Markovsky, G. Salitra, Y. Gofer, U. Heider, R. Oesten and M. Schmidt, *J. Electrochem. Soc.*, 2000, **147**, 1322.
94. D. R. Gallus, R. Schmitz, R. Wagner, B. Hoffmann, S. Nowak, I. Cekic-Laskovic, R. W. Schmitz and M. Winter, *Electrochim. Acta*, 2014, **134**, 393–398.
95. J. R. Nair, M. Destro, F. Bella, G. B. Appetecchi and C. Gerbaldi, *J. Power Sources*, 2016, **306**, 258–267.
96. J. R. Nair, C. Gerbaldi, M. Destro, R. Bongiovanni and N. Penazzi, *React. Funct. Polym.*, 2011, **71**, 409–416.
97. C. Gerbaldi, J. R. Nair, M. A. Kulandainathan, R. S. Kumar, C. Ferrara, P. Mustarelli and A. M. Stephan, *J. Mater. Chem. A*, 2014, **2**, 9948.
98. W. Liu, N. Liu, J. Sun, P.-C. Hsu, Y. Li, H.-W. Lee and Y. Cui, *Nano Lett.*, 2015, **15**, 2740–2745.
99. K. W. Chew and K. W. Tan, *Int. J. Electrochem. Sci.*, 2011, **6**, 5792–5801.
100. J. Xi, X. Qiu, J. Li, X. Tang, W. Zhu and L. Chen, *J. Power Sources*, 2006, **157**, 501–506.
101. C.-G. Wu, M.-I. Lu, C.-C. Tsai and H.-J. Chuang, *J. Power Sources*, 2006, **159**, 295–300.
102. L. Long, S. Wang, M. Xiao and Y. Meng, *J. Mater. Chem. A*, 2016, **4**, 10038–10039.

103. R. Hooper, L. J. Lyons, D. A. Moline and R. West, *Organometallics*, 1999, **18**, 3249–3251.
104. Y. Tominaga, *Polym. J.*, 2017, **49**, 291–299.
105. S. K. Chaurasia, A. L. Saroj, Shalu, V. K. Singh, A. K. Tripathi, A. K. Gupta, Y. L. Verma and R. K. Singh, *AIP Adv.*, 2015, **5**, 077178.
106. R. J. Hobson and A. H. Windle, *Macromolecules*, 1993, **26**, 6903–6907.
107. J. E. Mark, *Physical Properties of Polymers Handbook*, Springer, New York, NY, 2nd edn., 2007.
108. J. da, S. Reinaldo, I. Z. Damasceno, M. M. Ueki and E. N. Ito, *Mater. Res.*, 2017, **20**, 694–700.
109. M. Gilbert, *J. Macromol. Sci., Part C: Polym. Rev.*, 1994, **34**, 77–135.
110. J. Liu, X. Lu and C. Wu, *Membranes*, 2013, 3, 389–405.
111. Y. Cao, M. Liang, Z. Liu, Y. Wu, X. Xiong, C. Li, X. Wang, N. Jiang, J. Yu and C.-T. Lin, *RSC Adv.*, 2016, **6**, 68357–68362.
112. L. F. Malmonge, J. A. Malmonge and W. K. Sakamoto, *Mater. Res.*, 2003, **6**, 469–473.
113. X. Tian and X. Jiang, *J. Hazard. Mater.*, 2008, **153**, 128–135.
114. L. Porcarelli, A. S. Shaplov, M. Salsamendi, J. R. Nair, Y. S. Vygodskii, D. Mecerreyes and C. Gerbaldi, *ACS Appl. Mater. Interfaces*, 2016, **8**, 10350–10359.
115. L. Porcarelli, M. A. Aboudzadeh, L. Rubatat, J. R. Nair, A. S. Shaplov, C. Gerbaldi and D. Mecerreyes, *J. Power Sources*, 2017, **364**, 191–199.
116. M. Forsyth, L. Porcarelli, X. Wang, N. Goujon and D. Mecerreyes, *Acc. Chem. Res.*, 2019, **52**, 686–694.
117. D. R. MacFarlane, M. Forsyth, P. C. Howlett, M. Kar, S. Passerini, J. M. Pringle, H. Ohno, M. Watanabe, F. Yan, W. Zheng, S. Zhang and J. Zhang, *Nat. Rev. Mater.*, 2016, **1**, 15005.
118. L. Porcarelli, A. S. Shaplov, F. Bella, J. R. Nair, D. Mecerreyes and C. Gerbaldi, *ACS Energy Lett.*, 2016, **1**, 678–682.
119. L. Porcarelli, P. S. Vlasov, D. O. Ponkratov, E. I. Lozinskaya, D. Y. Antonov, J. R. Nair, C. Gerbaldi, D. Mecerreyes and A. S. Shaplov, *Eur. Polym. J.*, 2018, **107**, 218–228.
120. A. S. Shaplov, R. Marcilla and D. Mecerreyes, *Electrochim. Acta*, 2015, **175**, 18–34.
121. L. Fan, S. Wei, S. Li, Q. Li and Y. Lu, *Adv. Energy Mater.*, 2018, **8**, 1702657.
122. I. Osada, H. de Vries, B. Scrosati and S. Passerini, *Angew. Chem. Int. Ed.*, 2016, **55**, 500–513.
123. M. J. Park, I. Choi, J. Hong and O. Kim, *J. Appl. Polym. Sci.*, 2013, **129**, 2363–2376.
124. Z. Tu, S. Choudhury, M. J. Zachman, S. Wei, K. Zhang, L. F. Kourkoutis and L. A. Archer, *Joule*, 2017, **1**, 394–406.
125. A. Manuel Stephan, *Eur. Polym. J.*, 2006, **42**, 21–42.
126. X. Cheng, J. Pan, Y. Zhao, M. Liao and H. Peng, *Adv. Energy Mater.*, 2018, **8**, 1702184.

127. M. Galiński, A. Lewandowski and I. Stępniak, *Electrochim. Acta*, 2006, **51**, 5567–5580.
128. J. R. Nair, L. Porcarelli, F. Bella and C. Gerbaldi, *ACS Appl. Mater. Interfaces*, 2015, **7**, 12961–12971.
129. C. Gerbaldi, J. R. Nair, S. Ahmad, G. Meligrana, R. Bongiovanni, S. Bodoardo and N. Penazzi, *J. Power Sources*, 2010, **195**, 1706–1713.
130. C. Gerbaldi, J. R. Nair, S. Ferrari, A. Chiappone, G. Meligrana, S. Zanarini, P. Mustarelli, N. Penazzi and R. Bongiovanni, *J. Membr. Sci.*, 2012, **423–424**, 459–467.
131. M. Armand, F. Endres, D. R. MacFarlane, H. Ohno and B. Scrosati, *Nat. Mater.*, 2009, **8**, 621–629.
132. B. Rupp, M. Schmuck, A. Balducci, M. Winter and W. Kern, *Eur. Polym. J.*, 2008, **44**, 2986–2990.
133. J. R. Nair, M. Destro, C. Gerbaldi, R. Bongiovanni and N. Penazzi, *J. Appl. Electrochem.*, 2013, **43**, 137–145.
134. V. Vijayakumar, D. Diddens, A. Heuer, S. Kurungot, M. Winter and J. R. Nair, *ACS Appl. Mater. Interfaces*, 2020, **12**, 567–579.
135. J. R. Nair, D. Cíntora-Juárez, C. Pérez-Vicente, J. L. Tirado, S. Ahmad and C. Gerbaldi, *Electrochim. Acta*, 2016, **199**, 172–179.
136. J. R. Nair, C. Gerbaldi, G. Meligrana, R. Bongiovanni, S. Bodoardo, N. Penazzi, P. Reale and V. Gentili, *J. Power Sources*, 2008, **178**, 751–757.
137. L. Porcarelli, C. Gerbaldi, F. Bella and J. R. Nair, *Sci. Rep.*, 2016, **6**, 19892.
138. C. Gerbaldi, J. R. Nair, G. Meligrana, R. Bongiovanni, S. Bodoardo and N. Penazzi, *Electrochim. Acta*, 2010, **55**, 1460–1467.
139. P. Isken, C. Dippel, R. Schmitz, R. W. Schmitz, M. Kunze, S. Passerini, M. Winter and A. Lex-Balducci, *Electrochim. Acta*, 2011, **56**, 7530–7535.
140. P. Hilbig, L. Ibing, B. Streipert, R. Wagner, M. Winter and I. Cekic-Laskovic, *Curr. Top. Electrochem.*, 2018, **20**, 1.
141. T. Böttcher, N. Kalinovich, O. Kazakova, M. Ponomarenko, K. Vlasov, M. Winter and G.-V. Röschenthaler, *Advanced Fluoride-Based Materials for Energy Conversion*, Elsevier, 2015, pp. 125–145.
142. C. Gerbaldi, M. Destro, J. R. Nair, S. Ferrari, I. Quinzeni and E. Quartarone, *Nano Energy*, 2013, **2**, 1279–1286.
143. J. Wang, J. Luo, S. Feng, H. Li, Y. Wan and X. Zhang, *Green Energy Environ.*, 2016, **1**, 43–61.
144. W. J. Work, K. Horie, M. Hess and R. F. T. Stepto, *Pure Appl. Chem.*, 2004, **76**, 1985–2007.
145. S. Thomas, G. Zaikov, Valsaraj and Meera, *Recent Advances in Polymer Nanocomposites: Synthesis and Characterisation*, CRC Press, 2010.
146. International Union of Pure and Applied Chemistry, Subcommittee on Polymer Terminology, R. G. Jones and International Union of Pure and Applied Chemistry, Commission on Macromolecular Nomenclature, *Compendium of polymer terminology and nomenclature: IUPAC recommendations, 2008*, 2009.

147. V. Mittal, *Synthesis Techniques for Polymer Nanocomposites*, Wiley-VCH Verlag GmbH & Co. KGaA, Weinheim, Germany, 2014.
148. J. Liu, W.-J. Boo, A. Clearfield and H.-J. Sue, *Mater. Manuf. Process.*, 2006, **21**, 143–151.
149. J. Fawaz and V. Mittal, *Synthesis Techniques for Polymer Nanocomposites*, Wiley-VCH Verlag GmbH & Co. KGaA, Weinheim, Germany, 2014, pp. 1–30.
150. S. Chua, R. Fang, Z. Sun, M. Wu, Z. Gu, Y. Wang, J. N. Hart, N. Sharma, F. Li and D. Wang, *Chem. – Eur. J.*, 2018, **24**, 18180–18203.
151. F. Croce, L. Persi, B. Scrosati, F. Serraino-Fiory, E. Plichta and M. Hendrickson, *Electrochim. Acta*, 2001, **46**, 2457–2461.
152. J. Naik, R. F. Bhajantri, T. Sheela and S. G. Rathod, *Polym. Compos.*, 2018, **39**, 1273–1282.
153. K. K. Fu, Y. Gong, B. Liu, Y. Zhu, S. Xu, Y. Yao, W. Luo, C. Wang, S. D. Lacey, J. Dai, Y. Chen, Y. Mo, E. Wachsman and L. Hu, *Sci. Adv.*, 2017, **3**, e1601659.
154. H. M. J. C. Pitawala, M. A. K. L. Dissanayake and V. A. Seneviratne, *Solid State Ionics*, 2007, **178**, 885–888.
155. P. A. R. Jayathilaka, M. A. K. Dissanayake, I. Albinsson and B.-E. Mellander, *Electrochim. Acta*, 2002, **47**, 3257–3268.
156. S. Ferrari, J. R. Nair, Y. Zhou and C. Wan, *Polymer-based Nanocomposites for Energy and Environmental Applications*, Woodhead Publishing, 2018, pp. 283–313.
157. C. W. Lin, C. L. Hung, M. Venkateswarlu and B. J. Hwang, *J. Power Sources*, 2005, **143**, 397–401.
158. W. Wang and P. Alexandridis, *Polymers*, 2016, **8**, 387.
159. L. Zhou, N. Wu, Q. Cao, B. Jing, X. Wang, Q. Wang and H. Kuang, *Solid State Ionics*, 2013, **249–250**, 93–97.
160. M. Popall, M. Andrei, J. Kappel, J. Kron, K. Olma and B. Olsowski, *Electrochim. Acta*, 1998, **43**, 1155–1161.
161. H. Y. Sun, Y. Takeda, N. Imanishi, O. Yamamoto and H.-J. Sohn, *J. Electrochem. Soc.*, 2000, **147**, 2462.
162. N. Angulakshmi, K. S. Nahm, J. R. Nair, C. Gerbaldi, R. Bongiovanni, N. Penazzi and A. M. Stephan, *Electrochim. Acta*, 2013, **90**, 179–185.
163. L. Zhu, P. Zhu, Q. Fang, M. Jing, X. Shen and L. Yang, *Electrochim. Acta*, 2018, **292**, 718–726.
164. T. Yang, J. Zheng, Q. Cheng, Y.-Y. Hu and C. K. Chan, *ACS Appl. Mater. Interfaces*, 2017, **9**, 21773–21780.
165. S. Cho, S. Kim, W. Kim, S. Kim and S. Ahn, *Polymers*, 2018, **10**, 1364.
166. J. Liu, J. Zhou, M. Wang, C. Niu, T. Qian and C. Yan, *J. Mater. Chem. A*, 2019, **7**, 24477–24485.
167. J. R. Nair, A. Chiappone, C. Gerbaldi, V. S. Ijeri, E. Zeno, R. Bongiovanni, S. Bodoardo and N. Penazzi, *Electrochim. Acta*, 2011, **57**, 104–111.
168. A. Chiappone, J. R. Nair, C. Gerbaldi, E. Zeno and R. Bongiovanni, *Eur. Polym. J.*, 2014, **57**, 22–29.

169. A. Chiappone, J. R. Nair, C. Gerbaldi, R. Bongiovanni and E. Zeno, *Electrochim. Acta*, 2015, **153**, 97–105.
170. D. Klemm, F. Kramer, S. Moritz, T. Lindström, M. Ankerfors, D. Gray and A. Dorris, *Angew. Chem. - Int. Ed.*, 2011, **50**, 5438–5466.
171. J. R. Nair, F. Bella, N. Angulakshmi, A. M. Stephan and C. Gerbaldi, *Energy Storage Mater.*, 2016, **3**, 69–76.
172. A. Chiappone, J. R. Nair, C. Gerbaldi, L. Jabbour, R. Bongiovanni, E. Zeno, D. Beneventi and N. Penazzi, *J. Power Sources*, 2011, **196**, 10280–10288.
173. N. N. Mobarak, F. N. Jumaah, M. A. Ghani, M. P. Abdullah and A. Ahmad, *Electrochim. Acta*, 2015, **175**, 224–231.
174. M. Rinaudo, *Prog. Polym. Sci.*, 2006, **31**, 603–632.
175. N. Angulakhsmi, S. Thomas, J. R. Nair, R. Bongiovanni, C. Gerbaldi and A. M. Stephan, *J. Power Sources*, 2013, **228**, 294–299.
176. D. Zhao, Q. Huo, J. Feng, B. F. Chmelka and G. D. Stucky, *J. Am. Chem. Soc.*, 1998, **120**, 6024–6036.
177. S. Kim, E.-J. Hwang and S.-J. Park, *Curr. Appl. Phys.*, 2008, **8**, 729–731.
178. M. Eddaoudi, H. Li and O. M. Yaghi, *J. Am. Chem. Soc.*, 2000, **122**, 1391–1397.
179. B. M. Wiers, M. L. Foo, N. P. Balsara and J. R. Long, *J. Am. Chem. Soc.*, 2011, **133**, 14522–14525.
180. S. Horike, D. Umeyama and S. Kitagawa, *Acc. Chem. Res.*, 2013, **46**, 2376–2384.
181. S. Suriyakumar, S. Gopi, M. Kathiresan, S. Bose, E. B. Gowd, J. R. Nair, N. Angulakshmi, G. Meligrana, F. Bella, C. Gerbaldi and A. M. Stephan, *Electrochim. Acta*, 2018, **285**, 355–364.
182. R. Raccichini, A. Varzi, S. Passerini and B. Scrosati, *Nat. Mater.*, 2015, **14**, 271–279.
183. B. Zhang and T. Cui, *Appl. Phys. Lett.*, 2011, **98**, 073116.
184. M. G. Chung, D. H. Kim, H. M. Lee, T. Kim, J. H. Choi, D. K. Seo, J.-B. Yoo, S.-H. Hong, T. J. Kang and Y. H. Kim, *Sens. Actuators B Chem.*, 2012, **166–167**, 172–176.
185. P. K. Samantaray, S. Baloda, G. Madras and S. Bose, *ACS Sustain. Chem. Eng.*, 2019, **7**, 18775–18784.
186. P. K. Samantaray, G. Madras and S. Bose, *ACS Sustain. Chem. Eng.*, 2019, **7**, 1580–1590.
187. Q. Xiang, J. Yu and M. Jaroniec, *Chem. Soc. Rev.*, 2012, **41**, 782–796.
188. A. A. Balandin, S. Ghosh, W. Bao, I. Calizo, D. Teweldebrhan, F. Miao and C. N. Lau, *Nano Lett.*, 2008, **8**, 902–907.
189. C. Wan and B. Chen, *Biomed. Mater.*, 2011, **6**, 055010.
190. M. Koshino and T. Ando, *Phys. Rev. B*, 2008, **77**, 115313.
191. H. Shen, L. Zhang, M. Liu and Z. Zhang, *Theranostics*, 2012, **2**, 283–294.
192. H. Fredriksson, D. Chakarov and B. Kasemo, *Carbon*, 2009, **47**, 1335–1342.

193. K. S. Novoselov, *Science*, 2004, **306**, 666–669.
194. X. Z. Yu, C. G. Hwang, C. M. Jozwiak, A. Köhl, A. K. Schmid and A. Lanzara, *J. Electron Spectros. Relat. Phenom.*, 2011, **184**, 100–106.
195. Y. Hernandez, V. Nicolosi, M. Lotya, F. M. Blighe, Z. Sun, S. De, I. T. McGovern, B. Holland, M. Byrne, Y. K. Gun'Ko, J. J. Boland, P. Niraj, G. Duesberg, S. Krishnamurthy, R. Goodhue, J. Hutchison, V. Scardaci, A. C. Ferrari and J. N. Coleman, *Nat. Nanotechnol.*, 2008, **3**, 563–568.
196. C. Berger, *Science*, 2006, **312**, 1191–1196.
197. H. An, W.-J. Lee and J. Jung, *Curr. Appl. Phys.*, 2011, **11**, S81–S85.
198. J. Wintterlin and M.-L. Bocquet, *Surf. Sci.*, 2009, **603**, 1841–1852.
199. D. Chen, H. Feng and J. Li, *Chem. Rev.*, 2012, **112**, 6027–6053.
200. B. C. Brodie, *Philos. Trans. R. Soc. London*, 1859, **149**, 249–259.
201. L. Staudenmaier, *Ber. Dtsch. Chem. Ges.*, 1898, **31**, 1481–1487.
202. C. Botas, P. Álvarez, P. Blanco, M. Granda, C. Blanco, R. Santamaría, L. J. Romasanta, R. Verdejo, M. A. López-Manchado and R. Menéndez, *Carbon*, 2013, **65**, 156–164.
203. J. Chen, Y. Li, L. Huang, C. Li and G. Shi, *Carbon*, 2015, **81**, 826–834.
204. A. M. Shanmugharaj, W. S. Choi, C. W. Lee and S. H. Ryu, *J. Power Sources*, 2011, **196**, 10249–10253.
205. P. Lian, X. Zhu, S. Liang, Z. Li, W. Yang and H. Wang, *Electrochim. Acta*, 2010, **55**, 3909–3914.
206. Y. Fang, Y. Lv, R. Che, H. Wu, X. Zhang, D. Gu, G. Zheng and D. Zhao, *J. Am. Chem. Soc.*, 2013, **135**, 1524–1530.
207. J. R. Nair, G. Rius, P. Jagadale, M. Destro, M. Tortello, M. Yoshimura, A. Tagliaferro and C. Gerbaldi, *Electrochim. Acta*, 2015, **182**, 500–506.
208. L. Lefrançois Perreault, F. Colò, G. Meligrana, K. Kim, S. Fiorilli, F. Bella, J. R. Nair, C. Vitale-Brovarone, J. Florek, F. Kleitz and C. Gerbaldi, *Adv. Energy Mater.*, 2018, **8**, 1802438.
209. S.-L. Chou, J.-Z. Wang, M. Choucair, H.-K. Liu, J. A. Stride and S.-X. Dou, *Electrochem. Commun.*, 2010, **12**, 303–306.
210. H. A. Becerril, J. Mao, Z. Liu, R. M. Stoltenberg, Z. Bao and Y. Chen, *ACS Nano*, 2008, **2**, 463–470.
211. S. Mikhailov, *Physics and Applications of Graphene - Experiments*, InTech, 2011.
212. S. Tan, Z. Yang, H. Yuan, J. Zhang, Y. Yang and H. Liu, *J. Alloys Compd.*, 2019, **791**, 483–489.
213. J. Shim, D.-G. Kim, H. J. Kim, J. H. Lee, J.-H. Baik and J.-C. Lee, *J. Mater. Chem. A*, 2014, **2**, 13873–13883.
214. Y.-S. Ye, H. Wang, S.-G. Bi, Y. Xue, Z.-G. Xue, X.-P. Zhou, X.-L. Xie and Y.-W. Mai, *J. Mater. Chem. A*, 2015, **3**, 18064–18073.
215. J. Liu, X. Wu, J. He, J. Li and Y. Lai, *Electrochim. Acta*, 2017, **235**, 500–507.
216. W. Jia, Z. Li, Z. Wu, L. Wang, B. Wu, Y. Wang, Y. Cao and J. Li, *Solid State Ionics*, 2018, **315**, 7–13.

217. F. J. Yang, Y. F. Huang, M. Q. Zhang and W. H. Ruan, *Polymer*, 2018, **153**, 438–444.
218. J. Wu, X. Qin, H. Zhang, Y.-B. He, B. Li, L. Ke, W. Lv, H. Du, Q.-H. Yang and F. Kang, *Carbon*, 2015, **84**, 434–443.
219. X.-B. Cheng, H.-J. Peng, J.-Q. Huang, R. Zhang, C.-Z. Zhao and Q. Zhang, *ACS Nano*, 2015, **9**, 6373–6382.
220. Y. Gao, Z. Yan, J. L. Gray, X. He, D. Wang, T. Chen, Q. Huang, Y. C. Li, H. Wang, S. H. Kim, T. E. Mallouk and D. Wang, *Nat. Mater.*, 2019, **18**, 384–389.
221. N. Alem, R. Erni, C. Kisielowski, M. D. Rossell, W. Gannett and A. Zettl, *Phys. Rev. B*, 2009, **80**, 155425.
222. R. Haubner, M. Wilhelm, R. Weissenbacher and B. Lux, in *High Performance Non-Oxide Ceramics II. Structure and Bonding*, ed. M. Jansen, Springer, Berlin, Heidelberg, 2002, pp. 1–45.
223. R. Wills, *Int. J. High Technol. Ceram.*, 1985, **1**, 139–153.
224. *Concise Encyclopedia of Self-Propagating High-Temperature Synthesis*, ed. I. P. Borovinskaya, A. A. Gromov, E. A. Levashov, Y. M. Maksimov, A. S. Mukasyan and A. S. Rogachev, Elsevier Science, 2017.
225. E. C. Anota, Y. Tlapale, M. S. Villanueva and J. A. R. Márquez, *J. Mol. Model.*, 2015, **21**, 215.
226. K. Zhang, Y. Feng, F. Wang, Z. Yang and J. Wang, *J. Mater. Chem. C*, 2017, **5**, 11992–12022.
227. P. L. de Andres, R. Ramírez and J. A. Vergés, *Phys. Rev. B*, 2008, **77**, 045403.
228. J. Wang, F. Ma and M. Sun, *RSC Adv.*, 2017, **7**, 16801–16822.
229. H. Zeng, C. Zhi, Z. Zhang, X. Wei, X. Wang, W. Guo, Y. Bando and D. Golberg, *Nano Lett.*, 2010, **10**, 5049–5055.
230. W.-L. Song, P. Wang, L. Cao, A. Anderson, M. J. Meziani, A. J. Farr and Y.-P. Sun, *Angew. Chem. Int. Ed.*, 2012, **51**, 6498–6501.
231. C. Yuan, B. Duan, L. Li, B. Xie, M. Huang and X. Luo, *ACS Appl. Mater. Interfaces*, 2015, **7**, 13000–13006.
232. H. Li, R. Y. Tay, S. H. Tsang, W. Liu and E. H. T. Teo, *Electrochim. Acta*, 2015, **166**, 197–205.
233. H. Aydın, S. Ü. Çelik and A. Bozkurt, *Solid State Ionics*, 2017, **309**, 71–76.
234. M.-T. F. Rodrigues, K. Kalaga, H. Gullapalli, G. Babu, A. L. M. Reddy and P. M. Ajayan, *Adv. Energy Mater.*, 2016, **6**, 1600218.
235. J. Shim, H. J. Kim, B. G. Kim, Y. S. Kim, D.-G. Kim and J.-C. Lee, *Energy Environ. Sci.*, 2017, **10**, 1911–1916.
236. K.-H. Oh, D. Lee, M.-J. Choo, K. H. Park, S. Jeon, S. H. Hong, J.-K. Park and J. W. Choi, *ACS Appl. Mater. Interfaces*, 2014, **6**, 7751–7758.
237. Z. Zhang, R. G. Antonio and K. L. Choy, *J. Power Sources*, 2019, **435**, 226736.
238. W. Luo, L. Zhou, K. Fu, Z. Yang, J. Wan, M. Manno, Y. Yao, H. Zhu, B. Yang and L. Hu, *Nano Lett.*, 2015, **15**, 6149–6154.
239. Z. Zhang, A. R. Gonzalez and K. L. Choy, *ACS Appl. Energy Mater.*, 2019, **2**, 7438–7448.

240. K. Yan, H. W. Lee, T. Gao, G. Zheng, H. Yao, H. Wang, Z. Lu, Y. Zhou, Z. Liang, Z. Liu, S. Chu and Y. Cui, *Nano Lett.*, 2014, **14**, 6016–6022.
241. B. Shen, T.-W. Zhang, Y.-C. Yin, Z.-X. Zhu, L.-L. Lu, C. Ma, F. Zhou and H.-B. Yao, *Chem. Commun.*, 2019, **55**, 7703–7706.
242. A. Gröbmeyer, J. Becking, P. M. Bieker, M. Winter and M. C. Stan, *Energy Technol.*, 2019, **7**, 1800789.
243. T. Placke, R. Kloepsch, S. Dühnen and M. Winter, *J. Solid State Electrochem.*, 2017, **21**, 1939–1964.
244. G. Bieker, D. Diddens, M. Kolek, O. Borodin, M. Winter, P. Bieker and K. Jalkanen, *J. Phys. Chem. C*, 2018, **122**, 21770–21783.
245. M. Li, J. E. Frerichs, M. Kolek, W. Sun, D. Zhou, C. J. Huang, B. J. Hwang, M. R. Hansen, M. Winter and P. Bieker, *Adv. Funct. Mater.*, 2020, **30**, 1910123.
246. X. Xu, L. Wang, H. Fei and L. Ci, *J. Mater. Sci.: Mater. Electron.*, 2019, **30**, 19119–19125.
247. Y. Fan, Z. Yang, W. Hua, D. Liu, T. Tao, M. M. Rahman, W. Lei, S. Huang and Y. Chen, *Adv. Energy Mater.*, 2017, **7**, 1602380.
248. A. Yaya, B. Agyei-Tuffour, D. Dodoo-Arhin, E. Nyankson, E. Annan, D. S. Konadu, E. Sinayobye, E. A. Baryeh and C. P. Ewels, *Global J. Eng. Des. Technol.*, 2012, **1**(2), 32–41.
249. I. Bibi, J. Icenhower, N. K. Niazi, T. Naz, M. Shahid and S. Bashir, *Environmental Materials and Waste*, Elsevier, 2016, pp. 543–567.
250. F. Guo, S. Aryana, Y. Han and Y. Jiao, *Appl. Sci.*, 2018, **8**, 1696.
251. M. Ghadiri, W. Chrzanowski and R. Rohanizadeh, *RSC Adv.*, 2015, **5**, 29467–29481.
252. S. C. Aboudi Mana, M. M. Hanafiah and A. J. K. Chowdhury, *Geol. Ecol. Landscapes*, 2017, **1**, 155–161.
253. S. W. Bailey, *Crystal Structures of Clay Minerals and their X-Ray Identification*, Mineralogical Society of Great Britain and Ireland, Colchester and London, 1980.
254. B. Sarkar, M. Singh, S. Mandal, G. J. Churchman and N. S. Bolan, in *The Future of Soil Carbon*, ed. C. Garcia, P. Nannipieri and T. Hernandez, Elsevier, 2018, pp. 71–86.
255. F. Uddin, *Current Topics in the Utilization of Clay in Industrial and Medical Applications*, IntechOpen, 2018, pp. 3–23.
256. C. D. Barton, in *Encyclopedia of Soil Science, Second Edition*, ed. R. Lal, Marcel Dekker, New York, 2002, pp. 187–192.
257. J.-C. Viennet, B. Bultel, L. Riu and S. C. Werner, *J. Geophys. Res. Planets*, 2017, **122**, 2328–2343.
258. I. Shainberg and G. J. Levy, *Encyclopedia of Soils in the Environment*, Elsevier, 2005, pp. 27–34.
259. F. Bergaya, G. Lagaly and M. Vayer, in *Developments in Clay Science*, 2013, ed. G. L. Faïza *Bergaya*, pp. 333–359.
260. F. Bergaya and G. Lagaly, in *Developments in Clay Science*, 2013, ed. B. Faïza and L. Gerhard, pp. 1–19.

261. G. E. Christidis, in *Developments in Clay Science*, ed. B. Faïza and L. Gerhard, Elsevier, 2013, pp. 425–449.
262. W. R. Schell and J. V Jordan, *Plant Soil*, 1959, **10**, 303–318.
263. T. R. Jones, *Clay Miner.*, 1983, **18**, 399–410.
264. J. L. McAtee, *Clays Clay Miner.*, 1961, **10**, 153–162.
265. P. Liu, *Appl. Clay Sci.*, 2007, **38**, 64–76.
266. F. Annabi-Bergaya, *Microporous Mesoporous Mater.*, 2008, **107**, 141–148.
267. B. Sarkar, R. Rusmin, U. C. Ugochukwu, R. Mukhopadhyay and K. M. Manjaiah *Modified Clay and Zeolite Nanocomposite Materials*, Elsevier, 2019, pp. 113–127.
268. D. Chimene, D. L. Alge and A. K. Gaharwar, *Adv. Mater.*, 2015, **27**, 7261–7284.
269. T. Mishra, A. K. Mohanty and S. K. Tiwari, *Key Eng. Mater.*, 2013, **571**, 93–109.
270. S. Ismadji, F. E. Soetaredjo and A. Ayucitra, *Clay Materials for Environmental Remediation*, Springer, Cham, 2015, pp. 5–37.
271. F. Gao, *Mater. Today*, 2004, **7**, 50–55.
272. J. Cardoso, G. Vázquez, P. R. García and I. González, *Macromol. Symp.*, 2013, **325–326**, 156–167.
273. B. Kumar and L. G. Scanlon, *J. Power Sources*, 1994, **52**, 261–268.
274. S. A. L. Arya Anil, *Environmental Nanotechnology Volume 4*, Springer International Publishing, 2020, pp. 265–315.
275. A. Dantas de Oliveira and C. Augusto Gonçalves Beatrice, in *Nanocomposites - Recent Evolutions*, ed. S. Sivasankaran, IntechOpen, 2019, pp. 103–128.
276. R. Prasanth, N. Shubha, H. H. Hng and M. Srinivasan, *Eur. Polym. J.*, 2013, **49**, 307–318.
277. B. Chen, J. R. G. Evans, H. C. Greenwell, P. Boulet, P. V. Coveney, A. A. Bowden and A. Whiting, *Chem. Soc. Rev.*, 2008, **37**, 568–594.
278. X. Fu, D. Yu, J. Zhou, S. Li, X. Gao, Y. Han, P. Qi, X. Feng and B. Wang, *CrystEngComm*, 2016, **18**, 4236–4258.
279. F. Uddin, *Current Topics in the Utilization of Clay in Industrial and Medical Applications*, InTech, 2018.
280. Y. Zhao and Y. Wang, *Nanoscale Res. Lett.*, 2019, **14**, 366.
281. C. H. Manoratne, R. M. G. Rajapakse and M. A. K. L. Dissanayake, *Int. J. Electrochem. Sci.*, 2006, **1**, 32–46.
282. S. Kitajima, M. Matsuda, M. Yamato and Y. Tominaga, *Polym. J.*, 2013, **45**, 738–743.
283. S. Kim and S.-J. Park, *Solid State Ionics*, 2007, **178**, 973–979.
284. Z. Wen, Z. Gu, T. Itoh, Z. Lin and O. Yamamoto, *J. Power Sources*, 2003, **119–121**, 427–431.
285. W. Tang, S. Tang, X. Guan, X. Zhang, Q. Xiang and J. Luo, *Adv. Funct. Mater.*, 2019, **29**, 1900648.
286. W. Tang, S. Tang, C. Zhang, Q. Ma, Q. Xiang, Y.-W. Yang and J. Luo, *Adv. Energy Mater.*, 2018, **8**, 1800866.

287. B. Wang, M. Tang, Y. Wu, Y. Chen, C. Jiang, S. Zhuo, S. Zhu and C. Wang, *ACS Appl. Energy Mater.*, 2019, **2**, 5909–5916.
288. M. Rajamathi, G. S. Thomas and P. V. Kamath, *J. Chem. Sci.*, 2001, **113**, 671–680.
289. A.-L. Yan, X.-C. Wang and J.-P. Cheng, *Nanomaterials*, 2018, **8**, 747.
290. M. Laipan, J. Yu, R. Zhu, J. Zhu, A. T. Smith, H. He, D. O'Hare and L. Sun, *Mater. Horiz.*, 2020, **7**, 715–745.
291. F. Leroux and J.-P. Besse, *Chem. Mater.*, 2001, **13**, 3507–3515.
292. H. A. Patel, R. S. Somani, H. C. Bajaj and R. V. Jasra, *Bull. Mater. Sci.*, 2006, **29**, 133–145.
293. G. Starukh, O. Rozovik and O. Oranska, *Nanoscale Res. Lett.*, 2016, **11**, 228.
294. C.-S. Liao and W.-B. Ye, *J. Polym. Res.*, 2003, **10**, 241–246.
295. C.-S. Liao and W.-B. Ye, *Mater. Chem. Phys.*, 2004, **88**, 84–89.
296. M. M. Borgohain, T. Joykumar and S. V. Bhat, *Solid State Ionics*, 2010, **181**, 964–970.
297. H.-J. Peng, Z.-W. Zhang, J.-Q. Huang, G. Zhang, J. Xie, W.-T. Xu, J.-L. Shi, X. Chen, X.-B. Cheng and Q. Zhang, *Adv. Mater.*, 2016, **28**, 9551–9558.
298. L. Zeng, T. S. Zhao and Y. S. Li, *Int. J. Hydrogen Energy*, 2012, **37**, 18425–18432.
299. N. Chen, C. Long, Y. Li, D. Wang and H. Zhu, *J. Membr. Sci.*, 2018, **552**, 51–60.
300. X. He, L. Cao, G. He, A. Zhao, X. Mao, T. Huang, Y. Li, H. Wu, J. Sun and Z. Jiang, *J. Mater. Chem. A*, 2018, **6**, 10277–10285.
301. M. Ahmadi, O. Zabihi, S. Jeon, M. Yoonessi, A. Dasari, S. Ramakrishna and M. Naebe, *J. Mater. Chem. A*, 2020, **8**, 845–883.
302. S. Manzeli, D. Ovchinnikov, D. Pasquier, O. V. Yazyev and A. Kis, *Nat. Rev. Mater.*, 2017, **2**, 17033.
303. A. Gupta, T. Sakthivel and S. Seal, *Prog. Mater. Sci.*, 2015, **73**, 44–126.
304. S. Z. Butler, S. M. Hollen, L. Cao, Y. Cui, J. A. Gupta, H. R. Gutiérrez, T. F. Heinz, S. S. Hong, J. Huang, A. F. Ismach, E. Johnston-Halperin, M. Kuno, V. V. Plashnitsa, R. D. Robinson, R. S. Ruoff, S. Salahuddin, J. Shan, L. Shi, M. G. Spencer, M. Terrones, W. Windl and J. E. Goldberger, *ACS Nano*, 2013, **7**, 2898–2926.
305. M. Pumera, Z. Sofer and A. Ambrosi, *J. Mater. Chem. A*, 2014, **2**, 8981–8987.
306. M. A. Santa Ana, E. Benavente, P. Gómez-Romero and G. González, *J. Mater. Chem.*, 2006, **16**, 3107–3113.
307. X. Wang, W. Xing, X. Feng, L. Song and Y. Hu, *Polym. Rev.*, 2017, **57**, 440–466.
308. A. Liang, D. Li, W. Zhou, Y. Wu, G. Ye, J. Wu, Y. Chang, R. Wang, J. Xu, G. Nie, J. Hou and Y. Du, *J. Electroanal. Chem.*, 2018, **824**, 136–146.
309. V. Mazánek, C. C. Mayorga-Martinez, D. Bouša, Z. Sofer and M. Pumera, *Nanoscale*, 2018, **10**, 23149–23156.

310. J. Zhu, H. N. Alshareef and U. Schwingenschlögl, *Appl. Phys. Lett.*, 2017, **111**, 043903.
311. X. Fang, X. Guo, Y. Mao, C. Hua, L. Shen, Y. Hu, Z. Wang, F. Wu and L. Chen, *Chem. – Asian J.*, 2012, **7**, 1013–1017.
312. D. Yang, P. Westreich and R. F. Frindt, *Nanostruct. Mater.*, 1999, **12**, 467–470.
313. H.-L. Tsai, J. L. Schindler, C. R. Kannewurf and M. G. Kanatzidis, *Chem. Mater.*, 1997, **9**, 875–878.
314. E. Cha, M. D. Patel, J. Park, J. Hwang, V. Prasad, K. Cho and W. Choi, *Nat. Nanotechnol.*, 2018, **13**, 337–344.
315. Y. Li, Z. Sun, D. Liu, Y. Gao, Y. Wang, H. Bu, M. Li, Y. Zhang, G. Gao and S. Ding, *J. Mater. Chem. A*, 2020, **8**, 2021–2032.
316. L. F. Nazar, H. Wu and W. P. Power, *J. Mater. Chem.*, 1995, **5**, 1985.
317. B. Anasori, M. R. Lukatskaya and Y. Gogotsi, *Nat. Rev. Mater.*, 2017, **2**, 16098.
318. Q. Pan, Y. Zheng, S. Kota, W. Huang, S. Wang, H. Qi, S. Kim, Y. Tu, M. W. Barsoum and C. Y. Li, *Nanoscale Adv.*, 2019, **1**, 395–402.
319. Y. Shi, B. Li, Q. Zhu, K. Shen, W. Tang, Q. Xiang, W. Chen, C. Liu, J. Luo and S. Yang, *Adv. Energy Mater.*, 2020, **10**, 1903534.
320. L. Li, Y. Yu, G. J. Ye, Q. Ge, X. Ou, H. Wu, D. Feng, X. H. Chen and Y. Zhang, *Nat. Nanotechnol.*, 2014, **9**, 372–377.
321. M. Peruzzini, R. Bini, M. Bolognesi, M. Caporali, M. Ceppatelli, F. Cicogna, S. Coiai, S. Heun, A. Ienco, I. I. Benito, A. Kumar, G. Manca, E. Passaglia, D. Scelta, M. Serrano-Ruiz, F. Telesio, S. Toffanin and M. Vanni, *Eur. J. Inorg. Chem.*, 2019, **2019**, 1476–1494.
322. S. Das, M. Demarteau and A. Roelofs, *ACS Nano*, 2014, **8**, 11730–11738.
323. M. B. Erande, M. S. Pawar and D. J. Late, *ACS Appl. Mater. Interfaces*, 2016, **8**, 11548–11556.
324. P. Yasaei, B. Kumar, T. Foroozan, C. Wang, M. Asadi, D. Tuschel, J. E. Indacochea, R. F. Klie and A. Salehi-Khojin, *Adv. Mater.*, 2015, **27**, 1887–1892.
325. G. Nazri, *Solid State Ionics*, 1989, **34**, 97–102.
326. Y. Kim, D. Koo, S. Ha, S. C. Jung, T. Yim, H. Kim, S. K. Oh, D.-M. Kim, A. Choi, Y. Kang, K. H. Ryu, M. Jang, Y.-K. Han, S. M. Oh and K. T. Lee, *ACS Nano*, 2018, **12**, 4419–4430.
327. C. Liu, X. Han, Y. Cao, S. Zhang, Y. Zhang and J. Sun, *Energy Storage Mater.*, 2019, **20**, 343–372.
328. M. Yao, L. Ai, Y. Zhou, J. Ma and M. Chen, *J. Electrochem. Soc.*, 2019, **166**, A3394–A3400.
329. A. C. Serino, J. S. Ko, M. T. Yeung, J. J. Schwartz, C. B. Kang, S. H. Tolbert, R. B. Kaner, B. S. Dunn and P. S. Weiss, *ACS Nano*, 2017, **11**, 7995–8001.
330. J. Huang and S. R. Turner, *Polymer*, 2017, **116**, 572–586.

Subject Index